AF352876

Bioprocess Systems Engineering Applications in Pharmaceutical Manufacturing

Bioprocess Systems Engineering Applications in Pharmaceutical Manufacturing

Editors

Ralf Pörtner
Johannes Möller

MDPI • Basel • Beijing • Wuhan • Barcelona • Belgrade • Manchester • Tokyo • Cluj • Tianjin

Editors
Ralf Pörtner
Bioprocess and Biosystems
Engineering
Hamburg University of
Technology
Hamburg
Germany

Johannes Möller
Bioprocess and Biosystems
Engineering
Hamburg University of
Technology
Hamburg
Germany

Editorial Office
MDPI
St. Alban-Anlage 66
4052 Basel, Switzerland

This is a reprint of articles from the Special Issue published online in the open access journal *Processes* (ISSN 2227-9717) (available at: www.mdpi.com/journal/processes/specialissues/ PharmaceuticalManufacturing).

For citation purposes, cite each article independently as indicated on the article page online and as indicated below:

LastName, A.A.; LastName, B.B.; LastName, C.C. Article Title. *Journal Name* **Year**, *Volume Number*, Page Range.

ISBN 978-3-0365-5210-1 (Hbk)
ISBN 9978-3-0365-5209-5 (PDF)

Contents

Editorial

Special Issue "Bioprocess Systems Engineering Applications in Pharmaceutical Manufacturing"

Ralf Pörtner * and Johannes Möller *

Institute of Bioprocess and Biosystems Engineering, Hamburg University of Technology,
21073 Hamburg, Germany
* Correspondence: poertner@tuhh.de (R.P.); johannes.moeller@tuhh.de (J.M.)

Citation: Pörtner, R.; Möller, J. Special Issue "Bioprocess Systems Engineering Applications in Pharmaceutical Manufacturing". *Processes* **2022**, *10*, 1634. https://doi.org/10.3390/pr10081634

Received: 8 August 2022
Accepted: 12 August 2022
Published: 18 August 2022

Publisher's Note: MDPI stays neutral with regard to jurisdictional claims in published maps and institutional affiliations.

Biopharmaceutical and pharmaceutical manufacturing are strongly influenced by the process analytical technology initiative (PAT) and quality by design (QbD) methodologies, which are designed to enhance the understanding of more integrated processes. The major aim of this effort can be summarized as developing a mechanistic understanding of a wide range of process steps, including the development of technologies to perform online measurements and real-time control and optimization. Furthermore, minimization of the number of empirical experiments and the model-assisted exploration of the process design space are targeted. Even if tremendous progress has been achieved so far, there is still work to be carried out in order to realize the full potential of the process systems engineering toolbox.

Within this Special Issue of *Processes*, an overview of cutting-edge developments of process systems engineering for biopharmaceutical and pharmaceutical manufacturing processes is given, including model-based process design, Digital Twins, computer-aided process understanding, process development and optimization, and monitoring and control of bioprocesses. The biopharmaceutical processes addressed focus on the manufacturing of biopharmaceuticals, mainly by Chinese hamster ovary (CHO) cells, as well as adeno-associated virus production and generation of cell spheroids for cell therapies.

Both model-based process designs and the Digital Twin concept have gained increasing interest for the development and optimization of biopharmaceutical production processes. Such methods are still not state-of-the-art for cell culture processes during development or manufacturing, although first approaches have been proposed. This highlights a need for improved methods and tools for optimal experimental design, optimal and robust process design, and process optimization for the purposes of monitoring and control during manufacturing. Three contributions within this Special Issue address this topic, which are highlighted as follows:

Bayer et al. [1] present a digital bioprocess twin used for a model-based design of experiment (DoE) to accelerate the design space exploration and thereby decrease the time needed to identify the optimum combination of critical process parameters (CPP) for the variables of interest. This Digital Twin simultaneously delivers additional process understanding while accelerating bioprocess development and optimization by applying in silico simulations and only perform the recommended experiments. A structured workflow is presented using different initial data sets to reduce experimental efforts, evaluate the results, and additionally to investigate the applicability of an intensified DoE (iDoE) for such a model-based DoE. By this, the best CPP combinations in a design space are identified with the highest space-time yield.

Strategies for multi-objective optimization of industrial biopharmaceutical processes are addressed by Hernández Rodríguez et al. [2]. In industrial applications, it is typically desirable to optimize several conflicting objectives at a time, leading to suitable trade-offs and compromises. However, multi-objective optimization is more complex and its application is still not state-of-the-art in the context of cell culture processes (probably due to a lack of related studies and instructions). The contribution presents a conceptual workflow which couples uncertainty-based upstream simulation and Bayes optimization using Gaussian processes. Its application is demonstrated in a simulation case for the design of a robust cell culture expansion process (seed train), meaning that despite uncertainties and variabilities concerning cell growth, low variations of viable cell density during the seed train are obtained. This approach provides the potential to be used in the form of a decision tool (e.g., for the choice of an optimal and robust seed train design or for further optimization tasks within process development).

The selection of appropriate clones of production cells is essential for the optimized manufacturing of complex biopharmaceuticals using bioreactors. A simulation study on model-assisted clone selection for CHO-cells is presented by Hernández Rodríguez et al. The authors of [3] address the question, if clonal cell populations showing high cell-specific growth rates are more favourable than cell lines with higher cell-specific productivities (or vice versa). A mechanistic cell culture model was adapted to the experimental data of such clonally-derived cell population. Uncertainties and prior knowledge concerning model parameters were considered using Bayesian parameter estimations. This model was then used to define an inoculum train protocol. It could be shown that growth rates have a higher impact on overall process productivity and for product output per year, whereas cells with higher productivity can potentially generate higher product concentrations in the production vessel.

Besides the use of mechanistic process models, computational fluid dynamic studies can support bioprocess design and optimization. One paper in particular focuses on *computer-aided process understanding*. Bioreactor design and scale-up in today's biopharma industry relies mostly on empirical correlations, experience, and engineering heuristics, which can hardly provide the link between hydrodynamics within the bioreactor and biological process behaviour. Freiberger et al. [4] investigated cellular effects and locally resolved hydrodynamics in stirred bioreactors for impellers with different spatial hydrodynamics. Therefore, the hydrodynamics, mainly flow velocity, shear rate, and power input in a single- and a three-impeller bioreactor setup were analyzed by means of CFD simulations, and cultivation experiments with antibody-producing CHO cells were performed at various agitation rates in both reactor setups. It could be shown that behaviour of the cells in the different reactor set-ups cannot be linked to parameters commonly used to describe shear effects on cells such as the mean energy dissipation rate or the Kolmogorov length scale, even if this concept is extended by locally resolved hydrodynamic parameters. Alternatively, the hydrodynamic heterogeneity was statistically quantified by means of variance coefficients of the hydrodynamic parameters fluid velocity, shear rate, and energy dissipation rate. The calculated variance coefficients of all hydrodynamic parameters were higher in the setup with three impellers than in the single impeller setup, which might explain the rather stable process behaviour in multiple impeller systems due to the reduced hydrodynamic heterogeneity.

Three contributions are dedicated to aspects of *process development and optimization*. Müller et al. [5] studied seed train intensification using an ultra-high cell density cell banking process. A frequently used approach for seed train intensification uses $N - 1$ perfusion, in which perfusion cultivation is carried out as the final step of inoculum production to generate ultra-high cell banks exceeding 100×10^6 cells mL^{-1}. These cells can subsequently be used to inoculate a production bioreactor. On the one hand, the inoculum production steps can be reduced, and on the other hand a continuous process or a high-seed fed-batch process can be directly implemented with these cells instead of the otherwise usual low-seed fed-batch process. Within the study, an ultra-high cell density working cell bank was established for an immunoglobulin G-producing CHO cell line. A comparison with the standard approach shows that cell growth and antibody production are comparable, but time savings of greater than 35% are possible for inoculum production.

Ladd et al. [6] developed a process for continuous transfection for adeno-associated virus production in microcarrier-based culture. Adeno-associated virus vectors have great potential for gene therapy. However, a major bottleneck for this kind of therapy is the limitation of production capacity. Higher specific AAV vector yield is often reported for adherent cell systems compared to cells in suspension, and a microcarrier-based culture is well established for the culture of anchored cells on a larger scale. The purpose of the present study was to explore how microcarrier cultures could provide a solution for the production of adeno-associated virus vectors based on the triple plasmid transfection of HEK293T cells in a continuously operated stirred tank bioreactor. The present investigation provided a proof-of-concept of a continuous process based on microcarriers in a stirred-tank bioreactor.

Petry and Salzig [7] developed a large-scale production process of size-adjusted β-cell spheroids. The large and growing number of patients living with diabetes has generated interest in the promise of β-cell therapy to restore lost β-cell mass. For β-cell replacement therapies, one challenge is the manufacturing of a sufficient number of functional β-cells manufactured as 3D constructs, known as spheroids with a controlled size. For this, a process in a fully controlled stirred bioreactor systems was established using the INS-1 β-cell line as a model for process development. Specifically, the dynamic agglomeration of β-cells to determine minimal seeding densities, spheroid strength, and the influence of turbulent shear stress was investigated in order to generate spheroids of a defined size. The process developed in shaking flasks was successfully transferred to a stirred bioreactor, and it could be shown that functional β-cell spheroids sufficient for β-cell therapy applications can be generated.

Two contributions are dedicated to the aspects of *monitoring and control of bioprocesses*. Reyes et al. [8] provide a review on modern sensor tools and techniques for monitoring and control, addressing especial technological innovation directed towards online in situ continuous monitoring of quality attributes that could previously only be estimated offline. These new sensing technologies when coupled with software models have shown promise for unique fingerprinting, smart process control, outcome improvement, and prediction. State-of-the-art sensing technologies and their applications in the context of cell culture monitoring are reviewed with an emphasis on the coming push towards industry 4.0 and smart manufacturing within the biopharmaceutical sector. Additionally, perspectives concerning how this can be leveraged to improve both understanding and outcomes of cell culture processes are discussed.

A new and promising biosensor technology is introduced by Gaudreault et al. [9], a surface plasmon resonance (SPR)-based biosensor. SPR-based biosensors can play a role in enabling the development of improved bioprocess monitoring and control strategies. In this review, the applications of SPR that are or could be related to bioprocess monitoring in three spheres are examined such as biotherapeutics production monitoring, vaccine monitoring, and bacteria and contaminant detection. These applications mainly exploit SPR's ability to measure solution species concentrations, but performing kinetic analyses is also possible and could prove useful for product quality assessments. SPR-based biosensors exhibit potential as product monitoring tools from early production to the end of downstream processing, paving the way for more efficient production control. However, more work needs to be carried out to facilitate or eliminate the need for sample pre-processing and to optimize the experimental protocols.

We thank the authors, reviewers, and editors who have contributed to the success of this Special Issue.

Author Contributions: Conceptualization, J.M. and R.P.; Writing—original draft preparation, J.M. and R.P.; All authors have read and agreed to the published version of the manuscript.

Conflicts of Interest: The authors declare no conflict of interest.

References

1. Bayer, B.; Dalmau, D.R.; Melcher, M.; Striedner, G.; Duerkop, M. Digital Twin Application for Model-Based DoE to Rapidly Identify Ideal Process Conditions for Space-Time Yield Optimization. *Processes* **2021**, *9*, 1109. [CrossRef]
2. Rodríguez, T.H.; Sekulic, A.; Lange-Hegermann, M.; Frahm, B. Designing Robust Biotechnological Processes Regarding Variabilities Using Multi-Objective Optimization Applied to a Biopharmaceutical Seed Train Design. *Processes* **2022**, *10*, 883. [CrossRef]
3. Rodríguez, T.H.; Morerod, S.; Pörtner, R.; Wurm, F.M.; Frahm, B. Considerations of the Impacts of Cell-Specific Growth and Production Rate on Clone Selection—A Simulation Study. *Processes* **2021**, *9*, 964. [CrossRef]
4. Freiberger, F.; Budde, J.; Ateş, E.; Schlüter, M.; Pörtner, R.; Möller, J. New Insights from Locally Resolved Hydrodynamics in Stirred Cell Culture Reactors. *Processes* **2022**, *10*, 107. [CrossRef]
5. Müller, J.; Ott, V.; Eibl, D.; Eibl, R. Seed Train Intensification Using an Ultra-High Cell Density Cell Banking Process. *Processes* **2022**, *10*, 911. [CrossRef]
6. Ladd, B.; Bowes, K.; Lundgren, M.; Gräslund, T.; Chotteau, V. Proof-of-Concept of Continuous Transfection for Adeno-Associated Virus Production in Microcarrier-Based Culture. *Processes* **2022**, *10*, 515. [CrossRef]
7. Petry, F.; Salzig, D. Large-Scale Production of Size-Adjusted β-Cell Spheroids in a Fully Controlled Stirred-Tank Reactor. *Processes* **2022**, *10*, 861. [CrossRef]
8. Reyes, S.; Durocher, Y.; Pham, P.L.; Henry, O. Modern Sensor Tools and Techniques for Monitoring, Controlling, and Improving Cell Culture Processes. *Processes* **2022**, *10*, 189. [CrossRef]
9. Gaudreault, J.; Forest-Nault, C.; Crescenzo, G.; Durocher, Y.; Henry, O. On the Use of Surface Plasmon Resonance-Based Biosensors for Advanced Bioprocess Monitoring. *Processes* **2021**, *9*, 1996. [CrossRef]

Article

Digital Twin Application for Model-Based DoE to Rapidly Identify Ideal Process Conditions for Space-Time Yield Optimization

Benjamin Bayer [1,2,†], Roger Dalmau Diaz [1,2,†], Michael Melcher [1], Gerald Striedner [1,2] and Mark Duerkop [1,2,*]

[1] Department of Biotechnology, University of Natural Resources and Life Sciences, 1190 Vienna, Austria; benjamin.bayer@novasign.at (B.B.); roger.dalmau-diaz@novasign.at (R.D.D.); m.melcher@boku.ac.at (M.M.); gerald.striedner@boku.ac.at (G.S.)

[2] Novasign GmbH, 1190 Vienna, Austria

* Correspondence: mark.duerkop@novasign.at; Tel.: +43-660-1017239

† Co-first author, these authors contributed equally to this work.

Abstract: The fast exploration of a design space and identification of the best process conditions facilitating the highest space-time yield are of great interest for manufacturers. To obtain this information, depending on the design space, a large number of practical experiments must be performed, analyzed, and evaluated. To reduce this experimental effort and increase the process understanding, we evaluated a model-based design of experiments to rapidly identify the optimum process conditions in a design space maximizing space-time yield. From a small initial dataset, hybrid models were implemented and used as digital bioprocess twins, thus obtaining the recommended optimal experiment. In cases where these optimum conditions were not covered by existing data, the experiment was carried out and added to the initial data set, re-training the hybrid model. The procedure was repeated until the model gained certainty about the best process conditions, i.e., no new recommendations. To evaluate this workflow, we utilized different initial data sets and assessed their respective performances. The fastest approach for optimizing the space-time yield in a three-dimensional design space was found with five initial experiments. The digital twin gained certainty after four recommendations, leading to a significantly reduced experimental effort compared to other state-of-the-art approaches. This highlights the benefits of in silico design space exploration for accelerating knowledge-based bioprocess development, and reducing the number of hands-on experiments, time, energy, and raw materials.

Keywords: *Escherichia coli*; hybrid modeling; machine learning; model-assisted DoE; quality by design; upstream bioprocessing

Citation: Bayer, B.; Dalmau Diaz, R.; Melcher, M.; Striedner, G.; Duerkop, M. Digital Twin Application for Model-Based DoE to Rapidly Identify Ideal Process Conditions for Space-Time Yield Optimization. *Processes* **2021**, *9*, 1109. https://doi.org/10.3390/pr9071109

Academic Editors: Ralf Pörtner and Johannes Möller

Received: 2 June 2021
Accepted: 23 June 2021
Published: 25 June 2021

1. Introduction

For the production of biopharmaceuticals, it is of high importance to guarantee a specified product quality for patient safety. Raw materials, process deviations, and unrecognized faults may result in altered quality, and finally in batch rejection [1]. Process characterization in the biopharmaceutical industry has long been known and emphasized by the authorities, thus, processes must be closely monitored and well understood to ensure robust and uniform product quality. The most prominent guidance is the process analytical technology (PAT) guide by the US federal drug administration (FDA). Additionally, the quality by design (QbD) initiative [2] greatly emphasizes process understanding during the development of a bioprocess to guarantee a stable and uniform product quality output and fewer rejected batches [3]. To achieve these objectives, the statistical design of experiments (DoE) and advanced online monitoring are highlighted. The herein experimentally investigated design space is built by different combinations of critical process parameters (CPP) and critical material attributes (CMA), which affect the target parameters and the critical quality attributes (CQA) [4]. For such a design space exploration, different DoEs can be applied, e.g., full factorial, fractional factorial, Box–Behnken, Doehlert, and hyper-cubes, differing in the number of required experiments and the amount of information generated [5]. Besides the increased process understanding, for process optimization of the target molecule, it is still important to quickly find the best CPP combination in the design space, at which the production process will be performed, e.g., biomass, product titer, or space-time yield [6]. Such DoE studies are combined with process modeling to generate added value and further accelerate these tasks [7].

The most common techniques for bioprocess modeling are data-driven (black box) and mechanistic (white box) approaches, each with their own characteristics, advantages, and disadvantages [8]. Since the parameters in data-driven models do not have a physical meaning, no further process knowledge is needed, enabling a fast and easy implementation of these model types. Currently, various regression algorithms are available and commonly used, e.g., partial least squares, random forests, support vector machines, artificial neural networks (ANN) [9], and many more. However, such models are based on correlation and do not mandatorily imply causality, which can lead to inaccurate or even incorrect model predictions and conclusions. Contrarily, mechanistic models are based on theoretical considerations, i.e., the parameters have a physical meaning, and therefore ensure causality. Since these model predictions follow a purely mechanistic trend, temporary process deviations and unknown CPP impacts are not considered, which also interferes with the model performance and accuracy. To exploit the advantages of each individual model structure, a combined approach can be considered, called hybrid modeling (grey box) [10]. Since both models can complement each other in this combined structure, more precise predictions are anticipated. Such a hybrid model can be built in a parallel or serial structure, e.g., first, the data-driven part estimates parameters used in the mechanistic part, which otherwise would have to be assumed. Thereby, it is possible to also incorporate the CPP's impact into the hybrid model, which significantly strengthens the explanatory power of the model [11]. Additionally, to have assurance about the model performance and the risk of model mispredictions, typically cross-validation is performed to reduce variance, avoid overfitting, and investigate how the model performs when applied to new data [12]. A similar approach with a higher degree of freedom for creating the final model is model averaging from a leave-one-batch-out cross-validation, i.e., several developed models are averaged to improve the model stability and accuracy [13]. Even though this hybrid modeling approach has been the state-of-the-art in other industries for many years, due to the higher complexity of biological processes, it has only gained interest during the last few years [14]. Even though hybrid modeling is increasingly adopted for downstream applications [15–17], the response surface modeling of process endpoints is still more commonly applied [18], and the full potential of hybrid process modeling applications in bioprocessing has not yet been realized.

The high added value and the benefits of hybrid modeling for upstream bioprocessing become tangible when considering three major aspects of progressing towards digital biomanufacturing, i.e., delivering an increased process understanding, accelerating bioprocess development, and enabling advanced process control [19]. For all these components, various tools with different levels of complexity can be considered. Herein, soft sensors are frequently used, i.e., advanced online sensor systems such as spectrometry [20] or spectroscopy [21] in combination with a software algorithm to estimate the variables of interest in real-time, without any sampling and analytical time delay [22]. Depending on the used model structure, such soft sensors can be descriptive or predictive. While the descriptive model type can only be used to get estimated values up to the current time point, predictive models can also predict future values with a degree of uncertainty and therefore can additionally be used for process control [23]. Along with process models for the variables of interest, model-based methods for the optimization of process parameters such as the gained process information, the maximum amount of cells, or productivity were also introduced [24]. A highly interesting concept for accelerating bioprocess development and optimization in combination with model-related DoE approaches is a digital bioprocess twin [25,26]. Based on a minimal number of experiments, a hybrid model can be developed and subsequently be applied as a digital bioprocess. This digital twin then enables the simulation of further experiments, i.e., in silico exploration of the design space to shed light on the process behavior, without any additional laboratory experiments. This can be used to investigate the impact of the CPPs on the desired output, and thereby recommend the best CPP combination that maximizes it. A validation experiment at the recommended CPPs can be performed and compared to the simulation [27]. Subsequently, this digital twin model can be re-trained with the new experimental data, improving its performance by gaining a higher understanding of the process, and allowing it to explore a potential new optimum [28]. Once the recommendation of the digital twin converges at the process optimum, no new CPP combination will be proposed. Such model-based DoE and process modeling to find the best CPP combination in a design space saves raw materials and additionally operates more quickly and is cheaper compared to approaches in which experiments are only performed in the laboratory [29].

To accelerate the design space exploration and thereby greatly decrease the time needed to identify the optimum CPP combination for the variables of interest, we present a digital bioprocess twin used for model-based DoE [30]. This digital twin simultaneously delivers additional process understanding, while accelerating bioprocess development and optimization by applying in silico simulations that only perform the recommended experiments. We were particularly interested in determining the minimal number of required experiments for developing an initial digital twin, recommending further experiments to rapidly identify the best CPP combination in the design space. Such an iterative approach towards digitalization leads to a reduced experimental effort and saves various propositions of economic value while tackling current shortcomings for the implementation of such novel and promising tools [31]. Therefore, we present our structured workflow using different initial data sets to reduce experimental effort, evaluate the results, and additionally to investigate the applicability of an intensified DoE (iDoE) [32] for such a model-based DoE, to rapidly find the best CPP combinations in a design space and obtain the highest space-time yield.

2. Materials and Methods

2.1. Experimental Design

The experimental data set was derived from *E. coli* (HMS174 (DE3)) (Novagen, Germany) fed-batch cultivations at 20 L scale. For the workflow and the evaluation, a design space with three CPPs, each at three levels, was considered: the feed controlled specific growth rate μ (0.10, 0.15, and 0.20 h^{-1}), the cultivation temperature T (30, 34, and 37 °C), and the induction strength I (0.2, 0.5, and 0.9 µmol IPTG g^{-1} cell dry mass), respectively. The variables of interest to be modeled were the biomass concentration (g L^{-1}) and the space-time yield (g L^{-1} h^{-1}) of the soluble fraction of the expressed protein, recombinant human superoxide dismutase. The biomass was analytically measured by thermogravimetric analysis [33] once before induction and then hourly, and the soluble product titer was measured every 2 h from the time point of induction to the last sampling at the end of the process by ELISA [34]. The fed-batch phase was carried out for four doubling times, and induction of the cells took place after the first doubling time, i.e., product formation took place for the remaining three doubling times. The values for the online measurements were available every minute and included the pH (controlled by the addition of 12.5% NaOH), off-gas (%), cultivation temperature (°C), inlet air (slpm), dissolved oxygen (%), stirrer speed (rpm), base consumption (L), accumulated feed (L), inducer (kg), and head pressure (bar). More details about the applied exponential feeding strategy for the fed-batch phase, the utilized *E. coli* strain, the expression vector system, the online monitoring, and the offline measurements have already been presented elsewhere [35–37].

To receive meaningful information about the performance of the different digital twins and model-based DoE approaches, the design space was completely characterized. Once by common static cultivations (one CPP combination per experiment, i.e., 27 cultivations to cover all CPP combinations) and by iDoE cultivations (three CPP combinations per experiment, i.e., nine cultivations covering all 27 CPP combinations).

The intra-experimental CPP shifts in the intensified fed-batch fermentations were performed after each theoretical cell doubling post-induction of the cells, with a temporarily increased sampling interval, and executed by adjusting the setpoint value of the feed controlled specific growth rate and cultivation temperature in the process control system. Additionally, the feasibility of these shifts and the exclusion of a potential memory effect on the cells is presented in detail elsewhere [38]. A list of all the performed experiments used for comprehensive comparison is given in Appendix A.1 (Tables A1 and A2). Moreover, for the static cultivations, the maximum experimental values of the variables to be modeled are indicated. For the intensified cultivations, the maximum values were not conclusive, due to the intra-experimental shifts and the resulting multiple characterized CPP combinations, and therefore are not displayed. The two complete DoE and iDoE data sets are presented extensively and available for download as supporting information for an earlier publication [38].

2.2. Data Sets

For the initial hybrid model building and the model-based DoE, different initial data sets were used, and the respective performances for identifying the best CPP combination, obtaining the highest space-time yield were compared. These data sets were assembled out of the presented static and intensified fed-batch fermentations:

1. Full factorial DoE: the fully characterized design space, used as a reference ($N = 27$)
2. Fractional factorial DoE: the center point and the eight corners of the design space ($N = 9$)
3. Fractional factorial DoE: the center point and four corners of the design space ($N = 5$)
4. Fractional factorial DoE: the center point and two corners of the design space ($N = 3$)
5. Complete iDoE: all iDoE cultivations, covering the entire design space ($N = 9$)
6. Fractional iDoEs: one iDoE cultivation per induction level ($N = 3$, three different assemblies)

2.3. Hybrid Model Development

2.3.1. Model Building

For initial model training, the different data sets were considered. To deal with the small initial data sets, avoid loss of information, and provide a more robust basis for the digital twin simulations, for each practically performed experiment, two additional in silico experiments were generated, i.e., each performed experiment was available in triplicate. For these in silico experiments, an appropriate level of analytical error was considered as random noise for the biomass (up to 5%) and the soluble product titer (up to 10%). As model inputs, the cultivation temperature (°C), the accumulated feed (L), and the accumulated inducer (kg) were chosen to estimate the two response variables: the biomass (g L^{-1}) and the space-time yield (g L^{-1} h^{-1}). Prior to model building, the input variables were standardized using the z-score. To predict the response variables, a serial hybrid model structure was implemented. The data-driven model, an ANN, embedded in the hybrid model, and applying a Levenberg–Marquardt regularization algorithm, was chosen to estimate the specific growth rate μ and the soluble product formation rate $v_{p/x}$ as propagated predictions for the mechanistic part. The ANN consisted of three layers. The nodes of the hidden layer used hyperbolic tangent transfer functions, while the output layer used linear transfer functions. The values derived from the ANN were subsequently used in the mechanistic model, as shown in Equations (1) and (2), where X is the biomass concentration (g L^{-1}), P is the soluble product titer (g L^{-1}), $I_{y/n}$ is the inducer switch (zero for no induction or one for induction), and D is the dilution rate (h^{-1}). Herein, D is used as the comprehensive term to describe the ratio between the flow of all volume additions into the reactor (L h^{-1}), i.e., substrate feed, inductor feed and base, and the overall reactor volume (L), which comprises the initial volume and all the added volumes. Consequently, in Equation (3), the space-time yield (*STY*) was calculated with the soluble product titer (g L^{-1}) divided by the current utilization time of the bioreactor (h). This Bioreactor Utilization Time comprised the duration of the sterilization in place, inoculum, batch, harvest, cleaning, and the respective feed time.

$$\frac{\mathrm{d}X}{\mathrm{d}t} = u \cdot X - D \cdot X \tag{1}$$

$$\frac{\mathrm{d}P}{\mathrm{d}t} = v_{p/x} \cdot X \cdot I_{y/n} - D \cdot P \tag{2}$$

$$STY = \frac{P}{Bioreactor\ Utilization\ Time} \tag{3}$$

2.3.2. Model Validation

For validation of the model performance, leave-one-batch-out cross-validation was performed, i.e., the initial model was built on all but one experiment, and the parameters were optimized by applying them to the experiment left out. Once no further improvement was observed, the model training stopped. To find the optimal setting to fit the experimental data, the number of neurons and hidden layers were varied. While the number of neurons was individually adapted for each data set, a single hidden layer delivered the best performance in all cases with respect to the normalized root mean square error (NRMSE) in Equation (4), where y is the analytical value, $\hat{y}$ is the estimated counterpart for each sampling point (t), $\bar{y}$ is the mean of the analytical values, and N the total number of observations.

$$\mathrm{NRMSE}\ [\%] = \frac{\sqrt{\frac{1}{N} \cdot \sum \left(y_{(t)} - \hat{y}_{(t)}\right)^2}}{\bar{y}} \cdot 100 \tag{4}$$

2.3.3. Model Averaging

To assess the risk of model misprediction, averaging of the individual models was performed. This averaging of the estimations from multiple models represents a robust way to deal with model uncertainties. This approach allows selecting a single model from each of the cross-validations. Depending on the initial data set, the averaged hybrid models consisted of three to five individual models. To validate this averaged model performance and its uncertainty, the NRMSE was taken into account, along with its standard deviation (SD) (Equation (5)) and the prediction interval (PI) (Equation (6)), where $\hat{y}_{\mathrm{average}}$ is the estimation of the averaged model, $\hat{y}_{\mathrm{model}}$ is the estimation of the respective model, i the index of these models, and n is the number of observations for each time point.

$$\mathrm{SD}_{(t)} = \sqrt{\frac{1}{n-1} \cdot \sum \left(\hat{y}_{\mathrm{average}(t)} - \hat{y}_{\mathrm{model}(i)_{(t)}}\right)^2} \tag{5}$$

$$\mathrm{PI}_{(t)} = \hat{y}_{\mathrm{average}} \pm \mathrm{SD}_{(t)} \tag{6}$$

Subsequently, the final averaged hybrid models were transferred to a digital twin environment.

2.4. Digital Twin Application

The developed hybrid models were implemented as digital twins to simulate all experiments in the given design space. Therefore, the accumulated feed, the inducer, and the inducer switch were simulated according to the feeding strategy and process time of the individual constant CPP levels, according to the desired design space boundaries. Once the simulations were performed by the digital twin, a lookup table could be used to individually evaluate the digital twin simulations. This lookup table provides the options for investigating the simulations, i.e., find the minimum or maximum values for the response variables and their respective associated CPP combination along with the process time duration. For this case study, the lookup table was used to find the optima (maximum value) for the space-time yield in all simulated experiments, i.e., recommending the CPPs to obtain this simulated value. To validate the derived recommendation of the digital twin, a laboratory experiment with the respective settings was performed. The new experiment was then added to the previous data set and the hybrid model was re-trained including the new setup and its findings. This model-based DoE for optimizing the space-time yield was repeated until the digital twin identified the best CPP combination and no new CPP combination was recommended. The entire workflow of the model-based DoE is presented in Figure 1. This workflow was carried out for all of the different initial data sets presented before, to evaluate the possible minimum number of required experiments for each case.

The hybrid model development, digital twin simulation, and model-based DoE were accomplished in the Novasign GmbH (Vienna, Austria) hybrid modeling toolbox.

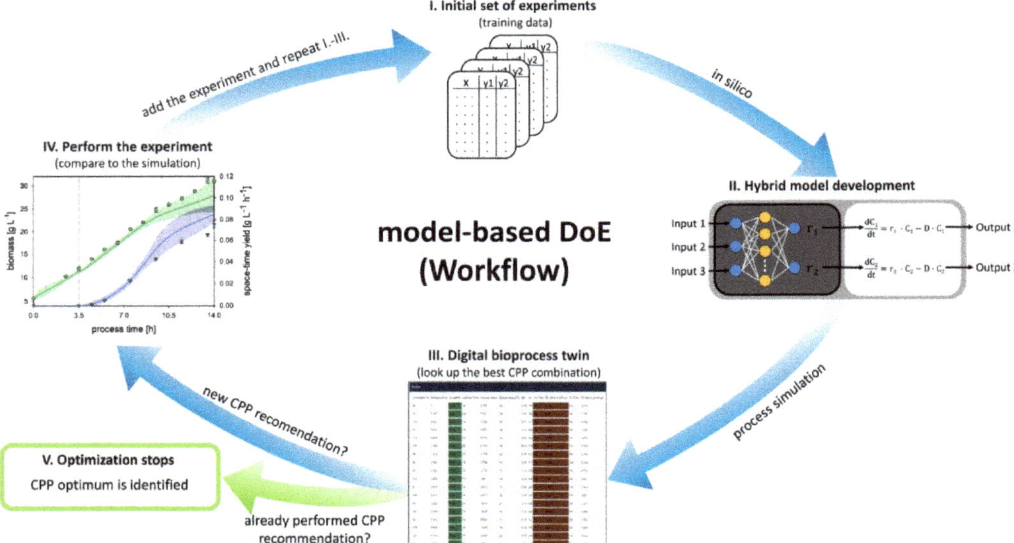

Figure 1. Schematic workflow of optimizing the space-time yield using model-based DoE. Starting with an initial set of experiments from a given design space (I), a hybrid model is developed (II) and transferred to a digital twin environment. Based on the hybrid model, the digital twin simulates all experiments of the design space and recommends the best CPP combination in the design space to obtain the maximum value of the variable of interest (space-time yield) (III). In the case of a new CPP recommendation, the experiment is performed, added to the training data, and utilized to re-train the hybrid model with the new process information (IV). Once no new CPP recommendation is obtained, the digital twin identifies the best CPP combination to maximize the space-time yield and the optimization stops (V).

3. Results

3.1. Analytical Space-Time Yield Maxima in the Design Space

To confirm the simulated values and correctness of the CPP recommendation by the digital twin, the space-time yield of each CPP combination was investigated. The analytical maximum space-time yield of each cultivation is presented as a response surface in Figure 2. For simpler visualization, the results are separated into the three levels of induction strength.

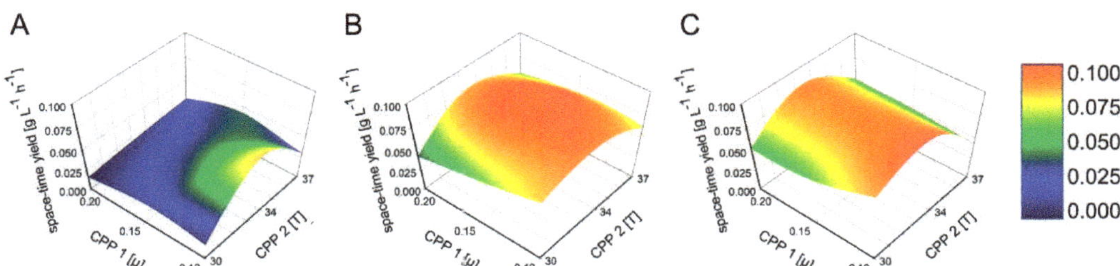

Figure 2. Response surfaces of the analytical space-time yield maxima in the design space. The maximum value of each CPP combination is displayed as a function of the specific growth rate and the cultivation temperature for each induction level: $I = 0.2$ (**A**), $I = 0.5$ (**B**), and $I = 0.9$ (**C**). The color indicates the values of the space-time yield from dark blue (lowest value) to red (highest value).

The graphical investigation of the analytical space-time yield of each CPP combination in Figure 2 reveals the local and global optima in the design space. While at induction level $I = 0.2$, the local maximum was found at 0.0726 g L^{-1} h^{-1} ($\mu = 0.10$ h^{-1} and $T = 34\,°$C), and the induction level $I = 0.5$ contained the global maximum at the center point ($\mu = 0.15$ h^{-1}, $T = 34\,°$C, and $I = 0.5$) with 0.0997 g L^{-1} h^{-1}. The local maximum at induction level $I = 0.9$ resulted in 0.0915 g L^{-1} h^{-1} ($\mu = 0.10$ h^{-1} and $T = 34\,°$C). This visualization demonstrates that a cultivation temperature of 34 °C seems to be highly favorable for product formation, along with a trend towards slower specific growth rates.

3.2. Initial Training Data for the Model-Based DoE

The objective for this model-based DoE for parameter optimization was to quickly identify the best CPP combination for the highest space-time yield in the design space. To determine the minimum number of required experiments to develop meaningful hybrid models, and applied as digital twins recommending the next experiments, different initial data sets were utilized (Section 2.2 Data sets). These comprised either static or intensified cultivations, as presented in Figure 3.

As presented in Figure 2 and Table A1, the best CPP combination in the design space to maximize the space-time yield was obtained at the center point. However, there was also a local maximum with a high space-time yield at the highest induction level, which is assumed to be challenging not to become trapped in. For the design space investigation and determination of this CPP combination, different approaches can be consulted, as presented in Figure 3. First, experiments at each CPP combination were performed, characterizing the entire space without comprehensive process modeling (Figure 3A). Using this approach, the optimum in the design space was found, but this was paired with a high experimental effort and therefore time and costs. This experimental effort can be reduced by selecting a fractional factorial design and process modeling, i.e., only certain CPP combinations are performed. For this comparison, three fractional factorial designs were performed with the center point and the corners of the design space, either using nine (Figure 3B), five (Figure 3C), or only three initial experiments to build the hybrid model (Figure 3D). Since the iDoE concept proved to be suitable for accelerating the process characterization, this approach was additionally considered. Therefore, a complete set of iDoE experiments (Figure 3E) and three fractional iDoE approaches (Figure 3F–H) were used. The initial experiments of these last seven approaches were used in combination with process modeling to find the optimal CPP combination for obtaining the highest space-time yield as fast as possible, and using the workflow presented in Figure 1.

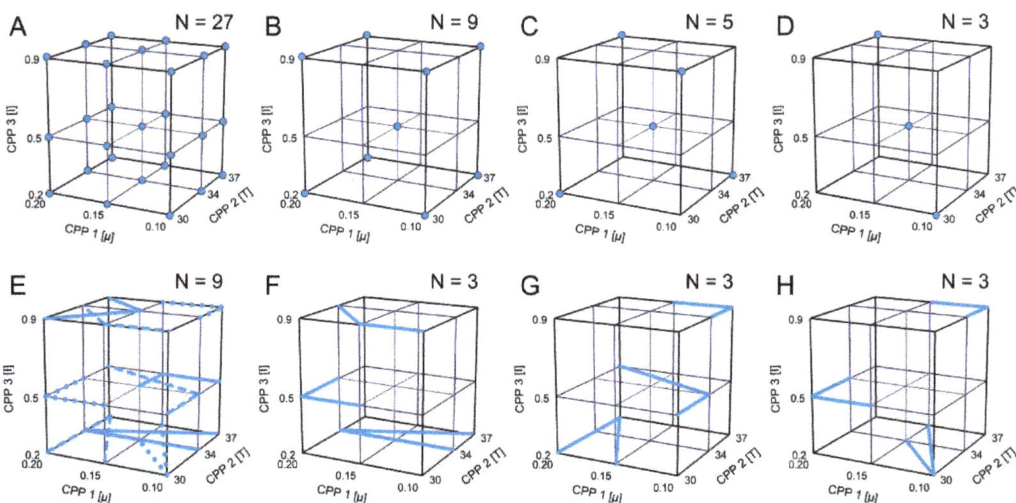

Figure 3. Approaches with varying initial data sets to set up the digital twin for model-based DoE. To find the best CPP combination in the given design space, different approaches with varying initial numbers of experiments were used (blue circles and lines). The full factorial DoE without the need for comprehensive modeling (**A**) was consulted in addition to fractional factorial DoEs with nine (**B**) and five (**C**), as well as a minimal approach using three (**D**), initial static cultivations. Model-based DoE approaches using iDoE cultivations were performed with the complete iDoE data set (**E**) and three fractional iDoEs (**F–H**).

3.3. Digital Twin Simulations of the Model-Based DoE

Out of all the presented initial data sets for the model-based DoE parameter optimization, the fractional factorial DoE with five initial static cultivations performed best, i.e., the fewest total experiments were needed by the digital twin to identify the CPP optimum for the space-time yield. A graphical presentation of this model-based DoE is presented in Figure 4. The step-by-step progression of the recommended experiments in the design space along with the simulated values compared to the analytical values for each re-trained digital twin are shown.

The model-based DoE quickly recommended the best CPP combination to obtain the highest space-time yield (Figure 4A). The correct induction level was already found after implementing the gained process knowledge from the first recommended experiment and the correct cultivation temperature after the second re-training of the digital twin. Even though the specific growth rate was the most difficult to properly assert, after two additional cultivations the optimum in the design space was found, identifying the center point CPPs as the optimum process conditions, which were already present in the initial training data. This resulted in nine performed experiments instead of twenty-seven, highlighting the advantages of knowledge-based bioprocess development. However, with this small initial data set, the simulated biomass of the first recommended experiment (Figure 4B) almost matched the analytical results, and the space-time yield was highly overestimated. Likewise, high overestimations were observed for the second (Figure 4C) and the third recommendation (Figure 4D). By adding these new recommended experiments to the initial data set, the resulting retrained hybrid model iteratively gained knowledge about the process for the next recommendation. Already, after only these three re-trainings, the fourth simulation almost converged on the analytical values (Figure 4E). The digital twin gained precision and certainty at the fifth and final recommendation (Figure 4F). Since this recommended experiment had already been performed, the model-based DoE stopped, i.e., the best CPP combination was identified, and the biomass and space-time yield of the process were accurately simulated.

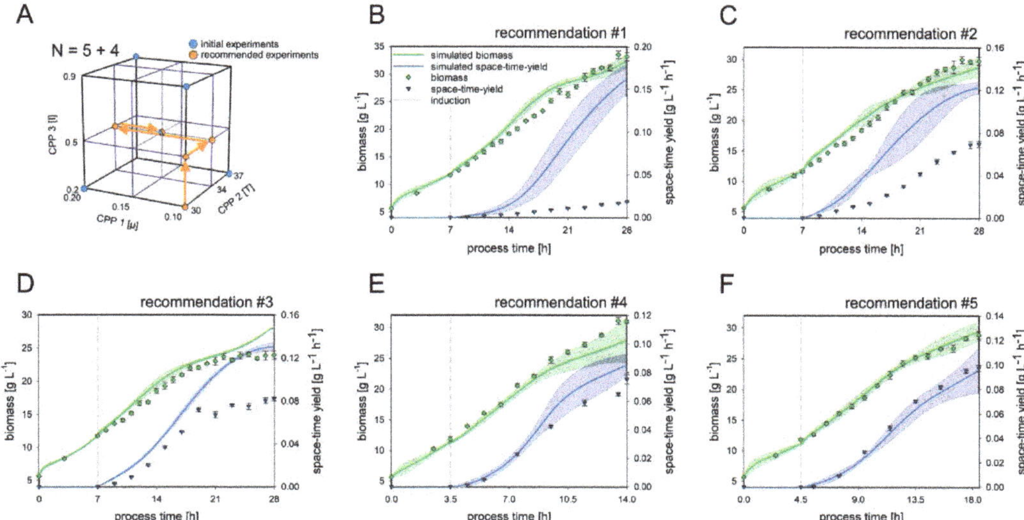

Figure 4. Step-by-step progression of the model-based DoE and the performance compared to experimental data. The initial data set (blue circles), the recommended experiments (orange circles), and the temporal order (orange arrows) are given (**A**). For each recommended experiment (**B–F**), the simulated biomass (green lines) and the simulated space-time yield (blue lines) are presented along with the PI (shaded area). The time point of induction (dashed grey line) and the mean analytical values for the biomass (green diamonds) and the space-time yield (blue triangles) are indicated along with the SD (error bars).

With five initial static experiments, the digital twin simulated the biomass concentration with an appropriate accuracy from the beginning, but highly overestimated the experimental values of the space-time yield. By consecutively adding the four recommended experiments, and extending the initial data set, precise simulations were obtained. This fast convergence of the simulated space-time yield on the analytical values, along with the SD, is displayed in Table 1.

Table 1. Progression of the model-based DoE until the optimum was found using five initial experiments.

Digital Twin Conversion	CPP I (μ)	CPP II (T)	CPP III (I)	Analytical Maximum $(g\ L^{-1}\ h^{-1})$	Simulated Maximum $(g\ L^{-1}\ h^{-1})$
1st recommendation	0.10	30	0.2	0.0185 ($\pm$0.0006)	0.1605 ($\pm$0.0185)
2nd recommendation	0.10	30	0.5	0.0696 ($\pm$0.0029)	0.1220 ($\pm$0.0058)
3rd recommendation	0.10	34	0.5	0.0820 ($\pm$0.0018)	0.1303 ($\pm$0.0040)
4th recommendation	0.20	34	0.5	0.0755 ($\pm$0.0032)	0.0848 ($\pm$0.0079)
5th recommendation	0.15	34	0.5	0.0976 ($\pm$0.0026)	0.0955 ($\pm$0.0186)

As seen in Table 1, the obtained recommendations of the digital twin, at which CPP combination the next experiment should be performed, converged at the best CPP combination in the design space after five recommended experiments, i.e., no new recommendation was derived. Moreover, a steep learning curve of the hybrid model was observed when the new experiments were added for re-training the digital twin. While the simulated space-time yield of the first recommended experiment, derived from the information gained from the initial five experiments, resulted in an 8.68-fold deviation compared to the analytical value, this factor quickly decreased after including the respective validation experiments in the training data and subsequent re-training of the hybrid model. For example, the simulation of the second recommendation already displayed a decreased deviation of only 1.75-fold compared to the analytical value, while the third simulation was down to a

1.59-fold deviation. The fourth simulation only displayed a deviation from the analytical value by 1.12-fold, and the final simulation of the fifth recommendation was highly precise, displaying a simulated maximum of 0.98-fold the analytical value. This demonstrates that with only five initial experiments to start the model-based DoE, the hybrid model promptly gained process knowledge and its digital twin was able to provide the best CPP combination to obtain the highest space-time yield.

A complete quantitative and qualitative performance comparison of all the presented approaches (Figure 3) is given in Table 2. Herein, the three different fractional iDoE approaches are summarized.

Table 2. Performance summary of the model-based DoE approaches. Capital letters in brackets represent the DoE conditions from Figure 3.

Initial Data Set	Initial Experiments	Recommended Experiments	Total Experiments	Optimum Found
full factorial DoE (A)	27	0	27	yes
fractional factorial DoE (B)	9	2	11	yes
fractional factorial DoE (C)	5	4	9	yes
fractional factorial DoE (D)	3	7	10	yes
complete iDoE (E)	9	2	11	no
fractional iDoEs (F–H)	3	1–4	4–7	no

Table 2 presents the quantitative effort and qualitative performance of each initial data set. With respect to the total required time for each presented approach, only the duration of the practical experiments (including pre- and post-processing) was taken into account for the evaluation, since using our setup, an entire experiment takes approximately one working week. However, the computational time for the hybrid model training and subsequent re-training can be neglected, since it ranges between half an hour and three hours, and depending highly on the performance of the utilized computer. While the number of required experiments remains unchanged, the needed experimental time can further be reduced by the utilization of multiple bioreactors or parallel bioreactor systems.

Since in the full factorial DoE all experiments are performed, comprehensive process modeling is redundant to find the best CPP combination for the highest space-time yield in the design space. By using this approach, the optimum was found, but paired with the highest experimental effort. For the other initial data sets, model-based DoE was applied to reduce the required number of experiments. For the fractional factorial DoEs, the number of recommended experiments increased until the optimum was found when decreasing the number of initial experiments. Herein, the fastest approach was the fractional factorial DoE with five initial experiments and four validation experiments required, i.e., only 9/27 experiments had to be performed. Moreover, in all cases, the optimum was identified. However, in this case study, the utilization of initial iDoE cultivations for model-based DoE did not lead to the identification of the best CPP combination in the design space. Regardless of selecting the entire iDoE data set or varying fractional iDoEs, the model-based DoE ended up at different locations in the design space than the optimum CPP combination. Herein, the final recommendations by the digital twin were all located at $\mu = 0.10$, $I = 0.9$ and either 30 °C or 34 °C, indicating a model bias towards slow specific growth rates and temperatures, apart from 37 °C, where a high value or local maximum of the space-time yield is located. A more detailed progression of the recommended experiments in the design space for each of the other six model-based DoEs is shown in Appendix A.2, Figure A1 (excluding the full factorial DoE).

4. Discussion

The prominent emerging concept of model-based DoE for parameter optimization is an interesting, and yet not completely explored, topic. To accelerate this identification of optimum process conditions is of great interest for manufacturers, to reduce bioprocess development timelines. Typically, by performing all experiments in a design space, these optimum process conditions can be found, but with high experimental effort. Herein, we challenged this approach by investigating the minimum requirements for such a model-based DoE workflow (Figure 1) to rapidly and properly discover the best CPP combinations in a design space (Figure 2), utilizing varying numbers of initial experiments (Figure 3). We demonstrated with our case study that the fastest approach to identifying the best process conditions for the highest space-time yield was an initial fractional factorial DoE with five static cultivations and four consecutively performed recommendations from the digital twin (Figure 4 and Table 1). In case scientists are limited to certain time slots for further experiments, the best x-recommendations from the digital twin can be used in the next campaign to obtain the maximum learning, according to the experimental possibilities. Interestingly, all model-based DoEs using initial iDoE cultivations failed to find the global maximum in the design space (Table 2), and recommending an incorrect optimal CPP combination after a few iterations (Figure A1). It has already been demonstrated that iDoE is favorable for accelerating process characterization. Here, a trade-off between decreased experimental effort and reduced process information can be accepted. This consideration must be handled with care when iDoE is used for process optimization, i.e., an increased model uncertainty due to decreased process information may result in divergent optima, as was the case herein. To the best of our knowledge, this iDoE concept has not been well investigated and little literature is available as a reference for microbial, and even less for mammalian, systems. Additionally, several degrees of freedom are introduced by iDoE, e.g., the number and duration of the intra-experimental CPP shifts, as well as how these should be performed. Therefore, before reliably applying iDoE for such model-based DoE approaches, more research should be performed on this subject.

Furthermore, the identification of optimum process conditions for the response to be optimized in design spaces with a higher dimensionality, as in our case study (>3 CPPs), could lead to new challenges, e.g., the occurrence of various local optima, which complicate the accurate identification of the global optimum. The robustness and applicability of digital twins to also perform reliably when confronted with this higher complexity must be further investigated. Moreover, our findings demonstrate that bioprocess modeling is not an all-in-one solution, eliminating all current limitations and obstacles; showing that it is important to consider many potentially influencing factors [39].

For instance, it is advisable for the initially used data set to introduce every CPP level to the hybrid model training, i.e., the minimal fractional factorial DoE with three initial cultivations in our case study. Otherwise, the hybrid model will be biased towards the included CPP levels in the training data and potentially would not recommend the missing setting, since the ability to correctly determine these causal relationships is lost. This bias towards CPP levels should be considered when initially investigating a design space, for which no prior process knowledge about process behavior and the responses is available, i.e., the CPPs and the appropriate levels should be well-considered and not too far apart. Hereby, the accidental generation of independent data sets, becoming missing and getting trapped in local optima, can be avoided at the start. Since this case study mainly focused on the practical application of digital twins, more detailed theoretical analysis should be performed in future studies. However, it might be desirable to re-define the CPP levels and look for new, more beneficial settings in the design space, e.g., with smaller intervals of the cultivation temperatures simulated by the digital twin. However, if a digital twin recommends an experiment next to the identified optimum CPP combination, but with a 0.5 °C decreased cultivation temperature and an increased space-time yield by 0.3%, the execution of this cultivation should be critically questioned. Additionally, for some CPPs, such simulated intervals are not always practically feasible, e.g., steps of 0.5 °C for the cultivation temperature, which might be adjustable but difficult to precisely control. This exemplary scenario demonstrates that such approaches must still be guided by human knowledge, rather than completely trusting an algorithm.

Herein, it has been demonstrated that such digital solutions enable a new knowledge-based perspective on bioprocess development and optimization, and to get more out of the available data. Even though several of these advantages have already been recognized and discussed, much more research will be required to fully implement and exploit the potential of digitalization in the biopharmaceutical industry [40]. For instance, an up-and-coming area for future application of model-based DoE, hybrid modeling, and digital twins is found in simulating new CPP combinations out of the design space, i.e., extrapolation where appropriate. However, this again poses new challenges, such as how to validate this new setting outside the design space, e.g., an additional smaller design space with the new CPP combination as the center point could potentially be performed. Besides the validation issue, the stability of the digital twin and the underlying hybrid model structure must also be ensured. Additionally, if the mechanistic relationships are known and understood, such digital twins could be used as a basis to initially simulate new bioprocesses with similar product properties without prior experiments, e.g., product size and cytotoxicity supporting platform approaches.

5. Conclusions

In silico design space exploration using a digital bioprocess twin increases the process understanding for QbD; the impact of the CPPs on the variables of interest can rapidly be investigated. The presented workflow enabled us to quickly find process optima in a design space despite using only a small initial experimental setup. Moreover, this approach to decreasing the number of required practical experiments for process optimization becomes even more advantageous for larger design spaces. Even though, herein the dimensionality and complexity increase, which will lead to new challenges, model-based DoE has the potential to significantly lower the experimental effort; saving money, time, raw materials, and other propositions of economic value for later stages.

Author Contributions: Conceptualization, B.B., M.D., R.D.D.; methodology, B.B., R.D.D.; software, R.D.D.; validation, B.B., R.D.D., M.M.; formal analysis, B.B., R.D.D.; investigation, B.B.; resources, G.S.; data curation, B.B., R.D.D.; writing—original draft preparation, B.B.; writing—review and editing, B.B., R.D.D., M.M., G.S., M.D.; visualization, B.B.; supervision, M.D.; project administration, M.D.; funding acquisition, G.S., M.D. All authors have read and agreed to the published version of the manuscript.

Funding: This research was funded by the Austrian Research Promotion Agency (FFG), grant number 859219.

Institutional Review Board Statement: Not applicable.

Informed Consent Statement: Not applicable.

Data Availability Statement: The data presented in this study are available in the supporting information of a previous publication (https://doi.org/10.1002/biot.202000121) (accessed on 24 June 2021).

Conflicts of Interest: Benjamin Bayer, Roger Dalmau Diaz, Mark Duerkop, and Gerald Striedner hold shares of Novasign GmbH.

Abbreviations

ANN	artificial neural network
CMA	critical material attribute
CPP	critical process parameter
CQA	critical quality attribute
DoE	design of experiments
iDoE	intensified design of experiments
FDA	US federal drug administration
NRMSE	normalized root mean square error
PAT	process analytical technology
PI	prediction interval
QbD	quality by design
SD	standard deviation

Appendix A

Appendix A.1 CPP Settings of All Experiments Used for Model-Based DoE

The design space, the herein investigated CPPs (and respective levels), and cultivation approaches are introduced in the Materials and Methods section of the main manuscript. A detailed list of all performed experiments of the comprehensive comparison for the applicability of the model-based DoE workflow (Figure 1, main manuscript) is given below. Table A1 provides information about the experiments performed with one CPP combination, and Table A2 contains the intensified experiments (three CPP combinations per cultivation) and the herein performed CPP shifts. For all static experiments, the maximum experimental values of the variables modeled (biomass and space-time yield) are provided for easier comparison. For the intensified experiments, these maximum experimental values are not indicated, because these quantities are not meaningful due to multiple characterized CPP combinations per experiment. The highest space-time yield in the entire design space was obtained at CPP combination #14 ($\mu = 0.15$ h^{-1}, T = 34 °C, and I = 0.5), reaching 0.0997 g L^{-1} h^{-1} in the performed cultivation. Subsequently, the different initial data sets were evaluated in the model-based DoE (Figure 3, main manuscript), considering the number of required recommendations by the digital twin until certainty about the best CPP combination is gained.

Table A1. CPP combinations of the static experiments for the model-based DoE approach.

CPP Combination	CPP 1 (μ)	CPP 2 (T)	CPP 3 (I)	Maximum Biomass (g L^{-1})	Maximum Space-Time Yield (g L^{-1} h^{-1})
1		30	0.2	33.18	0.0193
2		34	0.2	31.12	0.0726
3		37	0.2	30.31	0.0311
4		30	0.5	29.88	0.0733
5	0.10	34	0.5	23.96	0.0837
6		37	0.5	20.6	0.0621
7		30	0.9	26.07	0.0800
8		34	0.9	20.69	0.0915
9		37	0.9	18.23	0.0432
10		30	0.2	34.28	0.0264
11		34	0.2	32.09	0.0415
12		37	0.2	29.7	0.0430
13		30	0.5	31.74	0.0564
14	0.15	34	0.5	28.66	0.0997
15		37	0.5	24.06	0.0663
16		30	0.9	26.89	0.0564
17		34	0.9	25.17	0.0815
18		37	0.9	21.62	0.0485
19		30	0.2	34.51	0.0157
20		34	0.2	33.68	0.0227
21		37	0.2	32.93	0.0274
22		30	0.5	31.49	0.0418
23	0.20	34	0.5	30.97	0.0783
24		37	0.5	28.85	0.0578
25		30	0.9	29.14	0.0518
26		34	0.9	29.25	0.0818
27		37	0.9	23.98	0.0513

Table A2. CPP combinations of the intensified experiments for the model-based DoE approach.

iDoE CPP Combination	CPP 1 (μ)	CPP 2 (T)	CPP 3 (I)	CPP Shift 1	CPP Shift 2
1		37	0.2	37 °C to 34 °C 0.10 h^{-1} to 0.20 h^{-1}	0.20 h^{-1} to 0.10 h^{-1}
2	0.10	30	0.5	30 °C to 34 °C	34 °C to 37 °C 0.10 h^{-1} to 0.20 h^{-1}
3		34	0.9	34 °C to 37 °C	0.10 h^{-1} to 0.15 h^{-1}
4		37	0.2	37 °C to 30 °C 0.15 h^{-1} to 0.10 h^{-1}	30 °C to 34 °C 0.10 h^{-1} to 0.15 h^{-1}
5	0.15	30	0.5	0.15 h^{-1} to 0.20 h^{-1}	30 °C to 34 °C
6		34	0.5	34 °C to 37 °C	0.15 h^{-1} to 0.10 h^{-1}
7		30	0.2	30 °C to 37 °C	37 °C to 30 °C 0.20 h^{-1} to 0.15 h^{-1}
8	0.20	37	0.9	37 °C to 34 °C 0.20 h^{-1} to 0.15 h^{-1}	34 °C to 30 °C 0.15 h^{-1} to 0.20 h^{-1}
9		34	0.9	34 °C to 30 °C 0.20 h^{-1} to 0.15 h^{-1}	0.15 h^{-1} to 0.10 h^{-1}

Appendix A.2 Progression of the Recommended Experiments by Each Model-Based DoE Approach

Out of all presented initial data sets for the model-based DoE in Figure 3 (Results section of the main manuscript), the fractional factorial DoE with five initial static experiments proved to be the fastest for identifying the best CPP combination for the highest space-time yield. This detailed progression until the optimum was found is presented in Figure 4 and Table 1 (Results section of the main manuscript). For the other six data sets used for the model-based DoE (excluding the full factorial DoE), Figure A1 presents an overview of the respective progressions, including the initially performed experiments, as well as the recommended experiments.

Besides the best performing model-based DoE with five initial static cultivations, the two other initial fractional factorial DoEs also performed well. The approach with nine initial static cultivations (Figure A1A) needed two recommendations, i.e., two further experiments to gain certainty about the optimum CPP combination, resulting in a total of 11/27 cultivations. Herein, the model quickly gained certainty about the correct induction level from the beginning, and after the second experiment also about the other two CPP levels. The model-based DoE using three initial static cultivations performed seven recommendations until the optimum was identified, i.e., 10/27 cultivations (Figure A1B). Interestingly, here the induction level was also the first CPP to be correctly recommended after two additional experiments, followed by the cultivation temperature and then the specific growth rate. However, the complete iDoE as the basis for model-based DoE (Figure A1C) did not identify the optimum, and after two recommendations by the digital twin ended up recommending CPP combination #7 ($\mu = 0.10$ h^{-1}, $T = 30$ °C, and $I = 0.9$). Moreover, the model-based DoE based on three different fractional iDoEs was also not able to find the optimum CPP combination. Depending on the initially selected three iDoE cultivations, it took one to four recommendations by the digital twin until these model-based DoEs also recommended CPP combination #7 (Figure A1D,F) or CPP combination #8 ($\mu = 0.10$ h^{-1}, $T = 34$ °C, and $I = 0.9$) (Figure A1E) as the best CPP combination for the highest space-time yield.

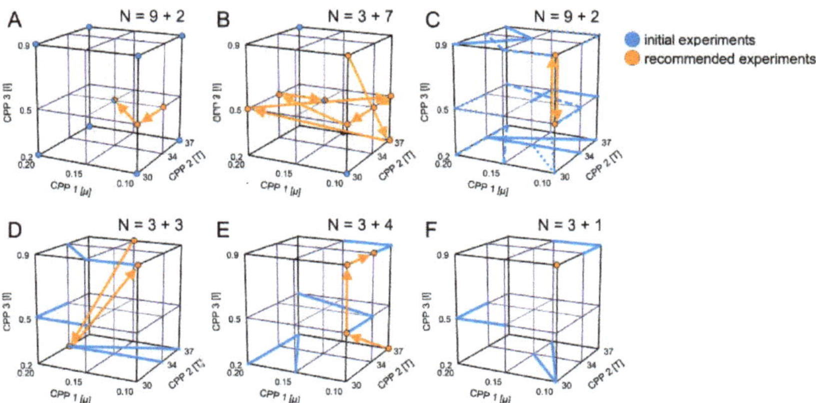

Figure A1. Step-by-step progressions of the recommended experiments by the model-based DoE, using varying initial data sets. The initial experiments (blue circles and lines) and the respective recommendations for the next experiment (orange dots), along with the temporal order (orange arrows) are given. The fractional factorial DoE with nine (**A**) and three (**B**) initial static cultivations, the complete iDoE (**C**), and the three fractional factorial iDoEs (**D**–**F**) are presented.

In conclusion, while every approach using static cultivations as a basis for model-based DoE could identify the optimum CPP combination in the design space, all the iDoE approaches failed to do so. However, it was already shown that the concept of iDoE is advantageous for reducing the experimental effort for process characterization but, in this particular case, it was not possible for model-based DoE to accurately identify the static process conditions optimizing a certain process output. The recommended CPP combinations by the model-based DoE with the initial iDoE cultivations were becoming trapped at high values or local maxima and were highly biased towards the highest induction level, slowest growth rate, and lower cultivation temperatures.

References

1. Pekarsky, A.; Konopek, V.; Spadiut, O. The impact of technical failures during cultivation of an inclusion body process. *Bioprocess Biosyst. Eng.* **2019**, *42*, 1611–1624. [CrossRef] [PubMed]
2. Guideline, I.H.T. International Conference on Harmonisation of Technical Requirements for Registration of Pharmaceuticals for Human Use Pharmaceutical development Q8(R2). *ICH Harmon. Tripart. Guidel.* **2009**, 1–24. [CrossRef]
3. Mandenius, C.-F.; Graumann, K.; Schultz, T.W.; Premstaller, A.; Olsson, I.M.; Petiot, E.; Clemens, C.; Welin, M. Quality-by-design for biotechnology-related pharmaceuticals. *Biotechnol. J.* **2009**, *4*, 600–609. [CrossRef]
4. Rathore, A.S.; Winkle, H. Quality by design for biopharmaceuticals. *Nat. Biotechnol.* **2009**, *27*, 26–34. [CrossRef] [PubMed]
5. Lundstedt, T.; Seifert, E.; Abramo, L.; Thelin, B.; Nyström, Å.; Pettersen, J.; Bergman, R. Experimental design and optimization. *Chemom. Intell. Lab. Syst.* **1998**, *42*, 3–40. [CrossRef]
6. Mandenius, C.-F.; Brundin, A. Bioprocess Optimization, Using Design-of-experiments Methodology. *Biotechnol. Progr.* **2008**, *24*, 1191–1203. [CrossRef] [PubMed]
7. Lee, K.-M.; Gilmore, D.F. Statistical Experimental Design for Bioprocess Modeling and Optimization Analysis. *Appl. Biochem. Biotechnol.* **2006**, *135*, 101–135. [CrossRef]
8. Hallow, D.M.; Mudryk, B.M.; Braem, A.D.; Tabora, J.E.; Lyngberg, O.K.; Bergum, J.S.; Rossano, L.T.; Tummala, S. An example of utilizing mechanistic and empirical modeling in quality by design. *J. Pharm. Innov.* **2010**, *5*, 193–203. [CrossRef]
9. Kadlec, P.; Gabrys, B.; Strandt, S. Data-driven Soft Sensors in the process industry. *Comput. Chem. Eng.* **2009**, *33*, 795–814. [CrossRef]
10. von Stosch, M.; Davy, S.; Francois, K.; Galvanauskas, V.; Hamelink, J.M.; Luebbert, A.; Mayer, M.; Oliveira, R.; O'Kennedy, R.; Rice, P.; et al. Hybrid modeling for quality by design and PAT-benefits and challenges of applications in biopharmaceutical industry. *Biotechnol. J.* **2014**, *9*, 719–726. [CrossRef]
11. Bayer, B.; Von Stosch, M.; Striedner, G.; Duerkop, M. Comparison of Modeling Methods for DoE-Based Holistic Upstream Process Characterization. *Biotechnol. J.* **2020**, *15*. [CrossRef]
12. Taylor, P.; Picard, R.R.; Cook, R.D. Cross-Validation of Regression Models. *J. Am. Stat. Assoc.* **1984**, *79*, 575–583. [CrossRef]

13. Mendes-Moreira, J.; Soares, C.; Jorge, A.M.; De Sousa, J.F. Ensemble approaches for regression: A survey. *ACM Comput. Surv.* **2012**, *45*. [CrossRef]

14. von Stosch, M.; Oliveira, R.; Peres, J.; Feyo de Azevedo, S. Hybrid semi-parametric modeling in process systems engineering: Past, present and future. *Comput. Chem. Eng.* **2014**, *60*, 86–101. [CrossRef]

15. Krippl, M.; Dürauer, A.; Duerkop, M. Hybrid modeling of cross-flow filtration: Predicting the flux evolution and duration of ultrafiltration processes. *Sep. Purif. Technol.* **2020**, *248*, 1–11. [CrossRef]

16. Krippl, M.; Bofarull-Manzano, I.; Duerkop, M.; Dürauer, A. Hybrid modeling for simultaneous prediction of flux, rejection factor and concentration in two-component crossflow ultrafiltration. *Processes* **2020**, *8*, 1625. [CrossRef]

17. Wang, G.; Briskot, T.; Hahn, T.; Baumann, P.; Hubbuch, J. Estimation of adsorption isotherm and mass transfer parameters in protein chromatography using artificial neural networks. *J. Chromatogr. A* **2017**, *1487*, 211–217. [CrossRef]

18. Kalil, S.J.; Maugeri, F.; Rodrigues, M.I. Response surface analysis and simulation as a tool for bioprocess design and optimization. *Process Biochem.* **2000**, *35*, 539–550. [CrossRef]

19. Sommeregger, W.; Sissolak, B.; Kandra, K.; von Stosch, M.; Mayer, M.; Striedner, G. Quality by control: Towards model predictive control of mammalian cell culture bioprocesses. *Biotechnol. J.* **2017**, *12*. [CrossRef]

20. Schmidberger, T.; Gutmann, R.; Bayer, K.; Kronthaler, J.; Huber, R. Advanced online monitoring of cell culture off-gas using proton transfer reaction mass spectrometry. *Biotechnol. Prog.* **2013**, *7*. [CrossRef]

21. Bayer, B.; Von Stosch, M.; Melcher, M.; Duerkop, M.; Striedner, G. Soft sensor based on 2D-fluorescence and process data enabling real-time estimation of biomass in Escherichia coli cultivations. *Eng. Life Sci.* **2020**, *20*, 26–35. [CrossRef]

22. Luttmann, R.; Bracewell, D.G.; Cornelissen, G.; Gernaey, K.V.; Glassey, J.; Hass, V.C.; Kaiser, C.; Preusse, C.; Striedner, G.; Mandenius, C.-F. Soft sensors in bioprocessing: A status report and recommendations. *Biotechnol. J.* **2012**, *7*, 1040–1048. [CrossRef]

23. Morari, M.; Lee, J.H. Model predictive control: Past, present and future. *Comput. Chem. Eng.* **1999**, *23*, 667–682. [CrossRef]

24. Kroll, P.; Hofer, A.; Ulonska, S.; Kager, J.; Herwig, C. Model-Based Methods in the Biopharmaceutical Process Lifecycle. *Pharm. Res.* **2017**, *34*, 2596–2613. [CrossRef]

25. Udugama, I.A.; Lopez, P.C.; Gargalo, C.L.; Li, X.; Bayer, C.; Gernaey, K.V. Digital Twin in biomanufacturing: Challenges and opportunities towards its implementation. *Syst. Microbiol. Biomanuf.* **2021**. [CrossRef]

26. Kritzinger, W.; Karner, M.; Traar, G.; Henjes, J.; Sihn, W. Digital Twin in manufacturing: A categorical literature review and classification. *IFAC-PapersOnLine* **2018**, *51*, 1016–1022. [CrossRef]

27. Shahmohammadi, A.; McAuley, K.B. Using prior parameter knowledge in model-based design of experiments for pharmaceutical production. *AIChE J.* **2020**, *66*. [CrossRef]

28. Abt, V.; Barz, T.; Cruz, N.; Herwig, C.; Kroll, P.; Möller, J.; Pörtner, R.; Schenkendorf, R. Model-based tools for optimal experiments in bioprocess engineering. *Curr. Opin. Chem. Eng.* **2018**, *22*, 244–252. [CrossRef]

29. Smiatek, J.; Jung, A.; Bluhmki, E. Towards a Digital Bioprocess Replica: Computational Approaches in Biopharmaceutical Development and Manufacturing. *Trends Biotechnol.* **2020**, *38*, 1141–1153. [CrossRef]

30. Möller, J.; Kuchemüller, K.B.; Steinmetz, T.; Koopmann, K.S.; Pörtner, R. Model-assisted Design of Experiments as a concept for knowledge-based bioprocess development. *Bioprocess Biosyst. Eng.* **2019**, *42*, 867–882. [CrossRef]

31. Narayanan, H.; Luna, M.F.; von Stosch, M.; Cruz Bournazou, M.N.; Polotti, G.; Morbidelli, M.; Butté, A.; Sokolov, M. Bioprocessing in the Digital Age: The Role of Process Models. *Biotechnol. J.* **2020**, *15*, 1–10. [CrossRef]

32. von Stosch, M.; Willis, M.J. Intensified Design of Experiments for upstream bioreactors. *Eng. Life Sci.* **2016**, *17*, 1173–1184. [CrossRef]

33. Cserjan-Puschmann, M.; Kramer, W.; Duerrschmid, E.; Striedner, G.; Bayer, K. Metabolic approaches for the optimisation of recombinant fermentation processes. *Appl. Microbiol. Biotechnol.* **1999**, *53*, 43–50. [CrossRef]

34. Porstmann, T.; Wietschke, R.; Schmechta, H.; Grunow, R.; Porstmann, B.; Bleiber, R.; Pergande, M.; Stachat, S.; von Baehr, R. A rapid and sensitive enzyme immunoassay for Cu/Zn superoxide dismutase with polyclonal and monoclonal antibodies. *Clin. Chim. Acta* **1988**, *171*, 1–10. [CrossRef]

35. Marisch, K.; Bayer, K.; Cserjan-Puschmann, M.; Luchner, M.; Striedner, G. Evaluation of three industrial Escherichia coli strains in fed-batch cultivations during high-level SOD protein production. *Microb. Cell Fact.* **2013**, *12*, 58. [CrossRef]

36. Luchner, M.; Striedner, G.; Cserjan-Puschmann, M.; Strobl, F.; Bayer, K. Online prediction of product titer and solubility of recombinant proteins in Escherichia coli fed-batch cultivations. *J. Chem. Technol. Biotechnol.* **2015**, *90*, 283–290. [CrossRef]

37. Melcher, M.; Scharl, T.; Spangl, B.; Luchner, M.; Cserjan, M.; Bayer, K.; Leisch, F.; Striedner, G. The potential of random forest and neural networks for biomass and recombinant protein modeling in Escherichia coli fed-batch fermentations. *Biotechnol. J.* **2015**, *10*, 1770–1782. [CrossRef] [PubMed]

38. Bayer, B.; Striedner, G.; Duerkop, M. Hybrid Modeling and Intensified DoE: An Approach to Accelerate Upstream Process Characterization. *Biotechnol. J.* **2020**, *15*. [CrossRef] [PubMed]

39. Mercier, S.M.; Diepenbroek, B.; Wijffels, R.H.; Streefland, M. Multivariate PAT solutions for biopharmaceutical cultivation: Current progress and limitations. *Trends Biotechnol.* **2014**, *32*, 329–336. [CrossRef] [PubMed]

40. Cardillo, A.G.; Castellanos, M.M.; Desailly, B.; Dessoy, S.; Mariti, M.; Portela, R.M.C.; Scutella, B.; von Stosch, M.; Tomba, E.; Varsakelis, C. Towards in silico Process Modeling for Vaccines. *Trends Biotechnol.* **2021**, 1–11. [CrossRef]

Article

Designing Robust Biotechnological Processes Regarding Variabilities Using Multi-Objective Optimization Applied to a Biopharmaceutical Seed Train Desig

Tanja Hernández Rodríguez [1], Anton Sekulic [2], Markus Lange-Hegermann [2] and Björn Frahm [1,*]

1   Biotechnology and Bioprocess Engineering, Ostwestfalen-Lippe University of Applied Sciences and Arts, 32657 Lemgo, Germany; tanja.hernandez@th-owl.de
2   inIT—Institute Industrial IT, Ostwestfalen-Lippe University of Applied Sciences and Arts, 32657 Lemgo, Germany; anton@sekulic.org (A.S.); markus.lange-hegermann@th-owl.de (M.L.-H.)
*   Correspondence: bjoern.frahm@th-owl.de

Abstract: Development and optimization of biopharmaceutical production processes with cell cultures is cost- and time-consuming and often performed rather empirically. Efficient optimization of multiple objectives such as process time, viable cell density, number of operating steps & cultivation scales, required medium, amount of product as well as product quality depicts a promising approach. This contribution presents a workflow which couples uncertainty-based upstream simulation and Bayes optimization using Gaussian processes. Its application is demonstrated in a simulation case study for a relevant industrial task in process development, the design of a robust cell culture expansion process (seed train), meaning that despite uncertainties and variabilities concerning cell growth, low variations of viable cell density during the seed train are obtained. Compared to a non-optimized reference seed train, the optimized process showed much lower deviation rates regarding viable cell densities (<10% instead of 41.7%) using five or four shake flask scales and seed train duration could be reduced by 56 h from 576 h to 520 h. Overall, it is shown that applying Bayes optimization allows for optimization of a multi-objective optimization function with several optimizable input variables and under a considerable amount of constraints with a low computational effort. This approach provides the potential to be used in the form of a decision tool, e.g., for the choice of an optimal and robust seed train design or for further optimization tasks within process development.

Keywords: Gaussian processes; Bayes optimization; Pareto optimization; multi-objective; cell culture; seed train

Citation: Hernández Rodríguez, T.; Sekulic, A.; Lange-Hegermann, M.; Frahm, B. Designing Robust Biotechnological Processes Regarding Variabilities Using Multi-Objective Optimization Applied to a Biopharmaceutical Seed Train Design. *Processes* **2022**, *10*, 883. https://doi.org/10.3390/pr10050883

Academic Editor: Alina Pyka-Pająk

Received: 6 April 2022
Accepted: 21 April 2022
Published: 29 April 2022

Publisher's Note: MDPI stays neutral with regard to jurisdictional claims in published maps and institutional affiliations.

1. Introduction

The development and optimization of biopharmaceutical production processes with cell cultures is cost- and time-consuming, requiring substantial lab work. This necessitates thorough planning of experiments and processes, taking into account existing process knowledge. The need for model-based decision support in biopharmaceutical manufacturing has been emphasized by the US Food and Drug Administration (FDA) [1,2], including taking into account available prior know-how and experience within the decision process and uncertainties [3]. Such methods are still not state-of-the-art for cell culture processes during development or manufacturing [3,4], although first approaches have been proposed, for example, in order to optimize the titer of a mammalian cell culture process [5]. This highlights a need for improved methods and tools for optimal experimental design, optimal and robust process design and process optimization for the purposes of monitoring and controlling during manufacturing.

But also in other engineering fields such as chemical engineering or mechanical engineering, process optimization plays an important role and is the subject of current research. Some application examples rely on dynamic models, an example is the optimization of sustainable algal production processes [6] or the improvement of the vibration performance of cold orbital forging machines [7]. Other approaches rely on machine-learning algorithms such as those reported in [8–10].

The optimization of one objective criterion (e.g., final titer) is relatively straight forward, i.e., building an objective function with a unique response variable and applying an appropriate optimization algorithm to maximize this function. However, in industry, it is typically desired to optimize several conflicting objectives at a time, leading to suitable trade-offs and compromises. For example, when trying to maximize final titer via viable cell density while minimizing cultivation time. Multi-objective optimization provides a decision-making tool for optimal decisions in the presence of trade-offs between two or more conflicting criteria.

However, multi-objective optimization is more challenging. Its application is still not state-of-the-art in the context of cell culture processes, probably due to a lack of related studies and instructions. Moreover, within the manufacturing life cycle of biopharmaceuticals, some phases are better investigated than others. Still very few investigations are reported concerning the cell expansion process (seed train). It consists of several consecutive cultivation and passaging (transfer) steps, starting with a small amount of cell suspension because cells are frozen in small vials until they are used for a production process. The goal is to expand the number of viable cells in order to reach the required amount to inoculate (start) the production bioreactor (e.g., 10,000 L at industrial scale) while keeping them in a healthy and growing state. A high amount of operational requirements and constraints have to be fulfilled and, as reported in literature [11,12], the cell expansion process critically effects product quality and the amount of product at production scale. In [12] for example, the passage duration, as well as the initial viable cell density for each passage are reported as important parameters with high impact on process time and productivity at production scale. A careful and optimal planning of a seed train is therefore essential. However, this is not a trivial task due to the inherent variability concerning cell growth (cell growth differs from cell line to cell line and also from cultivation run to cultivation run) and uncertainty about the real state of the process due to considerable measurement uncertainties. This requires the design of a reproducible process which is robust regarding viable cell density, meaning that despite (initial) variabilities concerning cell growth, low variations of viable cell density at the end of the seed train are obtained. The goal of this paper is to close the gap between state-of-the-art optimization techniques and modern techniques from machine learning to improve the biopharmaceutical production by allowing easy to use yet powerful multi-objective optimization.

In most multi-objective optimization problems, no single best (unique optimal) solution exists, instead there is a set of optimal solutions (also called Pareto optimal solutions or non-dominated solutions), meaning for each solution that one criterion cannot be improved without degrading at least one of the other criteria. So, the decision maker has to choose from the set of non-dominated solutions according to the most preferred or important objective criterion. A promising approach to optimize objective functions, which are expensive to evaluate, is Bayes optimization. The methodology of Bayes optimization dates back to the work of Harold Kushner in 1964 [13] and gained impact through the work of Jones et al. in 1998 [14]. It is a probabilistic global optimization method for finding the maximum of objective functions that are expensive to evaluate or unknown (black-box) objective functions that are approximated using simulations [15].

In practice, the objective function could be the outcome of interest of a process, for example, process productivity or control metrics to describe the quality of a product. Input parameters can be process parameters needed to be optimized. Bayesian optimization [16] creates a quick to evaluate model, the so-called surrogate model of the objective function. In order to reduce the objective function evaluations, the surrogate model is iteratively trained and updated on new data. The positions of this new data are chosen by finding a trade-off between exploration (improving the surrogate model) and exploitation (finding optimal points). Typical surrogate models are Gaussian processes.

Gaussian processes (GP's) are popular machine learning models [17] because, due to their Bayesian nature, they work well with few data points [18]. Furthermore, they allow the inclusion of expert knowledge [19,20] and can be used in dynamic systems [6,21]. GP's are very flexible non-parametric models, hence, they can approximate any function and do not assume a predefined set of modeling functions.

Bayes optimization is successfully applied in many fields of research and economics [22]. Moreover, applications of Bayes optimization in the field of bioprocess engineering were published during the last decade [6,9,23,24]. Furthermore, this methodology was shown to be efficient in solving multi-objective optimization problems [25] and has also been applied for parameter estimation of kinetic parameters [26]. However, no applications are reported so far applying model-based multi-objective Bayes optimization within biopharmaceutical process development.

This contribution aims to present the concept of a workflow which couples uncertainty-based upstream simulation and Bayes optimization using Gaussian processes and its application in the form of a simulation case study to illustrate its applicability to a relevant industrial task in process development.

This simulation case study addresses the question if a reference seed train setup comprising five shake flask scales can be optimized through varying shake flask volumes and how many shake flask scales, three, four or five, are recommendable in terms of two objective criteria, seed train duration and deviation rate. Moreover it is investigated how the results change if cells grow with 5% lower or 5% higher maximum cell-specific growth rate.

Afterwards, two more objective criteria, titer (product concentration) and viability after 8 days in the production bioreactor, are added and seed train optimization is performed regarding four objective criteria simultaneously.

Furthermore, the suitability of the proposed method and the required number of iterations is evaluated with respect to the obtained information gain.

2. Methods

The main components of the applied methodology and the corresponding tools are described.

2.1. Upstream Simulation

Upstream simulation comprises a simulation of the cell expansion process (seed train) and simulation of the production scale. The reference upstream process taken as an application example for the here presented simulation case study comprises five consecutive shake flask scales followed by three bioreactor scales and one production scale, similar to the upstream process investigated in [27]. Further specifications are listed in Table 1.

A mathematical model is required, describing cell growth and interactions with the main limiting substrates and eventually inhibiting metabolites over time. A cell growth model, a system of ordinary differential equations (ode), already adapted to an industrial cell culture upstream process using a CHO cell line [27] has been used, which describes the dynamic behavior of viable and total cell density, X_v and X_t, concentrations of glucose c_{Glc}, glutamine c_{Gln}, lactate c_{Lac}, ammonia c_{Amm} and product (volumetric titer) c_{titer} (see Table A1 in the Appendix A).

Moreover, such an upstream process includes several constraints, operation steps and process parameters (e.g., concerning passaging intervals, substrate/nutrient concentrations, initial viable cell densities and viable cell densities before transferring cells into the next cultivation vessel, as well as the amount of cell suspension and fresh medium), which have to be considered in the simulation workflow. A detailed description of the required components and calculation routines are described in [28,29].

Besides these requirements, several passaging strategies can be applied, helping to decide at which point in time cells should be transferred from one cultivation vessel into the next larger one and how to perform these passaging steps (e.g., which amount of cell suspension should be mixed with how much fresh cell culture medium).

For the here presented simulation study, the passaging strategy for robust seed train design was chosen, where robustness refers to the reproducibility of the seed train regarding viable cell density, meaning that despite initial uncertainties and variabilities concerning cell growth, low variations of viable cell density at the end of the seed train are obtained. This strategy grounds on the objective of reaching the previously determined threshold of viable cell density and corresponding probability distributions of viable cell density at different points in time. These distributions are used in combination with a utility function following the mean-variance principle, which grounds on the Markowitz mean-variance portfolio optimization theory [30,31]: The utility function $U(t)$ is defined as a function of viable cell density X_v including the expected value $E(X_v)$ and the variance $\mathrm{Var}(X_v)$ of viable cell density, as well as a risk aversion parameter α which controls the amount of risk (amount of uncertainty) the user is willing to bear. In the here presented example, the risk refers to the probability that viable cell density differs from the expected value (predicted mean). A risk aversion value of $\alpha = 1$ would mean that the expected time profile minus one time the standard deviation of X_v is considered.

The utility function is defined through:

$$U(t) = E(X_v(t)) - \alpha\sqrt{\mathrm{Var}(X_v(t))} \tag{1}$$

Based on the simulated time profiles of the current cultivation scale (by solving the corresponding ode system), Equation (1) is used to calculate the utility function value $U(t)$ per hour and to check if this value reaches or exceeds the required transfer viable cell density $X_{v,\mathrm{transfer}}$ which is necessary to inoculate (start) the next cultivation scale fulfilling the required seeding (initial) viable cell density and the filling volume.

In the next step, it is evaluated whether the calculated point in time lies within the range of practically feasible points in time for cell passaging, T_p. Thus, the objective is to find the minimum point in time out of the set of practically feasible points in time for passaging, T_p, which fulfills:

$$U(t) \geq X_{v,\mathrm{transfer}}, \tag{2}$$
$$\text{subject to:} \quad t \in T_p. \tag{3}$$

Based on the obtained point in time and the corresponding concentrations of viable cells, total cells, substrates and metabolites at this point in time, starting concentrations (=initial values of the system of ordinary differential equations) of the next cultivation scale are calculated based on the defined configurations and constraints (e.g., working volumes, acceptable range of seeding viable cell density and medium concentrations). This calculation has to be performed for every cultivation scale and passaging step. For more details refer to [27–29].

2.2. Bayes Optimization

A typical mathematical optimization problem is the following: Given an objective function $f : \mathcal{X} \to \mathbb{R}$ over input space $\mathcal{X} \subseteq \mathbb{R}^d$, the aim is to find an argument $x^* \in \mathcal{X}$, which optimizes (minimizes or maximizes) f.

The idea behind Bayes Optimization consists of creating a simple, probabilistic and cheap to evaluate model, a so-called surrogate model (substitute model), of the objective function f [15,17,32]. Bayesian optimization reduces the number of evaluations of the objective f via the following iterative approach: Before sampling f at another point, we take into account a trade-off between exploration (i.e., sampling of areas of high uncertainties) and exploitation (sampling from areas which are likely to move towards the optimum), which is encoded in a so-called acquisition function. We can find such points quickly from evaluation of the surrogate model.

Within Bayes optimization the following steps are performed:

1. Generate a set of initial points and evaluate the objective function at these points.
2. Train the surrogate model based on all evaluated points.
3. Optimize the acquisition function, which determines the next candidate point x_c to be evaluated.
4. Compute $f(x_c)$, the objective function f at the candidate point x_c.
5. Repeat steps 2–4 for N iterations

The key of Bayesian optimization is not to rely on local approximations as many other optimization algorithms and instead to have a global viewpoint of also evaluating the function at unknown positions.

The acquisition function is used to propose the next candidate point to be evaluated based on specific criteria, for example the expected improvement of the optimization criteria, and on the reduction in predictive uncertainty. As in the case of the kernels, there is also a wide variety of possible acquisition functions to choose from. In this study, the Expected Improvement (EI) acquisition function is used [33,34].

Gaussian processes (GPs) are well suited surrogate models when making few assumptions [15]. Just like a Gaussian distribution (a normal probability distribution) is fully described by its mean m and variance σ^2, a GP is fully described by a mean function $m(x)$ and a covariance function $k(x; x')$ [17]. A GP is an extension of a multivariate Gaussian (or normal) distribution to distributions of functions in the sense that if a function y follows a GP distribution, i.e., $y \sim \mathcal{GP}(m, k)$, then every evaluation of the function follows a Gaussian distribution $y(x) \sim \mathcal{N}(m(x), k(x, x))$. In particular, a GP returns mean and variance of the possible function values (instead of just returning a scalar), and hence also provides information about the uncertainty of a prediction. Moreover, GPs can take into account uncertainty in the form of noise, the class of Gaussian processes is closed under Bayesian updates, and such updates are computationally tractable [35].

The covariance function describes the assumed characteristics such as smoothness or periodicity of the objective function f [16]. They are so-called positive-definite functions, often also called kernels [17,36]. It specifies the relationship between two 'points' (vector of the input space) x and x' and the corresponding changes in f at these points. A covariance function is described by a set of parameters, also called hyperparameters, describing a specific behavior. This is how prior information is embedded in the Bayes optimization procedure. Also in this work, the most commonly used covariance function, the Squared Exponential (SE) kernel (often also referred to as Gaussian kernel) is used [32].

2.3. Problem Definition and Computational Procedure

The goal of the presented application example is to propose a concept and a numerical procedure for optimal robust seed train design, where robustness refers to the reproducibility of the seed train regarding viable cell density, meaning that despite initial uncertainties and variabilities concerning cell growth, low variations of viable cell density at the end of the seed train are obtained.

First, seed train constraints are defined based on a chosen cell line and its characteristics concerning optimal cultivation conditions and based on the operative possibilities (e.g., feasible points in time for cell passaging). Second, the optimizable input parameters and objective criteria (objective response variables) applied in this study are defined (as also illustrated in Figure 1), followed by the formulation of the mathematical optimization problem. Thereafter, the optimization problem is solved using a workflow which connects seed train simulation and Bayes optimization.

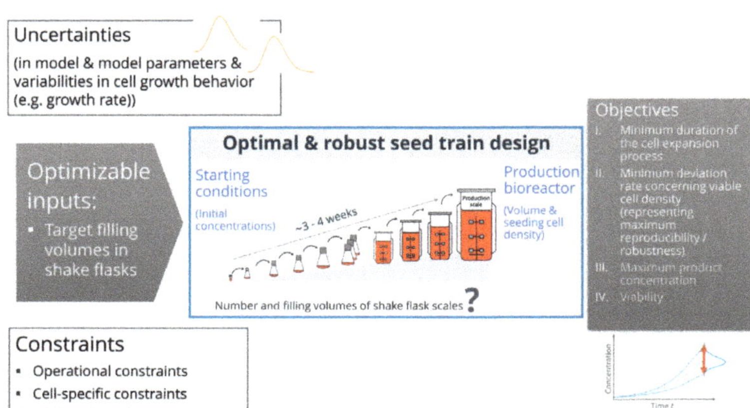

Figure 1. Goal of the study is to propose a concept and a numerical framework for optimal robust seed train design (blue box in the middle), including optimizable inputs (first gray box) as well as objectives (objective criteria) used in this study (right gray box).

The following objective criteria were chosen to represent an optimal seed train: (I) a minimum duration (d) (=required cultivation time) of the seed train and (II) a minimum deviation rate (D) regarding viable cell density, i.e., the probability that the seed train will run outside predefined ranges of viable cell density (for both, seeding viable cell density and transfer viable cell density) (This is important to consider because in the case that specific constraints are not fulfilled, the performance of the cells could decrease. The growth rate could decrease and, furthermore, it has been observed that the violation of constraints could also cause less viability of the cells in the production phase [12]) (compare to Figure 1 right gray box). These two attributes shall enable an optimal start of the production scale. Note that, in addition to these criteria, the growth rate is another important parameter affecting an optimal start of the production scale and the growth rate should be high until the end of the seed train. However, in this first optimization study it is not set as optimization criterion because the here defined seed train setup (in terms of medium concentrations and possible cultivation volumes per scale) together with the aim to reduce cultivation time already supports good growth during the entire cultivation. However, for other seed train setups, it might be advisable to include growth rate at the end of the seed train into the optimization problem.

After consideration of the two mentioned objective criteria, a third and fourth objective criterion, the product concentration (titer) and the viability at the end of the cultivation in the production bioreactor (in this simulation study: after 8 days in batch mode, i.e., without addition of nutrient feeds) are added to the optimization problem (see Figure 1 right gray box (III)). Note: The authors are aware of the fact that cultivation in the production vessel itself, which is often performed in fed-batch mode, is also influenced by several process parameters having an impact on product quantity and quality. Moreover, data of further attributes would be necessary to describe product quality (e.g., of a recombinant therapeutic protein or antibody) but these are not provided and therewith not considered in this study.

The input variables that can be varied to optimize the recently mentioned objective criteria, and thus the optimizable input variables, are the filling volumes in the first five shake flask scales, $V_1, \ldots, V_5$ (compare to Figure 1, the part of the seed train between thawing cells from a small vial and inoculation of the first biorector). These target values are important inputs of the seed train simulation process because they are used to calculate points in time for cell passaging. Volumes in the finally proposed seed train protocol (output of the seed train simulation) may vary within allowed working volume ranges and these are also presented in this work.

Formulation of the Mathematical Optimization Problem

The optimizable variables and therewith inputs of the optimization problem are the filling volumes of the n shake flask scales, $V_1, \ldots, V_n$ which are included in the input vector:

$$x = (V_1, \ldots, V_n)^T. \tag{4}$$

Outputs of the optimization problem are the defined objective criteria. These are seed train duration d and deviation rate D for the first optimization example. Thus, the unknown objective function (which should be minimized) can be written as follows:

$$f(x) = (f_1(x), f_2(x))^T \tag{5}$$

with $f_1(x) \triangleq d$ and $f_2(x) \triangleq D$.

The second optimization example includes a third and fourth optimization criterion, product concentration and viability at the end of the production scale (here after 8 days in the production vessel). Thus $f(x)$ expands to:

$$f(x) = (f_1(x), f_2(x), f_3(x), f_4(x))^T \tag{6}$$

with $f_1(x) \triangleq d$, $f_2(x) \triangleq D$, $f_3(x) \triangleq c_{\text{titer,end}}$ and $f_4(x) \triangleq \text{Viability}_{\text{end}}$.

2.4. Connecting Seed Train Simulation and Bayes Optimization

Uncertainty-based seed train simulation as described in Section 2.1 was coupled with algorithms for Bayes optimization as described in Section 2.2. The workflow integrating both components is illustrated in Figure 2. The inputs of the combined framework are the input variables: Boundaries for the optimizable variables (here filling volumes) and objective criteria (here seed train duration, deviation rate and in the second example also product concentration at the end of production scale) given all required seed train configuration settings and constraints (e.g., initial concentrations, practically feasible points in time for cell passaging, acceptable ranges for viable cell density, ...).

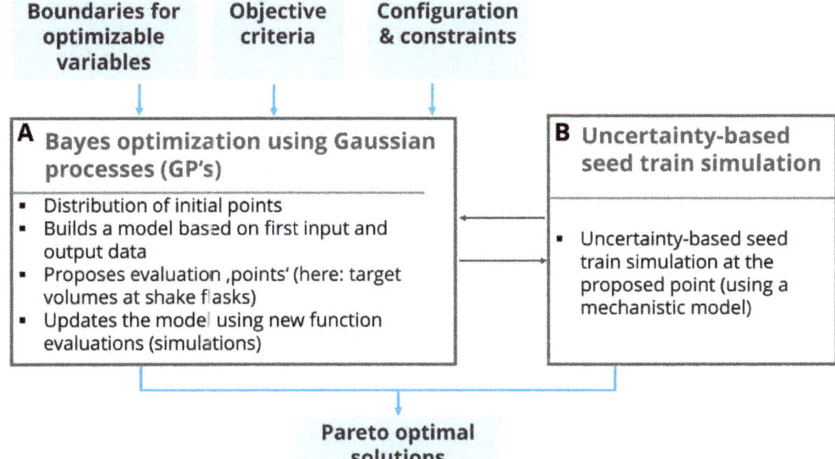

Figure 2. Scheme showing the applied computational workflow comprising: (**A**) a Bayes optimization algorithm which is coupled with (**B**) a seed train simulation routine. Input and output values are shown in the blue boxes above and below.

First points (=combinations of optimizable variables) are determined using a Latin Hypercube design distributing these points within the design space (see Figure 2, Box A). Seed train simulations are performed at these points in order to obtain the corresponding objective criteria values. Input values together with output values form a data set. An unknown model describing the relationship between inputs and outputs is approximated through a Gaussian process (GP) which has to be trained (see Figure 2, Box A) based on the given data set. Therefore, the Gaussian process proposes a point that has to be evaluated next (see Figure 2, Box B).

A robust seed train is simulated, using a mechanistic process model, and the objective criteria are calculated. This output is then returned to the Bayes optimization (Box A) to update the GP. Usually, experiments are performed to return the experimental output. The present approach instead exploits the advantages of the model-based upstream simulation in order to reduce the experimental effort to a minimum.

These steps are repeated various times, e.g., until a previously defined number of maximum iteration steps is reached. The latter depends on the resources (human and financial resources in case of laboratory experiments or computational resources in case of in silico experiments). In every iteration the Gaussian process chooses a new point aiming to move to the optimum and at the same time to reduce model uncertainty.

Results of this optimization framework are the set of Pareto optimal setups (also called Pareto front) and their corresponding response values.

2.5. Numerical Solvers and Tools

The programming language and numeric computing environment MATLAB [37] was used for the seed train simulations. The code for the optimization workflow was written in Python [38] using the MATLAB Engine API for Python to call MATLAB as a computational engine from Python code. To perform Bayes optimization within this workflow, the library GPflow [39] was used.

3. Results and Discussion

3.1. Optimization of Cultivation Vessels Regarding Number of Shake Flask Scales and Filling Volumes for Five, Four and Three Shake Flask Scales

In this section, it is investigated which cultivation filling volumes should be used for the flask scales in order to obtain optimal results in terms of seed train duration and deviation rate, here defined as the probability that the seed train will run outside the predefined acceptable ranges for initial viable cell density (VCD) and transfer VCD (final VCD before transfer into the next cultivation vessel) per scale. The latter is a measure for the robustness of the seed train regarding viable cell density.

For assessment of the optimization results, a conventional reference seed train comprising five shake flask scales was simulated based on a non-optimized design. Therefore, a common passaging interval of 3 days per cultivation scale was fixed and filling volumes were determined following a conservative layout (i.e., choosing not too huge differences between one cultivation scale and the next to ensure that enough viable cells are generated even if they grow a little bit slower than expected).

In the first step, the optimal combination of filling volumes for five shake flask scales is investigated and the results are compared to the reference seed train. Afterwards, it is investigated if a reduction in shake flask scales from five to four or three shake flask scales leads to similar or even better results in terms of seed train duration and deviation rate. The number of bioreactor scales was kept fixed. Three bioreactors with filling volumes of 40 L, 320 L and 2100 L were used as pre-stages before inoculation of the production bioreactor with 9600 L. The assumed seed train setup is given in Table 1.

Table 1. Specification of the exemplary seed train setup providing information concerning cultivation vessels, required viable cell densities and the transfer of cells from one cultivation vessel into the next larger one, assumed in this work.

Seed Train Setup	
Flask scales:	3, 4 or 5 flask scales between 0.014 L and 8 L filling volume
Bioreactor scales:	3 bioreactor scales, 38 L, 302 L and 2054 L filling volume
Production bioreactor:	9500 L filling volume
Optimal range for viable seeding cell density:	3×10^8–3.5×10^8 cells L^{-1} (3×10^5–3.5×10^5 cells mL^{-1})
Optimal range for transfer viable cell density:	0.1×10^{10}–1×10^{10} cells L^{-1} (0.1×10^7–1×10^7 cells mL^{-1})
Target seeding (initial) viable cell density:	3.15×10^8 cells L^{-1} (3.15×10^5 cells mL^{-1}) (=minimum viable seeding VCD + 5%)
Strategy concerning point in time for cell passaging:	'Xv transfer', i.e., passaging as soon as the calculated required viable transfer cell density is reached
Practically feasible points in time for passaging:	Passaging between 48 and 120 h possible (flexible ranges)
Strategy concerning current and new volume:	Discard cell suspension during the passaging step, if required to start within an optimal seeding cell density range

To find the optimal solution, multi-objective Bayesian optimization coupled with uncertainty-based seed train simulation, as described in Section 2.1, was applied. First, a Latin hypercube design for n_{lhs} design 'points' (combinations of filling volumes, here $n_{lhs} = 10$) was initiated and seed train simulation was applied to calculate the objective criteria values, here, deviation rate D and seed train duration d (replacing the normally required experimental cultivation runs) at each point. Within the Bayes optimization procedure, Gaussian processes were trained based on the simulation outcomes and an acquisition function was calculated in each iteration step in order to propose which point should be evaluated next. The input space for shake flask filling volumes (here the optimizable variables) was defined as described in Table 2, assuming the possibility of using several shake flasks in parallel for one shake flask scale and also considering their working volumes ranges.

Table 2. Input space for the shake flask filling volumes, containing the possible filling volumes per scale, given for optimization runs with 5, 4 or 3 shake flask scales.

	Filling Volumes		
	Range for 5 Shake Flask Scale [L]	Range for 4 Shake Flask Scales [L]	Range for 3 Shake Flask Scales [L]
V1	0.014–0.015	0.014–0.015	0.014–0.015
V2	0.05–0.15	0.1–1	0.1–2
V3	0.15–1.5	1.5–4	4–8
V4	1.5–4	4–8	-
V5	4–8	-	-

3.1.1. Optimization of Five Shake Flask Scales

The first optimization was performed for a seed train comprising five shake flask scales. Figure 3 shows the objective criteria values for each evaluated point, whereby the outcomes based on the initial Latin hypercube space are illustrated by blue dots and the outcomes for the proposed points based on the trained Gaussian processes are illustrated through yellow crosses. The optimal solutions are those near to the lower left corner aiming to minimize seed train duration and the deviation rate. The Pareto optimal solutions, also called non-dominated solutions, are illustrated through green circles. A solution (seed train setup/combination of filling volumes) is called non-dominated if no solution exists leading to better (here lower) objective criteria values. As described previously, several Pareto optimal solutions can be obtained because when considering two or more objective criteria then for two different solutions one criterion might have better (here lower) value then the other solution for the same objective, while the other criterion has worse (here higher) values. The set of all Pareto optimal solutions is called Pareto front.

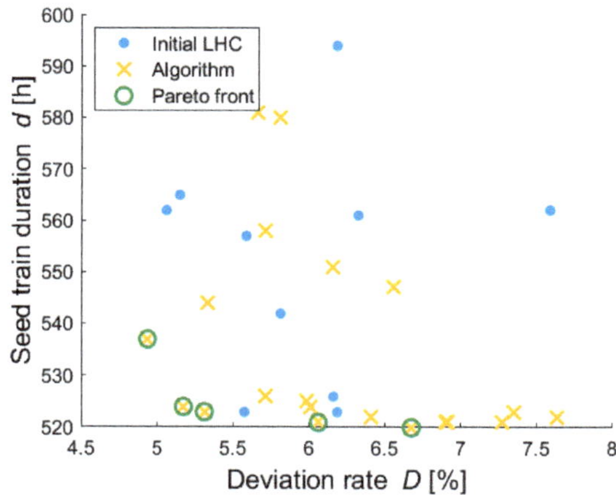

Figure 3. Algorthmically determined solutions and Pareto front of the two objective criteria seed train duration and deviation rate. (optimizable variables, here combinations of 5 shake flask filling volumes). Blue dots show an initial Latin hypercube design (LHC); yellow crosses are the points proposed by the algorithm; green circles are Pareto optimal solutions (=Pareto front).

For the investigated scenario (five shake flask scales and the seed train configuration according to Table 1) five Pareto optimal solutions were obtained (see green circles in Figure 3). It can be seen that comparing two of these solutions (green circles) each, one solution has a lower (here better) seed train duration value than the other solution and the opposite holds for the deviation rate.

The corresponding values for the optimizable variables, here shake flask filling volumes (V_1, V_2, V_3, V_4 and V_5), and the corresponding objective criteria values, here deviation rate D and seed train duration d, are listed in Table 3.

Table 3. Pareto optimal solutions concerning the choice of filling volumes in shake flask scales, for three scenarios for 5 flask scales. The following bioreactor filling volumes are 40 L, 320 L and 2210 L. The averaged filling volumes in L and the resulting deviation rate (D) in % and seed train duration (d) in h are listed for each Pareto optimal solution.

	Filling Volumes						
Solution	Vol. 1 [L]	Vol. 2 [L]	Vol. 3 [L]	Vol. 4 [L]	Vol. 5 [L]	D [%]	d [h]
1	0.015	0.065	0.904	2.355	7.78	537	4.9
2	0.015	0.115	0.451	1.672	7.89	521	6.1
3	0.015	0.104	0.340	1.614	6.85	524	5.2
4	0.014	0.103	0.369	1.582	7.87	523	5.3
5	0.015	0.114	0.431	2.026	7.97	520	6.7
	Filling volumes of reference seed train						
Reference	0.015	0.08	0.30	2	4	41.7	576

The filling volume of the first scale was limited to a very narrow range (14–15 mL) (A higher variation after cell thawing was not expected). Most obtained solutions start with the maximum value of this range (see Table 3, first column). The filling volume of flask scale 2 varies between 0.065 and 0.115 L, the filling volume of flask scale 3 between 0.340 and 0.904 L, the filling volume of flask scale 4 between 1.582 and 2.355 L and of flask scale 5 between 6.85 and 7.97 L. All five combinations lead to a deviation rate D of less than 7% and to a seed train duration between 520 to 537 h.

A more detailed illustration of the obtained results is presented in Figures 4 and 5. For two optimizable variables and one objective criterion each (deviation rate in Figure 4 and seed train duration in Figure 5), a contour plot is shown which illustrates the objective value for each calculated point (combination of the two variables), using the trained Gaussian processes, through colored isolines.

For example, the diagram in the top left of Figure 4 shows the deviation rate for each combination of V_1 (filling volume in flask scale 1) and V_2 (filling volume in flask scale 2) through colors representing the corresponding values in %, as indicated in the color bar. The results obtained through seed train simulations are shown by dots. The red dots represent the non-dominated (optimal solutions), optimal with respect to the defined multi-objective optimization problem. The dark blue area indicates combinations of V_1 and V_2 leading to a lower deviation rate. It can be seen that values above 0.1 for V_2 combined with any value of V_1 (within the given range) lead to the lowest deviation rates (below 6.2%, see dark blue area). Moreover, the optimal solutions (red dots) are mostly located in the area with higher filling volumes for shake flask 2, V_2, except one (red dot at $V_2 \approx 0.065$).

For some combinations, a closer delimitation is possible. For example, the middle diagram in the second row (V_3 over V_2) shows a limited region (dark blue area) and therewith a specific combination of V_3 and V_2 that leads to the lowest deviation rates (<5.6%). These are around 0.3 L for V_3 and around 0.105 L for V_2. Furthermore, two optimal solutions (red dots) out of the set of Pareto optimal solutions (considering both objective criteria, seed train duration and deviation rate) are located in this region. The remaining red dots are located outside of the dark blue regions (see turquoise regions in the same diagram), meaning that they have higher deviation rates. Analogously, Figure 5 shows the contour plots for the second objective criterion, seed train duration. The dark blue areas show the combinations with the lowest seed train durations (approximately below 528 h). It can be seen in these diagrams that most red dots are located in the dark blue regions. For some combinations the dark blue areas are wider, distributed over several possible values for one variable, e.g., the diagram in the top center, top left, center, and center right.

Other combinations show narrower regions with low seed train durations as can be seen in the diagram showing V_4 over V_3. The lowest seed train duration is obtained for filling volumes between 1.5 and 2.5 L for shake flask 4 in combination with filling volumes between 0.2 and 0.8 L for shake flask 3.

Overall, these diagrams give an overview of the impact of two combined optimizable variables each on a specific objective criterion.

In addition to this information, simulated time profiles (predictive mean in green, 90% prediction bands in blue) of viable and total cell density as well as concentrations of glucose, glutamine, lactate and ammonium (see Figure 6) can be obtained for each solution, as well as a seed train protocol containing information about the calculated passaging intervals, amount of medium, etc.

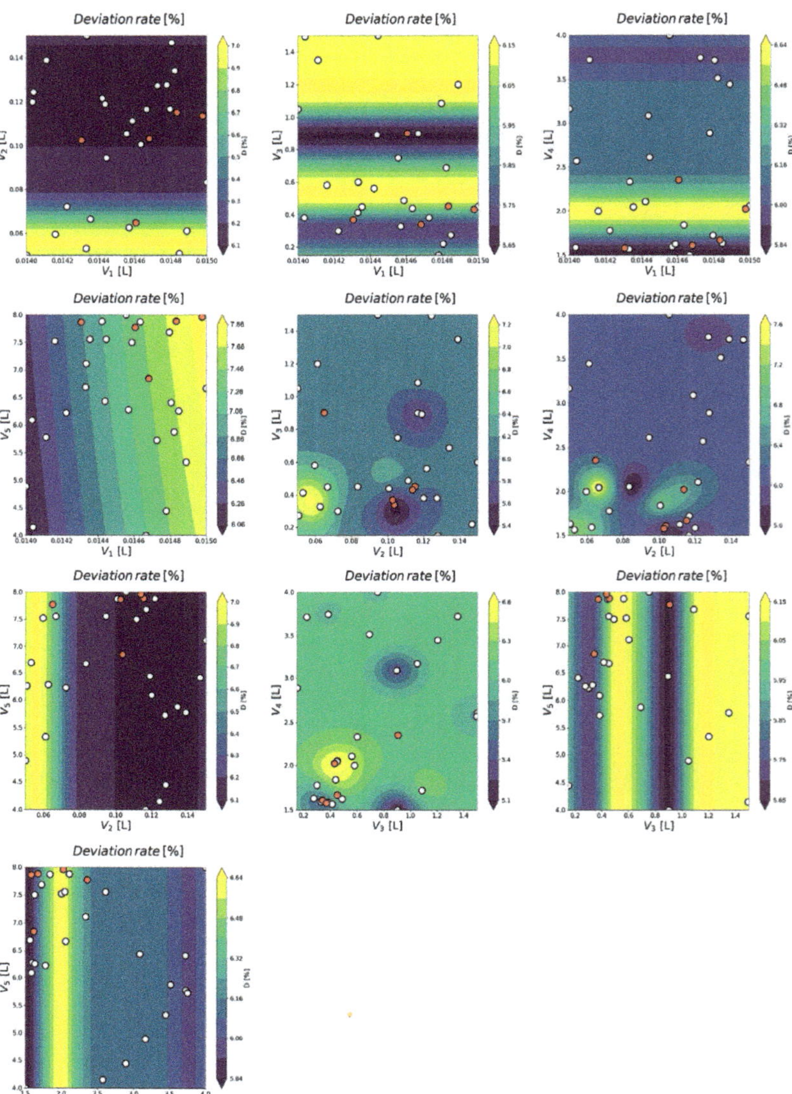

Figure 4. Contour plots showing two optimizable variables on x and y-axis and one objective (here Deviation rate D in %), assigned to each combination of the two variables, through colored isolines. For example, the diagram in the top left shows the deviation rate for each combination of V_1 (filling volume in flask scale 1) and V_2 (filling volume in flask scale 2) through colors representing the corresponding values in %, as indicated on the color bar. Moreover, the results obtained through seed train simulations are shown by dots. The red dots represent the non-dominated (optimal solutions).

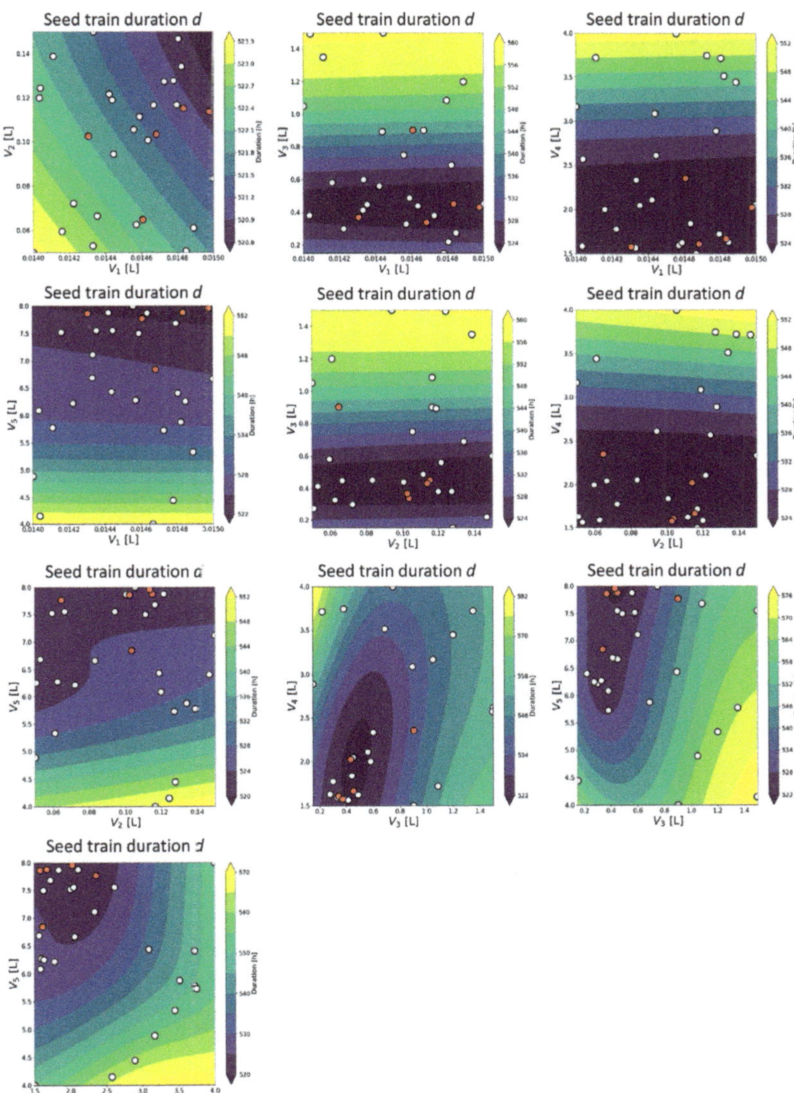

Figure 5. Contour plots showing two optimizable variables on x and y-axis and one objective (here seed train duration (d) in h), assigned to each combination of the two variables, through colored isolines. For example, the diagram top left shows the deviation rate for each combination of V_1 (filling volume in flask scale 1) and V_2 (filling volume in flask scale 2) through colors representing the corresponding values in %, as indicated on the color bar. Moreover, the results obtained through seed train simulations are shown by dots. The red dots represent the non-dominated (optimal solutions).

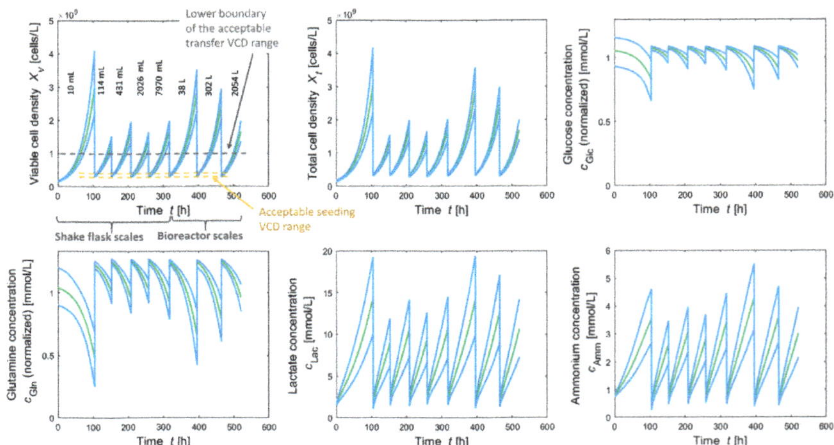

Figure 6. Seed train showing viable cell density (VCD) and total cell density, as well as substrate (glucose and glutamine) and metabolite (lactate and ammonium) concentrations over time and over the whole seed train (5 shake flask scales and three bioreactor scales), based on the shake flask filling volumes according to solution 1. The green lines represent the mean time course and the blue lines show the corresponding 90%-prediction band (5%- and 95%-quantiles). The plot (top left) also includes the filling volumes and the acceptable ranges for seeding VCD and transfer VCD, illustrated through dashed lines.

It can be seen in the top left of Figure 6 that based on the given filling volumes in addition to the flexibility to choose individual points in time for cell passaging in each scale, it is possible to set the seeding viable cell density at the beginning of each cultivation scale on the desired value with low variability, allowing to stay within the corresponding acceptable ranges for seeding VCD (see yellow dashed lines). Moreover, transfer VCDs lie within the corresponding acceptable range with high probability (see lower boundary, gray dashed line). Moreover, it can be seen that substrate concentrations are not depleted and according to [27], values of 20 mmol/L lactate and 5 mmol/L ammonium are not yet inhibiting concentrations for this cell line.

For a better assessment, the obtained results are compared to the reference seed train which is also defined in this work for five shake flask scales and illustrated in Figure 7. It grounds on a (non-optimized) configuration setup for five shake flask scales using fixed passaging intervals of 72 h each (common practice) and filling volumes of 15 mL (flask scale 1), 80 mL (flask scale 2), 300 mL (flask scale 3), 2000 mL (flask scale 4) and 4000 mL (flask scale 5). This choice grounds on a rather conservative approach aiming to avoid the risk of reaching too low transfer cell densities at the end of a cultivation scale but without the inclusion of probabilistic simulations.

The proposed method instead includes risk calculations and a passaging strategy aiming to minimize this risk but at the same time identifying a seed train configuration which is optimal regarding further objectives such as seed train duration in the present case.

A comparison of the seed train solutions obtained after optimization and the reference seed train shows that deviation rate is much lower after optimization (4.9–6.7% instead of 41.7%) and seed train duration could be reduced by 56 h from 576 h to 520 h. Figure 7, diagram top left shows where seeding or transfer viable cell density do not lie fully within the acceptable ranges (see red circles). This is different for the optimized solutions, e.g., solution 5, as illustrated in Figure 6, where seeding VCD lies within the acceptable range and also transfer VCD lies above the lower bound of the acceptable range for transfer VCD. This significant reduction in time ($\approx$2 days per seed train) would contribute to a meaningful acceleration of the production process.

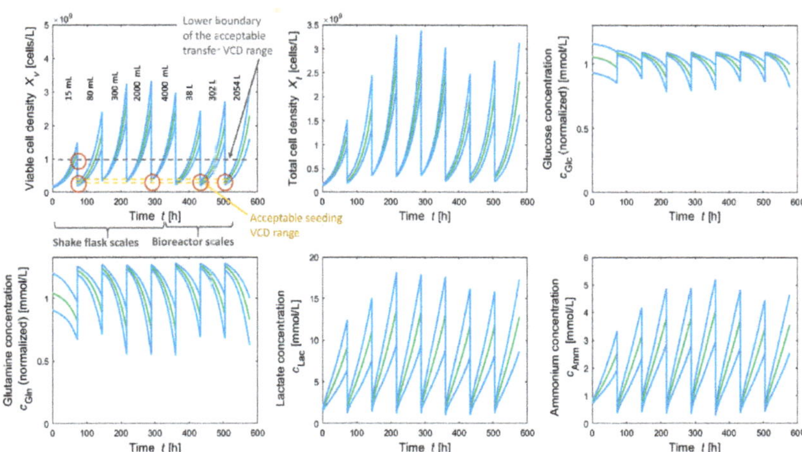

Figure 7. Reference (non-optimized) seed train showing viable and total cell density, as well as substrate (glucose and glutamine) and metabolite (lactate and ammonium) concentrations over time, based on a reference configuration setup for 5 shake flask scales using passaging intervals of 72 h each. The green lines represent the mean time course and the blue lines show the corresponding 90%-prediction band (5%- and 95%-quantiles).

3.1.2. Optimization of Three and Four Shake Flask Scales

In the next step, the number of shake flask scales was reduced from five to four and then to three shake flask scales and the same optimization procedure was applied. The aim was to investigate if less cultivation vessels would lead to comparable results and if so, which target and filling volumes should be chosen. This is of interest because less operations (such as transferring cells from one scale into another one) signify less risk of failure and deviations.

Figure 8 shows the obtained values for the objective criteria deviation rate and seed train duration for different combinations of filling volumes for three (left) and for four shake flask scales (right). Furthermore, here, the solutions based on the initial Latin hypercube design are shown by blue dots and Pareto optimal solutions are highlighted through green circles.

It can be seen that for both scenarios, combinations of filling volumes could be found leading to an overall seed train cultivation time between 519 and 530 h. However, the scenario of using four shake flask scales, leads to lower deviation rates ($D < 10\%$) compared to the scenario of using three shake flask scales ($23\% < D < 26\%$).

The corresponding filling volumes and the obtained filling volumes (based on the underlying passaging strategy) of the Pareto optimal solutions are listed in Table 4 together with the results for five shake flask scales from Table 3. The results are sorted as discovered by the optimization algorithm. The obtained filling volumes for four shake flask scales are very similar, except for shake flask scale 4 ($V_1 = 15$ mL, $V_2 = 158$–200 mL, $V_3 = 1.51$–1.60 L and $V_4 = 4.81$–7.58 L). Some of the obtained solutions would be seen or treated as equal in practice, because the differences are rather small. For example it would not be distinguished between 0.190 and 0.195 L. Probably 200 mL would be used instead. However, the applied optimization algorithm works on a continuous input space and differentiates between the solutions listed in the Table 4, even though the differences are very low. The obtained optimal filling volumes for three shake flask scales also look similar, but with a bit more variation for shake flask 3 (4.45–5.58 L).

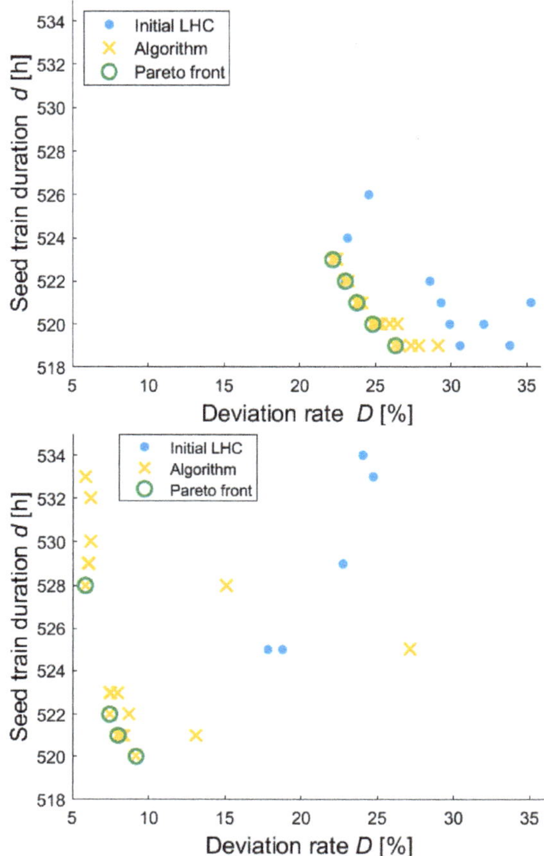

Figure 8. Algorithmically determined solutions and Pareto front regarding seed train duration and deviation rate for 3 resp. 4 shake flask scales on top resp. bottom. Blue dots show an initial Latin hypercube design (LHC); yellow crosses are the points proposed by the algorithm; green circles are Pareto optimal solutions (=Pareto front).

Comparing the results for the three scenarios (three, four and five shake flask scales) endorses a decision against the three flask scales-scenario due to the higher deviation rates (>20%), which stands for less process robustness. Between the other two scenarios (four or five shake flask scales) only little differences with respect to deviation rates are observed for the determined optimal solutions (4.9–6.7% for five shake flasks, 5.9–9.2% for four shake flasks). Using five shake flask scales would lead to more or less similar cultivations times (520–537 h) but one operational step more would be required.

This information, together with the corresponding seed train protocol, provides a solid basis to take a decision for one of the proposed optimal seed trains designs, taking into account seed train duration, robustness (expressed through deviation rates) and operational steps.

Table 4. Pareto optimal solutions concerning the choice of filling volumes in shake flask scales, for 3, 4 and 5 shake flask scales. The following bioreactor filling volumes are 40 L, 320 L and 2210 L. The averaged filling volumes in L, the resulting deviation rate (D) in % and seed train duration d in h are listed for each solution.

	Filling Volumes						
Solution	Vol. 1 [L]	Vol. 2 [L]	Vol. 3 [L]	Vol. 4 [L]	Vol. 5 [L]	D [%]	d [h]
	5 flask scales						
1	0.015	0.065	0.904	2.355	7.78	4.9	537
2	0.015	0.115	0.451	1.672	7.89	6.1	521
3	0.015	0.104	0.340	1.614	6.85	5.2	524
4	0.014	0.103	0.369	1.582	7.87	5.3	523
5	0.0015	0.114	0.431	2.026	7.97	6.7	520
	4 flask scales						
1	0.015	0.195	1.60	7.58		9.2	520
2	0.015	0.190	1.51	5.39		8.0	521
3	0.015	0.169	1.52	6.33		7.5	522
4	0.015	0.158	1.59	4.81		5.9	528
	3 flask scales						
1	0.015	0.733	4.45			23.0	522
2	0.015	1.046	4.85			23.8	521
3	0.015	1.103	5.26			24.8	520
4	0.015	0.934	4.77			23.0	522
5	0.015	1.110	4.65			22.2	523
6	0.015	1.306	5.58			26.4	519

3.2. Application to Further Cell Lines with Potentially Different Growth Rates

The optimization examples presented in the previous subsection were applied to a specific CHO cell line with growth characteristics described by a set of model parameters derived from an industrial cell culture process which was investigated in [27]. If a different cell line or a clonal cell population with potentially differing growth behavior is used, then the optimization has to be performed for this specific cell line. In the following simulation study, a cell line having a 5% lower and a cell line having a 5% higher maximum cell-specific growth rate compared to the reference maximum growth rate ($\mu_{max} = 0.028$ h^{-1} for the first bioreactor scale and $\mu_{max} = 0.029$ h^{-1} for the remaining seed train scales) are assumed and the optimization is applied for both scenarios.

The results for the obtained/proposed filling volumes, as well as the corresponding seed train duration and deviation rate are listed in Table 5.

As expected, cells which grow faster (higher maximum growth rate μ_{max}) would require less time until reaching a specific target cell density. This can be seen in the right column of Table 5. Using five flask scales, the optimal required seed train duration would lie between 494 and 503 h for a cell line with a 5% higher growth rate compared to the reference cell line which would need 520–537 h (see Table 3). Correspondingly, cells with a 5% lower growth rate would need more time (550–568 h). The same is observed when using four or three shake flasks.

With respect to the deviation rates which represent the robustness of the seed train design regarding variability of viable cells, it can be seen that low deviation rates of between 4.1% and 11.6% can be reached when using five or four flask scales, even if the maximum growth rate varies ±5%. A critical limit was identified for the combination of using three shake flask scales for a slower growing cell line. The corresponding optimal solution shows a comparatively higher deviation rate (19.2–29.1%) together with a high seed train duration (548–552 h).

Table 5. Pareto optimal solutions concerning the choice of filling volumes in shake flask scales, for 3, 4 and 5 shake flask scales, for two different scenarios. Scenario 1 assumes a 5% lower and scenario 2 a 5% higher cell-specific maximum growth rate compared to the reference maximum growth rate. The bioreactor filling volumes which follow after the shake flask scales are 40 L, 320 L and 2210 L. The filling volumes in L, the resulting deviation rate (D) in % and seed train duration (d) in h are listed for each solution (several Pareto optimal solutions can be obtained per setup).

			Filling Volumes				
Solution	Vol. 1 [L]	Vol. 2 [L]	Vol. 3 [L]	Vol. 4 [L]	Vol. 5 [L]	D [%]	d [h]
5 flask scales							
5% lower growth rate							
1	0.015	0.083	0.45	2.06	6.67	7.3	550
2	0.014	0.072	0.30	2.78	6.23	7.0	551
3	0.014	0.105	0.45	2.53	6.77	6.1	553
4	0.015	0.119	0.84	2.96	6.21	5.5	568
5	0.014	0.122	0.56	2.64	6.62	5.8	559
5% higher growth rate							
6	0.015	0.08	0.354	1.52	7.24	6.2	495
7	0.014	0.09	0.560	1.89	7.19	4.6	502
8	0.014	0.14	0.545	1.89	7.16	4.1	503
9	0.015	0.06	0.313	1.63	7.72	6.5	494
10	0.014	0.09	0.312	1.82	7.08	5.3	496
11	0.014	0.13	0.564	2.16	7.41	4.8	501
4 flask scales							
5% lower growth rate							
12	0.015	0.158	1.99	7.6		8.8	551
13	0.015	0.147	1.59	7.7		7.7	552
14	0.015	0.132	2.00	7.8		7.6	553
15	0.015	0.167	1.56	7.5		9.8	550
16	0.015	0.180	1.58	7.4		11.6	548
5% higher growth rate							
17	0.015	0.199	1.59	5.2		5.3	501
18	0.015	0.246	1.53	7.9		10.5	493
19	0.015	0.215	1.53	7.2		7.7	494
20	0.015	0.193	1.56	6.8		6.8	496
21	0.015	0.191	1.60	5.6		5.8	498
22	0.015	0.210	1.56	7.5		7.3	495
23	0.014	0.183	1.71	6.1		5.4	499
3 flask scales							
5% lower growth rate							
24	0.015	0.174	4.14			20.0	550
25	0.015	1.011	4.03			25.7	549
26	0.015	0.151	4.02			19.2	552
27	0.015	0.929	4.34			29.1	548
5% higher growth rate							
28	0.015	0.235	4.277			11.13	496
29	0.015	0.987	7.683			25.5	493
30	0.015	0.267	4.589			14.9	495

3.3. Optimization Regarding Four Objectives Including Product Concentration

To show the applicability of the proposed method to more than two objectives, a third and a fourth objective criterion, titer concentration and viability at the end of the production vessel (after 8 days) was added. Whereas the first two objective criteria (seed train duration and deviation rate) are related to the seed train itself, the third and fourth criterion refer to the generated product in the production vessel and to the viability of the cells in the production vessel. Product concentration, as well as product quality can be influenced by many factors (seeding cell density, substrate concentrations and nutrient feeds, metabolite production, temperature, pH, dissolved oxygen and carbon dioxide concentration, osmolality and more) and also by the amount and the state of the cells at the end of the seed train. Since no data describing product quality are available, product concentration and viability are considered in this study. A further simplification that was made is the assumption that the production vessel is performed in batch-mode (meaning without any addition of nutrient feeds or medium renewals). The reason for this simplification is to avoid confounding effects. The authors are aware of the fact that many factors affect product concentration and product quality and when data of other critical process parameters or quality attributes are available, these could also be considered in the same manner. The main purpose of the present simulation example is to demonstrate how the proposed method can be applied to more than two objectives and how the corresponding results can be illustrated and interpreted.

To obtain a visual overview for multiple objective criteria in one figure, a so-called spider plot (or net plot) can be used, which is shown in Figure 9.

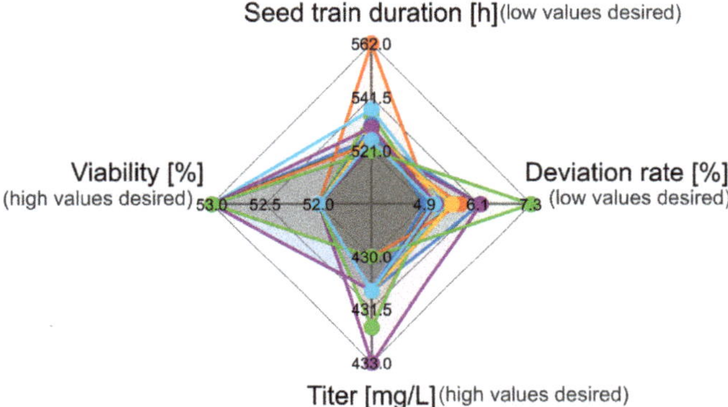

Figure 9. Spider plot showing the objective criteria values (seed train duration, deviation rate, titer and viability after 8 days in the production vessel) for the Pareto optimal solutions for 5 shake flask scales.

The horizontal axis shows the values of the deviation rate (on the right) and of the viability (on the left). The vertical axis shows the values of the seed train duration (above) and of the titer (below). The aim of the optimization was to minimize seed train duration and deviation rate and to maximize viability and titer. Each color (hyperplane) represents one of the Pareto optimal seed train configurations (based on the optimal combinations of filling volumes in shake flask scales). Since seed train duration and deviation rate should be minimal and titer and viability should be maximal, hyperplanes covering the lower left area would be desired. However, no such solution (hyperplane) was obtained. The reason is that the optimization problem contains conflicting objective criteria, meaning that an improvement of one criterion leads to a degradation of another criterion. The here presented solutions are all non-dominated (see the green circles in the figures for two objective criteria). For all shown solutions, the deviation rate is rather low (4.9–7.3%), the seed train duration lies between 521 and 562 h and a titer of approximately 430–433 mg/L (assuming here a cell-specific production rate of $q_{\text{titer,max}} = 3.9 \times 10^{-10}$ mg cell^{-1} h^{-1}, as reported in [40]) and a viability of 52–53% is reached after 8 days in the production vessel (here via batch-mode). Of course, the obtained values depend a lot on the real process conditions (production bioreactor probably performed in fed-batch model) and the model parameter values obtained after model validation. However, the presented simulation example shall illustrate how the proposed approach can be applied for risk-based decision making under consideration of several criteria that should be optimal.

3.4. Impact of Performed Iterations during Bayes Optimization

For the example of three shake flask scales, (followed by three bioreactor scales) and optimizing filling volumes for all shake flask scales with respect to the two objective criteria: seed train duration and deviation rate, the number of performed iterations during the optimization procedure was varied. First, 10 initial points (combinations of filling volumes) distributed based on a Latin hypercube design were evaluated, followed by 10 Bayes iterations, which means that 10 times the algorithm updates the black box model (the Gaussian process), calculates the acquisition function and proposes the next point based on the outcome of this calculation. Then, the optimization was performed again for the same seed train setup but using 20 and then 30 Bayes iterations. The obtained solutions are illustrated in Figure 10.

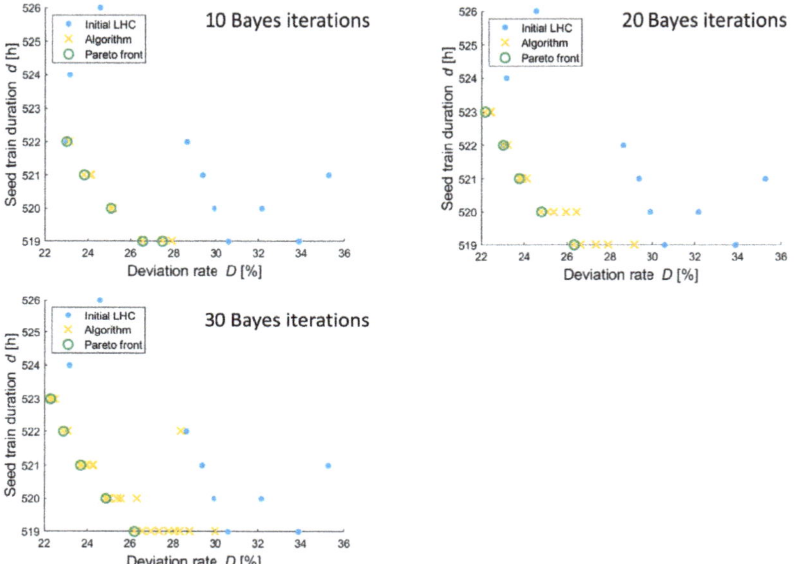

Figure 10. Algorthmically determined solutions and Pareto front of the two objective criteria seed train duration and deviation rate (optimizable variables, here combinations of 3 shake flask filling volumes) for 10 (**top left**), 20 (**top right**) and 30 (**bottom left**) Bayes iterations; blue dots show the initial Latin hypercube design (LHC); yellow crosses show the points proposed by the algorithm; green circles are the Pareto optimal solutions (=Pareto front).

Increasing the number of iterations from 10 to 20 helped to identify one solution that has not been discovered when running only 10 iterations. This can be seen when comparing the green circles in the diagram top left and the green circles in the diagram top right. The solution with $D \approx 22$ and $d = 523$ cannot been found in the diagram top left.

Increasing the number of iterations from 20 to 30 did not lead to an improved optimum as can be seen when comparing the green circles in the Figure 10 top right diagram and bottom left diagram. This underlines the efficiency of the Bayes optimization. In the present example, only 10 initial points (distributed randomly according to a Latin hypercube design) and 20 Bayes optimization iteration steps were required to obtain the results which were confirmed when applying 30 iteration steps.

3.5. Summary

The objective of the first optimization problem was to design a robust seed train (cell expansion process), which means a seed train layout (including the number of cultivation scales, filling volumes and passaging intervals) leading to a reproducible seed train with low variability regarding viable cell density and with a minimum seed train duration. The obtained solutions were compared to a non-optimized reference seed train and a comparison showed that the deviation rate is much lower after optimization (<10% instead of 41.7%) and seed train duration could be reduced by 56 h from 576 h to 520 h, which means a significant reduction of more than 2 days.

Addressing the question of if variation of the number of shake flask scales (and therewith the number of passaging steps) would lead to similar results in terms of deviation rates and seed train duration, it turned out that a reduction to three shake flask scales, would mean an increase in deviation rate and is therefore not recommended, at least under the assumed working volume ranges.

In industrial practice, typically more than one cell line is in use (different cell lines may be used to produce different molecules/products). Since growth rates of different cell lines differ, it was investigated how optimal seed train designs would differ for cell lines with 5% higher or lower growth rates. It turned out that the same optimization procedure could be easily adapted (by modification of the model parameter maximum growth rate) and applied to the modified setup revealing critical limits, e.g., for the combination of using three shake flask scales for a slower growing cell line. The latter shows comparatively high deviation rates (19.2–29.1%) together with high seed train durations (548–552 h instead of 519–523 h for the reference growth rate).

To show the applicability of the proposed method to more than two objective criteria, a third and fourth objective criterion, product concentration (titer) and viability after 8 days in the production phase, were added and the optimization was performed regarding four objective criteria in total. These are seed train duration, deviation rate (i.e., the probability that the seed train will run outside the predefined criteria), titer and viability at the end of the production phase.

Moreover, it was investigated for one seed train configuration (three shake flask scales and two objectives using the reference cell growth rate) if increasing the number of Bayes iterations would identify different optima. A number of 20 Bayes iterations turned out to be sufficient, because running 20 or 30 Bayes iterations showed similar results, which underlines the efficiency of the Bayes optimization approach.

In the present case study, the volumes are considered as fixed after optimization. If the production process allows for more flexibility in terms of adapting the volume within a specific range in the case that cells grow slower than the expected mean, then a reduction in the deviation rate can be achieved because varying the volumes allows for regulation of the inoculum viable cell density. However, this flexibility is not always given due to regulatory requirements and therefore not considered in the present study.

4. Conclusions

A concept has been developed to use process models in combination with algorithms for Bayes optimization using Gaussian processes to solve multi-objective optimization problems in the context of biopharmaceutical production processes. To illustrate this approach, a relevant exemplary optimization problem was chosen and solved using the proposed method.

The goal was to find optimal combinations of filling volumes for the shake flask scales of a seed train leading to a minimum deviation rate regarding viable cell densities and a minimum process duration. Compared to a non-optimized reference seed train, the optimized process showed much lower deviation rates regarding viable cell densities (<10% instead of 41.7%) using five or four shake flask scales and seed train duration could be reduced by 56 h from 576 h to 520 h.

Overall, it is shown that applying Bayes optimization to a multi-objective optimization function with several optimizable input variables and under a considerable amount of constraints, lead to revealing results with a low computational effort. This approach provides the potential to be used in form of a decision tool, e.g., for the choice of an optimal and robust seed train design but also to further optimization tasks within process development.

It should be noted that Bayes optimization and the corresponding computational modules could also be applied, even if no mechanistic process model is available, following a slightly different workflow. Instead of performing model-based in silico experiments (process simulations), real lab experiments would be performed and fed back to update the black box model (here the Gaussian process). This adaptive procedure (also called Bayesian experimental design or experimental design with Bayesian optimization [41]) or further related optimization methods might be promising tools to support experimental planning, process characterization, process transfer or optimization of cell culture processes but they still require further research and being embedded in software solutions that are easy to use for operators.

Author Contributions: Conceptualization, T.H.R., A.S., B.F. and M.L.-H.; methodology, T.H.R., A.S. and M.L.-H.; software, A.S. and T.H.R.; validation, A.S. and T.H.R.; formal analysis, T.H.R.; investigation, T.H, A.S., M.L.-H. and B.F.; resources, M.L.-H. and B.F.; data curation, T.H.R.; writing—original draft preparation, T.H.R.; writing—review and editing, A.S., B.F. and M.L.-H.; visualization, T.H.R. and A.S.; supervision, M.L.-H. and B.F.; project administration, T.H.R., B.F. and M.L.-H.; funding acquisition, B.F. and T.H.R. All authors have read and agreed to the published version of the manuscript.

Funding: The article processing charge (APC) was funded partially by Ostwestfalen-Lippe University of Applied Sciences and Arts (TH OWL).

Acknowledgments: The authors would like to express special thanks to Christoph Posch (Novartis Technical Research and Development) for the fruitful scientific exchange regarding the here presented case study. Moreover, we acknowledge support for the Open Access fees by Ostwestfalen-Lippe University of Applied Sciences and Arts (TH OWL) in the funding program Open Access Publishing.

Conflicts of Interest: All authors T.H.R., A.S., M.L.-H. and B.F. do not have any conflict of interest.

Abbreviations

The following abbreviations are used in this manuscript:

CHO	Chinese hamster ovary
EI	Expected improvement
FDA	Food and Drug Administration
GP	Gaussian process
LHS	Latin hypercube sampling
ode	Ordinary differential equations
SE	Squared exponential
VCD	Viable cell density

List of symbols

α	Risk aversion parameter (-)
μ	Cell-specific growth rate (h^{-1})
μ_d	Cell-specific death rate (h^{-1})
$\mu_{d,max}$	Maximum cell-specific death rate (h^{-1})
$\mu_{d,min}$	Minimum cell-specific death rate (h^{-1})
μ_{max}	Maximum cell-specific growth rate (h^{-1})
μ_{ref}	Reference maximum cell-specific growth rate (h^{-1})
σ^2	Variance
c_{Amm} ($c_{Amm,0}$)	(Initial) ammonia concentration (mmol L^{-1})

c_{Glc} $(c_{Glc,0})$	(Initial) glucose concentration (mmol L^{-1})
c_{Gln} $(c_{Gln,0})$	(Initial) glutamine concentration (mmol L^{-1})
c_{Lac} $(c_{Lac,0})$	(Initial) lactate concentration (mmol L^{-1})
c_{titer} $(c_{titer,0})$	(Initial) volumetric titer (product concentration) (mg L^{-1})
d	Dimension of the input space, number of optimizable variables = seed train duration (h)
D	Data, Deviation rate
$E(\cdot)$	Expectation value
f	Objective function (-)
f_i	Component i of a multidimensional objective function (-)
F_{sample}	Change of volume due to sampling (L h^{-1})
i	Running index (-)
k	Covariance function
K_{Amm}	Correction factor for ammonia uptake (-)
K_{Lys}	Cell lysis constant (h^{-1})
$K_{S,Glc}$	Monod kinetic constant for glucose (mmol L^{-1})
$K_{S,Gln}$	Monod kinetic constant for glutamine (mmol L^{-1})
k_{Glc}	Monod kinetic constant for glucose uptake (mmol L^{-1})
k_{Gln}	Monod kinetic constant for glutamine uptake (mmol L^{-1})
m $(m(\cdot))$	Mean (mean function)
n	Number of shake flasks (-)
N	Number of iterations (-)
$\mathcal{N}$	Normal distribution (-)
n_{lhs}	Number of latin hypercube points (-)
q_{Amm} $(q_{Amm,uptake,max})$	(Maximum) cell-specific ammonia uptake rate (mmol cell^{-1} h^{-1})
q_{Glc} $(q_{Glc,max})$	(Maximum) cell-specific glucose uptake rate (mmol cell^{-1} h^{-1})
q_{Gln} $(q_{Gln,max})$	(Maximum) cell-specific glutamine uptake rate (mmol cell^{-1} h^{-1})
q_{Lac} $(q_{Lac,uptake,max})$	(Maximum) cell-specific lactate uptake rate (mmol cell^{-1} h^{-1})
q_{titer} $(q_{titer,max})$	(Maximum) cell-specific product production rate (mg cell^{-1} h^{-1})
$\mathbb{R}$	Set of real number
t	Time (h)
T_p	Set of feasible points in time for passaging
$U(\cdot)$	Utility function
V	Volume (L)
V_i	Volume in shake flask scale i
$Var(\cdot)$	Variance
x_c	Candidate point
x, x'	Multidimensional points (vectors) of the input space
$x*$	Argument that maximizes $f(s)$
X_t	Total cell density (cells L^{-1})
X_v	Viable cell density (cells L^{-1})
$X_{v,i}$	Viable cell density at point in time with index i (cells L^{-1})
$\mathcal{X}$	Input space
y	Arbitrary function (-)
Y	Arbitrary random variable (-)
$Y_{Amm/Gln}$	Kinetic production constant for ammonia (mmol mmol^{-1})
$Y_{Lac/Glc}$	Kinetic production constant for lactate (mmol mmol^{-1})

Appendix A

Table A1. Mechanistic model [27,42–45] for description of cell growth, cell death, substrate uptake, metabolite production and antibody production applicable to batch and fed-batch mode.

Balance Equations	Kinetic Equations

Biomass

$$\frac{dX_v}{dt} = X_v \cdot (\mu - \mu_d) - \frac{F_{Glc} + F_{Gln} + F_{Medium}}{V} \cdot X_v$$

$$\mu = \mu_{max} \cdot \frac{c_{Glc}}{c_{Glc} + K_{S,Glc}} \cdot \frac{c_{Gln}}{c_{Gln} + K_{S,Gln}}, \text{if } t > t_{Lag}$$

$$\mu = \mu_{max} \cdot \frac{c_{Glc}}{c_{Glc} + K_{S,Glc}} \cdot \frac{c_{Gln}}{c_{Gln} + K_{S,Gln}} - \left(1 - \frac{t}{t_{Lag}}\right) \cdot a_{Lag} \cdot \mu_{max},$$
$$\text{if } t \leq t_{Lag}$$

$$\frac{dX_t}{dt} = X_v \cdot \mu - K_{Lys} \cdot (X_t - X_v)$$
$$- \frac{F_{Glc} + F_{Gln} + F_{Medium}}{V} \cdot X_t$$

$$\mu_d = \mu_{d,min} + \mu_{d,max} \cdot \frac{K_{S,Glc}}{K_{S,Glc} + c_{Glc}} \cdot \frac{K_{S,Gln}}{K_{S,Gln} + c_{Gln}}$$

Substrates

$$\frac{dc_{Glc}}{dt} = -X_v \cdot q_{Glc} + \frac{F_{Glc}}{V} \cdot c_{Glc,F} + \frac{F_{Medium}}{V} \cdot c_{Glc,Medium}$$
$$- \frac{F_{Glc} + F_{Gln} + F_{Medium}}{V} \cdot c_{Glc}$$

$$q_{Glc} = q_{Glc,max} \cdot \frac{c_{Glc}}{c_{Glc} + k_{Glc}}$$

$$\frac{dc_{Gln}}{dt} = -X_v \cdot q_{Gln} + \frac{F_{Gln}}{V} \cdot c_{Gln,F} + \frac{F_{Medium}}{V} \cdot c_{Gln,Medium}$$
$$- \frac{F_{Glc} + F_{Gln} + F_{Medium}}{V} \cdot c_{Gln}$$

$$q_{Gln} = q_{Gln,max} \cdot \frac{c_{Gln}}{c_{Gln} + k_{Gln}}$$

Metabolites

$$\frac{dc_{Lac}}{dt} = X_v \cdot q_{Lac} - \frac{F_{Glc} + F_{Gln} + F_{Medium}}{V} \cdot c_{Lac}$$

$$q_{Lac} = Y_{Lac/Glc} \cdot q_{Glc} \cdot \frac{c_{Glc}}{c_{Lac}} - q_{Lac,uptake} \cdot \frac{\mu_{max} - \mu}{\mu_{max}}$$
$$\text{with } q_{Lac,uptake} = 0, \text{if } c_{Glc} > 0.5 \text{ mmol L}^{-1}$$
$$\text{with } q_{Lac,uptake} = q_{Lac,uptake,max}, \text{if } c_{Glc} \leq 0.5 \text{ mmol L}^{-1}$$

$$\frac{dc_{Amm}}{dt} = X_v \cdot q_{Amm} - \frac{F_{Glc} + F_{Gln} + F_{Medium}}{V} \cdot c_{Amm}$$

$$q_{Amm} = Y_{Amm/Gln} \cdot q_{Gln} \cdot \frac{c_{Gln}}{c_{Amm}}$$
$$- K_{Amm} \cdot q_{Amm,uptake,max} \cdot \frac{\mu_{max} - \mu}{\mu_{max}}$$
$$\text{with } K_{Amm} = 0, \quad \text{if} \quad (c_{Gln} > c_{Amm})$$
$$\text{with } K_{Amm} = 1, \quad \text{if} \quad (c_{Gln} \leq c_{Amm}) \text{ and } (\mu > \mu_d)$$
$$\text{with } K_{Amm} = -k_{Amm} \quad \text{(constant), if } (\mu \leq \mu_d)$$

Product titer and volume

$$\frac{dc_{titer}}{dt} = X_v \cdot q_{titer} - \frac{F_{Glc} + F_{Gln} + F_{Medium}}{V} \cdot c_{titer}$$

$$q_{titer} = q_{titer,max}$$

$$\frac{dV}{dt} = -F_{Sample} + F_{Glc} + F_{Gln} + F_{Medium}$$

Appendix B

Application to Other Cell Lines with Potentially Higher and Lower Maximum Growth Rates

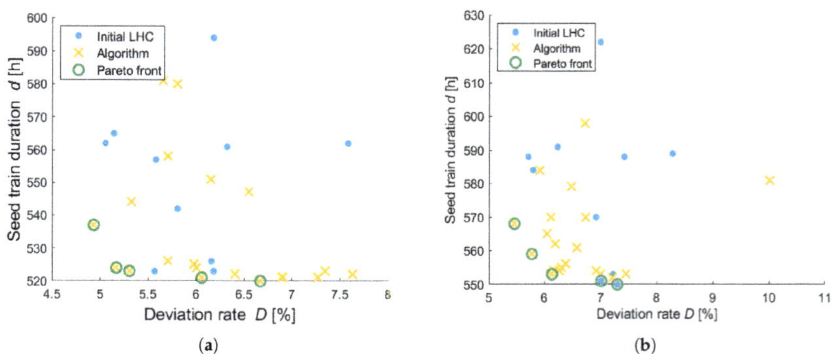

Figure A1. *Cont.*

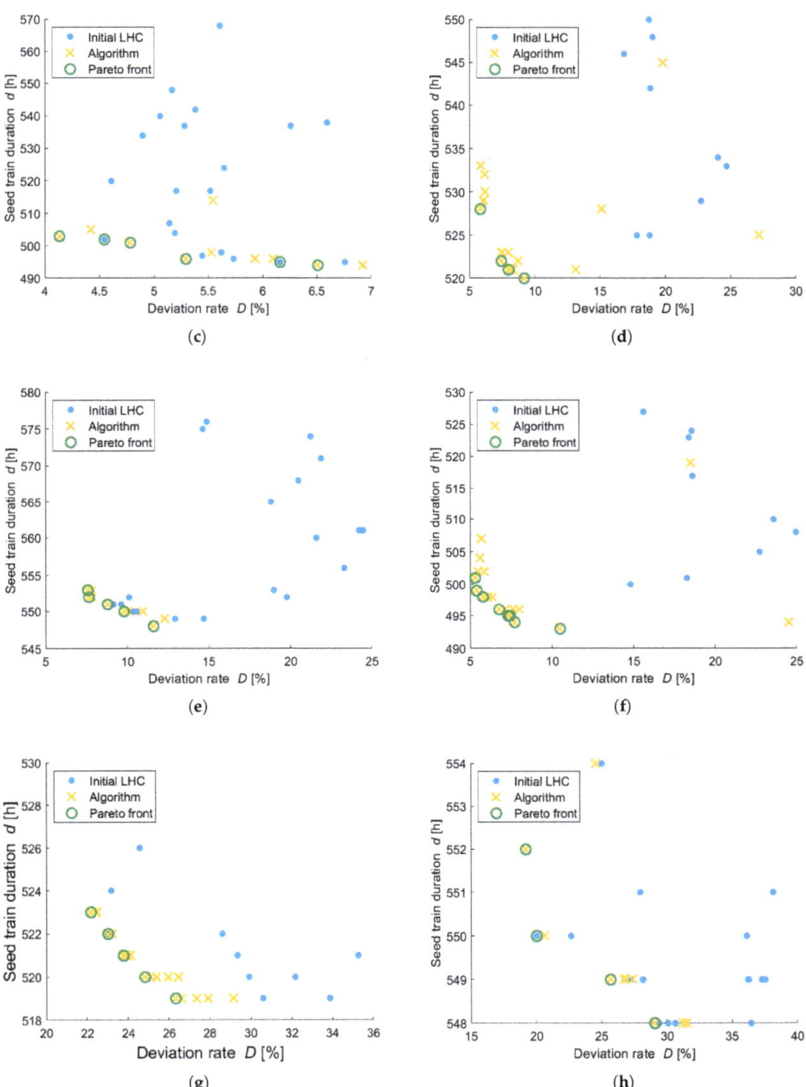

Figure A1. *Cont.*

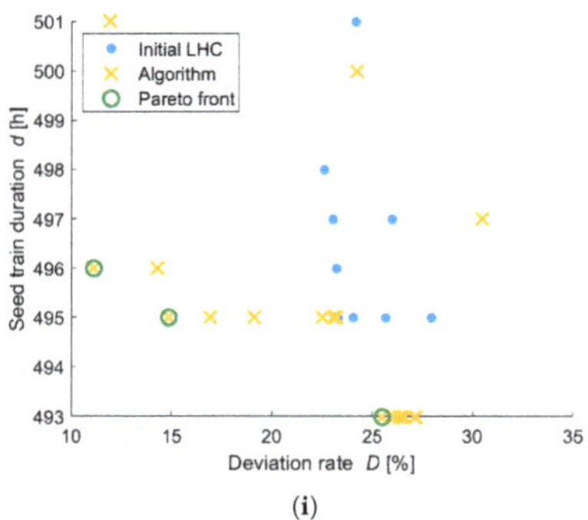

Figure A1. Pareto solutions for 3, 4 and 5 shake flask scales and for three different growth rates, reference maximum growth rate (left column), a 5% lower (middle column) and a 5% higher growth rates (right column) showing the objective criterion seed train duration over objective criterion deviation rate, using 20 optimization iterations; Blue dots: based on the initial Latin hypercube (LHC) design; Yellow crosses: based on the proposed points (by the algorithm); Green circles: Pareto optimal solutions. (**a**) 5 sf, $\mu_{,max,ref}$. (**b**) 5 sf, $\mu_{,max,95\%}$. (**c**) 5 sf, $\mu_{,max,105\%}$. (**d**) 4 sf, $\mu_{,max,ref}$. (**e**) 4 sf, $\mu_{,max,95\%}$. (**f**) 4 sf, $\mu_{,max,105\%}$. (**g**) 3 sf, $\mu_{,max,ref}$. (**h**) 3 sf, $\mu_{,max,95\%}$. (**i**) 3 sf, $\mu_{,max,105\%}$.

References

1. Herwig, C.; Garcia-Aponte, O.F.; Golabgir, A.; Rathore, A.S. Knowledge management in the QbD paradigm: Manufacturing of biotech therapeutics. *Trends Biotechnol.* **2015**, *33*, 381–387. [CrossRef] [PubMed]
2. U.S. Department of Health and Human Services; Food and Drug Administration. Guidance for Industry PAT—A Framework for Innovative Pharmaceutical Development, manufacturing, and Quality Assurance: U.S. Department of Health and Human Services, Food and Drug Administration: Guidance for industry: PAT—A Framework for Innovative Pharmaceutical Development, Manufacturing and Quality Assurance. Available online: https://www.fda.gov/regulatory-information/search-fda-guidance-documents/pat-framework-innovative-pharmaceutical-development-manufacturing-and-quality-assurance (accessed on 6 April 2022).
3. Sokolov, M. Decision Making and Risk Management in Biopharmaceutical Engineering—Opportunities in the Age of COVID-19 and Digitalization. *Ind. Eng. Chem. Res.* **2020**, *59*, 17587–17592. [CrossRef]
4. Xie, X.; Schenkendorf, R. Robust Process Design in Pharmaceutical Manufacturing under Batch-to-Batch Variation. *Processes* **2019**, *7*, 509. [CrossRef]
5. Liu, Y.; Gunawan, R. Bioprocess optimization under uncertainty using ensemble modeling. *J. Biotechnol.* **2017**, *244*, 34–44. [CrossRef] [PubMed]
6. Bradford, E.; Schweidtmann, A.M.; Zhang, D.; Jing, K.; del Rio-Chanona, E.A. Dynamic modeling and optimization of sustainable algal production with uncertainty using multivariate Gaussian processes. *Comput. Chem. Eng.* **2018**, *118*, 143–158. [CrossRef]
7. Hua, L.; Chen, M.; Han, X.; Zhang, X.; Zheng, F.; Zhuang, W. Research on the vibration model and vibration performance of cold orbital forging machines. *J. Eng. Manuf.* **2022**, *236*, 828–843. [CrossRef]
8. Schweidtmann, A.M.; Clayton, A.D.; Holmes, N.; Bradford, E.; Bourne, R.A.; Lapkin, A.A. Machine learning meets continuous flow chemistry: Automated optimization towards the Pareto front of multiple objectives. *Chem. Eng. J.* **2018**, *352*, 277–282. [CrossRef]
9. Clayton, A.D.; Schweidtmann, A.M.; Clemens, G.; Manson, J.A.; Taylor, C.J.; Niño, C.G.; Chamberlain, T.W.; Kapur, N.; Blacker, A.J.; Lapkin, A.A.; et al. Automated self-optimisation of multi-step reaction and separation processes using machine learning. *Chem. Eng. J.* **2020**, *384*, 123340. [CrossRef]
10. Rangaiah, G.P.; Feng, Z.; Hoadley, A.F. Multi-Objective Optimization Applications in Chemical Process Engineering: Tutorial and Review. *Processes* **2020**, *8*, 508. [CrossRef]
11. Le, H.; Kabbur, S.; Pollastrini, L.; Sun, Z.; Mills, K.; Johnson, K.; Karypis, G.; Hu, W.S. Multivariate analysis of cell culture bioprocess data–lactate consumption as process indicator. *J. Biotechnol.* **2012**, *162*, 210–223. [CrossRef]
12. Böhl, O.J.; Schellenberg, J.; Bahnemann, J.; Hitzmann, B.; Scheper, T.; Solle, D. Implementation of QbD strategies in the inoculum expansion of a mAb production process. *Eng. Life Sci.* **2020**, *27*, 196–207. [CrossRef] [PubMed]
13. Kushner, H.J. A New Method of Locating the Maximum Point of an Arbitrary Multipeak Curve in the Presence of Noise. *J. Basic Eng.* **1964**, *86*, 97–106. [CrossRef]
14. Jones, D.R.; Schonlau, M.; Welch, W.J. Efficient Global Optimization of Expensive Black-Box Functions. *J. Glob. Optim.* **1998**, *13*, 455–492. [CrossRef]

15. Brochu, E.; Cora, V.M.; Freitas, N.d. A Tutorial on Bayesian Optimization of Expensive Cost Functions, with Application to Active User Modeling and Hierarchical Reinforcement Learning. *arXiv* **2010**, arXiv:1012.2599.

16. Shahriari, B.; Swersky, K.; Wang, Z.; Adams, R.P.; Freitas, N.d. Taking the Human Out of the Loop: A Review of Bayesian Optimization. *Proc. IEEE* **2016**, *104*, 148–175. [CrossRef]

17. Rasmussen, C.E.; Williams, C.K.I. *Gaussian Processes for Machine Learning*, 3rd ed.; MIT Press: Cambridge, UK, 2008.

18. Tulsyan, A.; Garvin, C.; Undey, C. Industrial batch process monitoring with limited data. *J. Process. Control.* **2019**, *77*, 114–133. [CrossRef]

19. Lange-Hegermann, M. Algorithmic Linearly Constrained Gaussian Processes. In *Advances in Neural Information Processing Systems 31 (NeurIPS 2018)*; Bengio, S., Wallach, H., Larochelle, H., Grauman, K., Cesa-Bianchi, N., Garnett, R., Eds.; Curran Associates, Inc.: Red Hook, NY, USA, 2018; pp. 2137–2148.

20. Lange-Hegermann, M. Linearly constrained gaussian processes with boundary conditions. In Proceedings of the 24th International Conference on Artificial Intelligence and Statistics, San Diego, CA USA, 13–15 April 2021; pp. 1090–1098.

21. Bradford, E.; Imsland, L.; Zhang, D.; del Rio Chanona, E.A. Stochastic data-driven model predictive control using gaussian processes. *Comput. Chem. Eng.* **2020**, *139*, 106844. [CrossRef]

22. Petsagkourakis, P.; Sandoval, I.O.; Bradford, E.; Zhang, D.; Chanona, E.A.d.R. Constrained Reinforcement Learning for Dynamic Optimization under Uncertainty. *PapersOnLine* **2020**, *53*, 11264–11270. [CrossRef]

23. Bradford, E.; Schweidtmann, A.M.; Lapkin, A. Efficient multiobjective optimization employing Gaussian processes, spectral sampling and a genetic algorithm. *J. Glob. Optim.* **2018**, *71*, 407–438. [CrossRef]

24. Narayanan, H.; Stosch, M.; Luna, M.F.; Cruz Bournazou, M.N.; Buttè, A.; Sokolov, M. Consistent value creation from bioprocess data with customized algorithms: Opportunities beyond multivariate analysis. In *Process Control, Intensification, and Digitalisation in Continuous Biomanufacturing*; Subramanian, G., Ed.; Wiley and Sons: Hoboken, NJ, USA, 2022; Volume 36, pp. 231–264. [CrossRef]

25. Yang, K.; Emmerich, M.; Deutz, A.; Bäck, T. Multi-Objective Bayesian Global Optimization using expected hypervolume improvement gradient. *Swarm Evol. Comput.* **2019**, *44*, 945–956. [CrossRef]

26. Manheim, D.C.; Detwiler, R.L. Accurate and reliable estimation of kinetic parameters for environmental engineering applications: A global, multi objective, Bayesian optimization approach. *Methods X* **2019**, *6*, 1398–1414. [CrossRef] [PubMed]

27. Hernández Rodríguez, T.; Posch, C.; Schmutzhard, J.; Stettner, J.; Weihs, C.; Pörtner, R.; Frahm, B. Predicting industrial-scale cell culture seed trains-A Bayesian framework for model fitting and parameter estimation, dealing with uncertainty in measurements and model parameters, applied to a nonlinear kinetic cell culture model, using an MCMC method. *Biotechnol. Bioeng.* **2019**, *116*, 2944–2959. [CrossRef] [PubMed]

28. Hernández Rodríguez, T.; Frahm, B. Design, optimization, and adaptive control of cell culture seed trains. In *Animal Cell Biotechnology*; Pörtner, R., Ed.; Humana Press: New York, NY, USA, 2020; Volume 2095, pp. 251–267. [CrossRef]

29. Hernández Rodríguez, T.; Frahm, B. Digital Seed Train Twins and Statistical Methods. *Adv. Biochem. Eng.* **2020**, *9*, 964.

30. Kellerer, B. Portfolio Optimization and Ambiguity Aversion. *Jr. Manag. Sci.* **2019**, *4*, 305–338. [CrossRef]

31. Guerard, J.B. *Handbook of Portfolio Construction*; Springer: Boston, MA, USA, 2010. [CrossRef]

32. Frazier, P.I. A Tutorial on Bayesian Optimization. *arXiv* **2018**, arXiv:1807.02811.

33. Couckuyt, I.; Deschrijver, D.; Dhaene, T. Fast calculation of multiobjective probability of improvement and expected improvement criteria for Pareto optimization. *J. Glob. Optim.* **2014**, *60*, 575–594. [CrossRef]

34. Sekulic, A. Bayes'sche Optimierung von Multikriteriellen Zielfunktionen bei Zellkultur-Seed-Trains. Ph.D. Thesis, Ostwestfalen-Lippe University of Applied Sciences and Arts, Lemgo, Germany, 2020.

35. Snoek, J.; Larochelle, H.; Adams, R.P. Practical Bayesian Optimization of Machine Learning Algorithms. *arXiv* **2012**, arXiv:1206.2944.

36. Duvenaud, D. Automatic Model Construction with Gaussian Processes. Ph.D. Thesis, Apollo–University of Cambridge Repository, Cambridge, UK, 2014. [CrossRef]

37. MathWorks Inc. *MATLAB*, Version 9.7.0 (R2019b); The MathWorks Inc.: Natick, MA, USA, 2019.

38. van Rossum, G. *The Python Language Reference. Documentation for Python. Python Software Foundation*; Release 3.0.1 [Repr.] Ed.; SoHo Press: Hampton, WA, USA; Redwood City, CA, USA, 2010.

39. Knudde, N.; van der Herten, J.; Dhaene, T.; Couckuyt, I. GPflowOpt: A Bayesian Optimization Library Using TensorFlow. *arXiv* **2017**, arXiv:1711.03845.

40. Hernández Rodríguez, T.; Morerod, S.; Pörtner, R.; Wurm, F.M.; Frahm, B. Considerations of the Impacts of Cell-Specific Growth and Production Rate on Clone Selection—A Simulation Study. *Processes* **2021**, *9*, 964. [CrossRef]

41. Greenhill, S.; Rana, S.; Gupta, S.; Vellanki, P.; Venkatesh, S. Bayesian Optimization for Adaptive Experimental Design: A Review. *IEEE Access* **2020**, *8*, 13937–13948. [CrossRef]

42. Frahm, B. Seed train optimization for cell culture. In *Animal Cell Biotechnology*; Pörtner, R., Ed.; Humana Press: New York, NY, USA, 2014; Volume 1104, pp. 355–367. [CrossRef]

43. Kern, S.; Platas-Barradas, O.; Pörtner, R.; Frahm, B. Model-based strategy for cell culture seed train layout verified at lab scale. *Cytotechnology* **2016**, *68*, 1019–1032. [CrossRef] [PubMed]

44. Möller, J.; Kuchemüller, K.B.; Steinmetz, T.; Koopmann, K.S.; Pörtner, R. Model-assisted Design of Experiments as a concept for knowledge-based bioprocess development. *Bioprocess Biosyst. Eng.* **2019**, *42*, 867–882. [CrossRef] [PubMed]
45. Pörtner, R.; Platas Barradas, O.; Frahm, B.; Hass, V.C. Advanced process and control strategies for bioreactors. In *Current Developments in Biotechnology and Bioengineering*; Larroche, C., Ángeles Sanromán, M., Du, G., Pandey, A., Eds.; Elsevier: Amsterdam, The Netherlands, 2017; Volume 105, pp. 463–493. [CrossRef]

Article

Considerations of the Impacts of Cell-Specific Growth and Production Rate on Clone Selection—A Simulation Study

Tanja Hernández Rodríguez [1], Sophie Morerod [2], Ralf Pörtner [3], Florian M. Wurm [2,4] and Björn Frahm [1,*]

[1]   Biotechnology & Bioprocess Engineering, Ostwestfalen-Lippe University of Applied Sciences and Arts, 32657 Lemgo, Germany; tanja.hernandez@th-owl.de
[2]   ExcellGene SA, 1870 Monthey, Switzerland; sophie.morerod@excellgene.com (S.M.); florian.wurm@excellgene.com (F.M.W.)
[3]   Institute for Bioprocess and Biosystems Engineering, Hamburg University of Technology, 21071 Hamburg, Germany; poertner@tuhh.de
[4]   Life Science Faculty, École Polytechnique de Lausanne (EPFL), 1015 Lausanne, Switzerland
*   Correspondence: bjoern.frahm@th-owl.de

Abstract: For the manufacturing of complex biopharmaceuticals using bioreactors with cultivated mammalian cells, high product concentration is an important objective. The phenotype of the cells in a reactor plays an important role. Are clonal cell populations showing high cell-specific growth rates more favorable than cell lines with higher cell-specific productivities or vice versa? Five clonal Chinese hamster ovary cell populations were analyzed based on the data of a 3-month-stability study. We adapted a mechanistic cell culture model to the experimental data of one such clonally derived cell population. Uncertainties and prior knowledge concerning model parameters were considered using Bayesian parameter estimations. This model was used then to define an inoculum train protocol. Based on this, we subsequently simulated the impacts of differences in growth rates ($\pm10\%$) and production rates ($\pm10\%$ and $\pm50\%$) on the overall cultivation time, including making the inoculum train cultures; the final production phase, the volumetric titer in that bioreactor and the ratio of both, defined as overall process productivity. We showed thus unequivocally that growth rates have a higher impact (up to three times) on overall process productivity and for product output per year, whereas cells with higher productivity can potentially generate higher product concentrations in the production vessel.

Keywords: clonal cell population; phenotypic diversity; inoculum train; uncertainty-based; cell culture model; biopharmaceutical manufacturing

Citation: Hernández Rodríguez, T.; Morerod, S.; Pörtner, R.; Wurm, F.M.; Frahm, B. Considerations of the Impacts of Cell-Specific Growth and Production Rate on Clone Selection—A Simulation Study. *Processes* **2021**, *9*, 964. https://doi.org/10.3390/pr9060964

Academic Editor: Ewa Kaczorek

Received: 30 April 2021
Accepted: 24 May 2021
Published: 28 May 2021

Publisher's Note: MDPI stays neutral with regard to jurisdictional claims in published maps and institutional affiliations.

1. Introduction

For the production of certain biopharmaceuticals, animal cells have to be expanded from a frozen vial. Today, Chinese hamster ovary (CHO) cells are by far the most popular system in use [1] because they are known to be easy to grow; safe as far as not carrying any infectious agents; and last but not least, highly productive with yields in the multiple grams per liter range [2,3]. Nevertheless, questions and issues remain to be solved to maximize their utility, particularly since CHO cells have a very wide range of genotypic diversity and thus corresponding phenotypic differences [4–6]. This is quite obvious when clonally derived cell populations from a single transfection are compared against each other. These phenotypic differences have impacts on growth-related characteristics, cell-specific productivity and the quality of the final product (e.g., glycosylation patterns) [6,7]. Screening and profiling methods have been introduced (amongst others by [5,7]) to assess cell growth rate, cell-specific productivity and glycosylation patterns, along with further quality attributes of the produced recombinant proteins. In addition, the phenotypic stability of cell populations is another important parameter. A factor to maintaining stability over a reasonable time frame is the use of environmental conditions of cells within narrow and favorable ranges [2,3]. For the testing of the genetic and production stability, clonally derived cell lines, also in the following referred to as "clones" or "clonally derived cell populations", undergo typically stability studies which can last up to six months.

Desired phenotypic parameters to be maintained in such generated cell lines (besides properties characterizing the quality of the produced recombinant protein) are high cell-specific productivity [3,4,8,9] and high growth rates in order to reduce overall cultivation times (including the duration of the cell expansion process). However, growth rates and specific productivity of recombinant cells are often inversely related to each other [3]. Thus, a frequent trade-off has to be weighed between clonal populations, with one showing faster growth but a lower cell-specific production rate and vice versa (see Figure 1a), assuming little or no quality differences in the product obtained.

Mathematical process models appear to be suitable tools for analysis, for the generation of process understanding and for simulation and prediction. Several examples of using such process models addressing biopharmaceutical manufacturing can be found in the literature [10–16]. Within this field, uncertainty-based methods gained attention because model uncertainty, uncertainty in measurements and batch-to-batch variability can be taken into consideration in this way.

This study aims to present a model-based investigation of the impacts of clonal differences concerning cell-specific growth rates and cell-specific production rates on the duration of an inoculum train; the volumetric titer in production; and the overall process productivity, defined by the ratio of volumetric titer in production to the overall cultivation time, including the duration of the cell expansion process (inoculum train). In this study, a batch process (for simulation of the inoculum train and also for the production bioreactor) has been used for simulation and evaluation because it is a good first step to obtain results regarding the impacts of phenotypic differences on the above-described response values. This can be further expanded—once a smaller number of clonally derived cell lines have been chosen—to also involve fed-batch processes and/or perfusion mode.

The investigation is divided into four main blocks (see also Figure 1b(I–IV) for orientation).

I: Growth rate and production rate were analyzed for five clonal populations based on the data of a stability study (Section 3.1).

II: A reference cell line was taken from one of these and a mechanistic cell culture model was adapted to the data obtained in the laboratory (Section 3.2). Uncertainties were considered and prior knowledge from previous studies concerning model parameters was integrated into the model using Bayesian parameter estimation.

III: Upstream simulations were performed for three different clonal cell lines under consideration of the variabilities observed in Section 3.1. For each clonal cell line a suitable inoculum train protocol is defined. Furthermore, these inoculum train protocols are compared to each other with respect to inoculum train duration and volumetric titer in production (see Section 3.3).

IV: Several combinations of maximum growth rate ($\pm$10%) and maximum production rate ($\pm$10% and $\pm$50%) within realistic ranges were considered (Section 3.4). First, production rate was varied $\pm$10% for a multiple regression. Second, a variation of $\pm$50% was applied to cover all three investigated clones with their growth and production rates and to illustrate them in a response surface plot. Based on the results, a decision criterion is provided which is expected to help in evaluating different clonal cell lines.

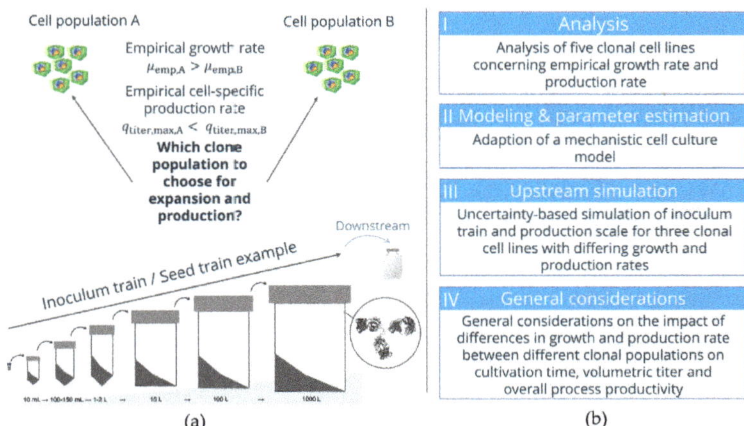

Figure 1. (**a**) Problem definition: Find the clone with a higher potential regarding inoculum train creation and final quantities of interest (volumetric titer in production and overall process productivity (=Space-Time-Yield; volumetric titer in production/overall cultivation time including inoculum train)). (**b**) Structure of the presented studies.

2. Materials and Methods

2.1. Data from a Stability Study

Experimental data from a stability study for analysis of cell-specific growth rates (in the following growth rates) and cell-specific production rates (in the following production rates) for five different clonal CHO populations (named clone 1, ..., clone 5) have been used for statistical analysis. The CHOExpress cells have been used as a host system, which are known to be moderate producers of ammonia. This is also based on the media formulations used in the work. Cells were cultivated with and without puromycin in duplicate runs in 50 mL OrbShake tubes (TubeSpin bioreactor 50™, TPP, Trasadingen, Switzerland, 30 mm diameter) with culture volumes of 5 mL. Each subcultivation was started with a viable cell density of $5 \cdot 10^5$ cells mL^{-1}. The cultures were shaken at 180 rpm in a Kühner SFX-1 incubator (Kühner AG, Birsfelden, Switzerland), set at a temperature of 37 °C and a CO_2 set-point of 5%. For each clonal population, cells have been cultivated and passaged (subcultivated) every 3 or 4 days during a time period of 13 weeks (25 subultivations in total). Measurements of volumetric product titer were taken 4 days after starting a new subcultivation for every second subcultivation. Viable cell densities were determined at the end of every subcultivation. The seeding density of $5 \cdot 10^5$ cells mL^{-1} was based on calculated dilutions into fresh medium. These data have been used for approximations of empirical growth rates and production rates according to Section 2.4.

2.2. Experimental Data and Set Up for Modeling Purposes

Batch culture experiments were performed by ExcellGene SA for clonal cell population 1 (clone 1) for modeling purposes. Cell expansion was carried out in volumes of 10 mL (TubeSpin bioreactor 50™, TPP, Trasadingen, Switzerland, 30 mm diameter), 50 mL (Erlenmeyer bottle—250 mL, 85 mm diameter) and 500 mL (TubeSpin bioreactor 600™, TPP, Trasadingen, Switzerland, 100 mm diameter) while shaking at 37 °C temperature, 180 rpm shaking speed, 80% humidity and 5% CO_2. Viable cell density and viability were determined using Guava easyCyte™ 5HT cytometry (Luminex Corporation, Austin, TX, USA) and glucose, glutamine, lactate and ammonia were measured using a NOVA Bioanalyzer (Nova Biomedical Corporation, Waltham, MA, US). At 10 mL and 50 mL scale, measurements of viable cell density and viability were performed on days 0, 1, 2, 3 and 4, and volumetric titer was measured on day 8. At the 500 mL scale of operation, measurements of viable cell density, viability, glutamine and ammonia were performed on days 0, 1, 2, 3, 4 and 8. Glucose and lactate were determined at days 0, 1, 2, 3 and 4 and by volumetric titer on days 4 and 8.

For clarity, in the following and throughout this paper, data presentation and discussions on cell cultures refer to experimental work with clonal cell lines by identifying these in numbers, i.e., clonal cell line # 1, 2, 3, or equivalent. In contrast, modelled cell lines are referred to as Clone A, B, etc.

2.3. Cultivation Systems for Inoculum Train Simulations

For inoculum train simulation, only vessel-types applicable for orbital shaking have been considered, with the expectation that the cultivation conditions were highly similar during cell expansion. These have been taken from the list reported in [17]. Based on the given working volumes, an inoculum train has been designed to include 5 scales from 10 mL to 100 L target volume and a production scale of 1000 L target volume (see Table 1).

Table 1. Cultivation systems, including working volumes per vessel.

Scale	Cylindrical Vessel	Working Volume per Scale [L]
1	TubeSpin bioreactor 50	0.001–0.035
2	Schott glass bottle (2 L)	0.4–1.8
3	Schott glass bottle (5 L)	0.5–4.5
4	OrbShake bioreactor prototype (50 L)	15
5	OrbShake bioreactor prototype (200 L)	100
6	OrbShake bioreactor prototype (2500 L)	1000

2.4. Approximations of Empirical Growth Rates and Production Rates

Based on data of viable cell densities of a clonal population, empirical (averaged) growth rates and empirical (averaged) production rates have been determined. The empirical growth rate μ_{emp} between two points in time t_i and t_{i+1} was calculated using the corresponding viable cell density values $X_{v,i}$ and $X_{v,i+1}$ according to

$$\mu_{\text{emp}}(t_i, t_{i+1}) = \frac{\ln X_{v,i+1} - \ln X_{v,i}}{t_{i+1} - t_i}. \tag{1}$$

The empirical production rate $q_{\text{titer,emp}}$ between two points in time t_i and t_{i+1} was calculated using the corresponding volumetric titer values $c_{\text{titer},i}$ and $c_{\text{titer},i+1}$ according to

$$q_{\text{titer,emp}}(t_i, t_{i+1}) = \frac{c_{\text{titer},i+1} - c_{\text{titer},i}}{(t_{i+1} - t_i) \cdot 0.5 \cdot (X_{v,i} + X_{v,i+1})}. \tag{2}$$

2.5. Statistical Testing of the Differences in Means between Clonal Cell Populations

Clonal populations have been analyzed regarding their growth rates and production rates by applying statistical tests to determine the differences between population means using the statistical software R [18]. Variance homogeneity was tested using the Bartlett test [19]. A global test on differences between population means was performed using the Brown and Forsythe F-test [20] (similar to the classical ANOVA but adapted for heterogeneous variances). To identify where the differences come from and to determine the differences between individual groups, post hoc tests have to be performed. When comparing more than two populations, a method for multiple testing containing an adjustment of the significance level is additionally required. Multiple testing methods exist for groups showing heterogeneous variances. In this work, a pairwise comparison was performed using the adjustment method by Benjamini and Yekutieli [21]. The applied statistical tests and R-commands are given in Figure 2.

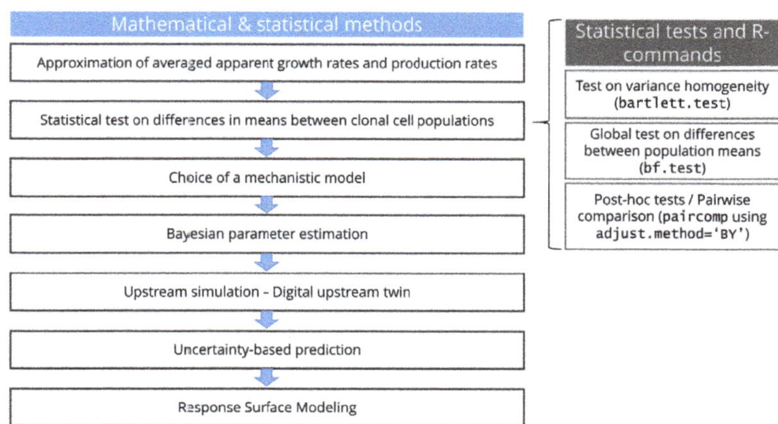

Figure 2. Summary of mathematical, statistical and computational methods applied in the presented study including the applied R-commands for statistical tests.

2.6. Mechanistic Model

The applied kinetic model is a modification of previous model variations published in [12,13,22,23]. Differential equations (see Table 2), consisting of nine mostly Monod-type algebraic equations (description of growth rate, death rate, substrate uptake, metabolite production kinetics and production rate) and 18 model parameters, describe the cell culture dynamics of total and viable cell density, X_t and X_v and concentrations of glucose c_{Glc}, glutamine c_{Gln}, limiting substrate c_{LS}, lactate c_{Lac}, ammonia c_{Amm} and volumetric titer c_{titer}. All these variables and model parameters are listed in Table A1, including units and descriptions.

Table 2. Mechanistic model (for batch and fed-batch mode) [12,13,22,23] for descriptions of cell growth, cell death, substrate uptake, metabolite production and antibody production.

Balance Equations	Kinetic Equations
Biomass $$\frac{dX_v}{dt} = X_v \cdot (\mu - \mu_d) - \frac{F_{Glc}+F_{Gln}+F_{Medium}}{V}$$ $$\frac{dX_t}{dt} = X_v \cdot \mu - K_{Lys} \cdot (X_t - X_v) - \frac{F_{Glc}+F_{Gln}+F_{Medium}}{V}$$ **Substrates and metabolites** $$\frac{dc_{Glc}}{dt} = -X_v \cdot q_{Glc} + \frac{F_{Glc} \cdot c_{Glc,F}}{V} + \frac{F_{Medium} \cdot c_{Glc,Medium}}{V} - \frac{F_{Glc}+F_{Gln}+F_{Medium}}{V}$$ $$\frac{dc_{Gln}}{dt} =$$ $$-X_v \cdot q_{Gln} + \frac{F_{Gln} \cdot c_{Gln,F}}{V} + \frac{F_{Medium} \cdot c_{Gln,Medium}}{V} - \frac{F_{Glc}+F_{Gln}+F_{Medium}}{V}$$ $$\frac{dc_{LS}}{dt} = -X_v \cdot q_{LS} - \frac{F_{Glc}+F_{Gln}+F_{Medium}}{V}$$ $$\frac{dc_{Lac}}{dt} = X_v \cdot q_{Lac} - \frac{F_{Glc}+F_{Gln}+F_{Medium}}{V}$$ $$\frac{dc_{Amm}}{dt} = X_v \cdot q_{Amm} - \frac{F_{Glc}+F_{Gln}+F_{Medium}}{V}$$ $$\frac{dc_{titer}}{dt} = X_v \cdot q_{titer} - \frac{F_{Glc}+F_{Gln}+F_{Medium}}{V}$$ $$\frac{dV}{dt} = -F_{Sample} + F_{Glc} + F_{Gln} + F_{Medium}$$	$$\mu = \mu_{max} \cdot \frac{c_{Glc}}{c_{Glc}+K_{S,Glc}} \cdot \frac{c_{Gln}}{c_{Gln}+K_{S,Gln}} \cdot \frac{c_{LS}}{c_{LS}+K_{S,LS}}$$ $$\mu_d = \mu_{d,min} + \mu_{d,max} \cdot \frac{K_{S,Glc}}{K_{S,Glc}+c_{Glc}}$$ $$q_{Gln} = q_{Glc,max} \cdot \frac{c_{Glc}}{c_{Glc}+k_{Glc}} \cdot \left(\frac{\mu}{\mu+\mu_{max}} + 0.5\right)$$ $$q_{Gln} = q_{Gln,max} \cdot \frac{c_{Gln}}{c_{Gln}+k_{Gln}}$$ $$q_{LS} = q_{LS,max} \cdot \frac{c_{LS}}{c_{LS}+k_{LS}}$$ $$q_{Lac} = Y_{Lac/Glc} \cdot q_{Glc} \cdot \frac{c_{Glc}}{c_{Lac}} - q_{Lac,uptake,max}$$ $$q_{Amm} = Y_{Amm/Gln} \cdot q_{Gln} \cdot \frac{c_{Gln}}{c_{Amm}} - K_{Amm} \cdot q_{Amm,uptake,max} \cdot \frac{\mu_{max}-\mu}{\mu_{max}}$$ $$q_{titer} = q_{titer,max}$$

The model can be applied for batch-mode and fed-batch mode. The presented model example contains extended fed-batch terms for a glucose feed, a glutamine feed and a medium feed containing specific glucose and glutamine concentrations. Therefore, the differential equations are extended by the terms (at the end of each differential equation) including feeding rates for glucose (F_{Glc}) and glutamine (F_{Gln}) and for the medium feed

(F_{Medium}). The specific glucose and glutamine concentrations are denoted by $c_{Glc,F}$, $c_{Gln,F}$, $c_{Glc,Medium}$ and $c_{Gln,Medium}$. When applying fed-batch mode, all considered concentrations (viable and total cells, substrates and metabolites) are diluted during addition of the feed. This is represented by the dilution term $-\frac{F_{Glc}+F_{Gln}+F_{Medium}}{V}$. At the same time, glucose or glutamine concentrations increase during the glucose or glutamine feeding and during the medium feeding. This is represented by the terms $+\frac{F_{Glc}\cdot c_{Glc,F}}{V}$, $+\frac{F_{Gln}\cdot c_{Gln,F}}{V}$ and $+\frac{F_{Medium}\cdot c_{Glc,Medium}}{V}$ or $+\frac{F_{Medium}\cdot c_{Gln,Medium}}{V}$, respectively. When applying this model to batch-mode, all these feeding terms are omitted.

2.7. Bayesian Parameter Estimation

One of the main differences between Bayesian statistics and frequentist statistical methods ("classical statistics" based on frequencies) is that Bayesian statistics provides a framework to integrate prior process knowledge (knowledge available before applying new data for analysis), including input uncertainty, and to calculate probabilities based on both, prior knowledge and new collected data. Applying this principle within the context of parameter estimation is called Bayesian parameter estimation. A very brief description of this procedure is given through the following steps: Step 1 quantifies the prior knowledge, including input uncertainties (e.g., measurement uncertainties of initial concentrations and uncertainties concerning model parameters). In this contribution a gamma distribution has been chosen to describe the probability distribution of model parameters (further details can be found in the Appendix A.1).

The second step is to determine the posterior parameter distributions using an appropriate algorithm. A Markov chain Monte Carlo method was applied based on a single-component metropolis algorithm, resulting in posterior distributions, including the maximum a posteriori (MAP) estimate and variance.

Step 3 is to evaluate the parameter estimation results, for example, based on the Monte Carlo error and the posterior parameter distributions. This method was implemented in the self-developed seed train-software tool developed at Ostwestfalen-Lippe University of Applied Sciences and Arts. For a more detailed description of this approach, refer to [13].

2.8. Upstream Simulation—Software Tool

The upstream process has been simulated and digitally displayed using the seed train-software tool [12–14] implemented in MATLAB [24]. To digitally display an upstream process, several inputs are required: The estimated model parameters, initial concentrations of cells of the first scale of operation, a passaging or subcultivation strategy for the cells (e.g., concerning the point in time for cell passaging), the inoculum train vessels and operating conditions and medium concentrations. For further details, see [12–14].

2.9. Uncertainty-Based Prediction

For simulation of the production scale, uncertainty in measurements of initial concentrations and in parameters was considered and propagated onto the output. Therefore, Monte Carlo samples were generated sampling initial values of state variables for scale 1 from a gamma distribution described above. The corresponding histograms can be found in the appendix; see Figure A1.

The obtained 90% prediction bands (credible bands) were used for comparisons of different clonal cell populations, and were calculated using the 5% and 95% quantiles of the obtained Monte Carlo sample at a specific point in time.

2.10. Response Surface Modeling

Response surface models (RSM) describe the relationships between individual explanatory variables (here μ_{max} and $q_{titer,max}$) on one or more process variables of interest. In this contribution, the first response variable was the volumetric titer in production. The second response variable was the overall process productivity (=Space-Time-Yield: volumetric titer in production/overall cultivation time, including inoculum train). To explain: This is very important when a given manufacturing plant can be used for production during one year—how much product can this facility deliver for the market? This methodology mostly consists of solving a multiple regression model, meaning to estimate the corresponding regression coefficients. First order and second order polynomials have been adapted (through estimation of regression coefficients) using MATLAB [24]. To evaluate the obtained model, the coefficient of determination was calculated. It is a measure used to explain how well differences in the response variable can be explained by its relationship to the considered independent factors.

The β-coefficients (here $\beta = (\beta_1, \beta_2)$) indicate how much the response value changes per each unit variation of the independent variable (or factor, here μ_{max}, or $q_{titer,max}$). Thus, a higher β-coefficient stands for a higher correlation between the factor and the response value (here volumetric titer or overall process productivity).

3. Results and Discussion

The following results provide insights into the roles of cell-specific growth rate (in the following growth rate) and cell-specific production rate (in the following production rate) in the cell expansion process (inoculum train) and the final production scale of operation using a model-based simulation approach. The following variables are considered: duration of the inoculum train; the volumetric titer in production; and the overall process productivity, defined by the ratio of volumetric titer in production to the overall cultivation time, including inoculum train.

In the first step (compare to Figure 2), data of a stability study of five clonal CHO cell lines were analyzed concerning growth rates and production rates. Can statistically significant differences can be observed between these five cell populations?

An implemented and tested mechanistic cell culture model was adapted to further exploit the experimental data of one of these populations. Modeling and parameter estimations based on new experiments at 10, 50 and 500 mL were performed. This model was then used for further theoretical considerations.

Uncertainty-based simulations of inoculum train and production scale were performed for three clonal cell lines with established differences of growth and production rates.

Finally, a study was performed for several combinations of growth rate and production rate, showing the impacts of these differences on cultivation time and overall process productivity.

3.1. Analysis of Variabilities in Growth Rate and Production Rate for Five Clonal Cell Lines

Experimental data from a 3-month stability study were used to calculate growth and production rates for each clonal population. The clonal populations were subcultivated every 3 or 4 days during a time period of 13 weeks (see Section 2.1). The averaged empirical growth rates for two measurements of viable cell density $X_{v,i}$ and $X_{v,i+1}$ (at the beginning and at the end of a subcultivation) have been calculated according to Equation (1). The averaged empirical production rate between the beginning of a subcultivation and 4 days later was calculated according to Equation (2) for every second subcultivation (volumetric titer were only determined for the 4-day subcultivations).

The obtained average growth and production rates are illustrated in Figure 3 over every second subcultivation. The corresponding distributions can be found in the appendix (see Figures A2 and A3).

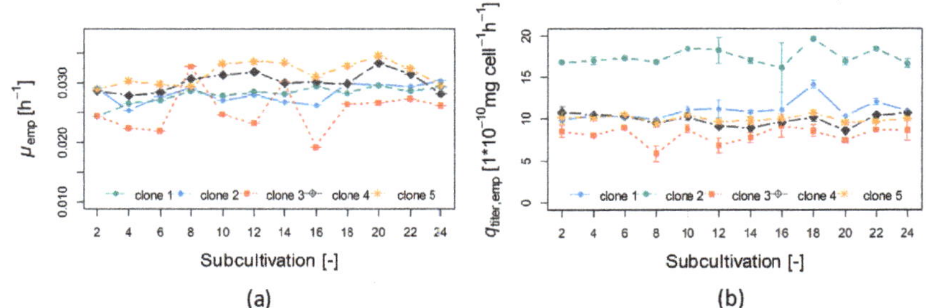

Figure 3. Growth rates μ_{emp} (**a**) and production rates $q_{titer,emp}$ (**b**) for clonal populations "clone 1" to "clone 5" for every second subcultivation in 50 mL OrbShake tubes.

Mean, standard deviation (sd), coefficient of variation (cv) and maximum (max) are listed for both quantities, growth rate and production rate, for all five populations in Table 3.

Table 3. Mean, standard deviation (sd), coefficient of variation (cv) and maximum of empirical growth rate μ_{emp} and mean, standard deviation, coefficient of variation and maximum of empirical production rate $q_{titer,emp}$ for 'clone 1' to 'clone 5' based on data from a 13 weeks-stability study.

Clone	Mean	sd	Max	cv	Mean	sd	Max	cv
		$[\,h^{-1}]$		[%]	$[1 \cdot 10^{-10}$ mg cell^{-1} h$^{-1}]$			[%]
1	0.028	0.0014	0.030	5	11.1	1.16	10.5	14.2
2	0.027	0.0019	0.030	7	**17.4**	1.01	19.6	**5.8**
3	0.026	0.0028	0.033	10	8.09	0.99	9.2	12.2
4	0.029	0.0018	0.033	6	9.92	0.76	10.8	7.6
5	**0.030**	0.0022	0.035	7	10.0	0.40	10.7	4.0

(header spans: μ_{emp} over Mean/sd/Max/cv; $q_{titer,emp}$ over Mean/sd/Max/cv)

Clone 5 showed the highest growth rate (a mean of 0.030 h^{-1}, a cv of 7% and a maximum value of 0.035 h^{-1}) and clone 2 the highest production rate (a mean of 17.4 $\cdot 10^{-10}$ mg cell^{-1} h^{-1}, a cv of 5.8% and a maximum value of 19.6 $\cdot 10^{-10}$ mg cell^{-1} h^{-1}) (see also Figure 3).

Overall, the growth rates (mean values) varied between 0.62 d^{-1} (=0.026 h^{-1}) and 0.72 d^{-1} (=0.030 h^{-1}), and production rates varied between 19 pg cell^{-1} d^{-1} (=8.09 $\cdot 10^{-10}$ mg cell^{-1} h^{-1}) and 42 pg cell^{-1} d^{-1} (=17.4 $\cdot 10^{-10}$ mg cell^{-1} h^{-1}). These growth rates are in the range of those reported recently in [6] (0.48–0.76 d^{-1}), where different CHO host cell lines were compared. The production rates found in the present study exceeded the production rates presented in [6], where production rates between 1.6 and 16.2 pg cell^{-1} d^{-1} were found, and [25], where averaged production rates ranged between 8 and 22 pg cell^{-1} d^{-1}, though the cultivation set ups in those studies may have differed in some aspects from ours, e.g., concerning cultivation vessels and volumes.

However, to identify which clones differ from each other in terms of averaged growth rates and production rates, the variations of the calculated rates have to be considered as well. To decide if several clonal populations have significant differences in terms of their means, an analysis of variance adapted for heterogeneous variances and post hoc tests (multiple comparison) has been performed according to the statistical procedure described in Section 2.5. To test on variance homogeneity, the Bartlett test was applied, and the result (p-value = 0.019 for $\mu_{emp,max}$, p-value = 0.021 for $q_{titer,emp,max}$) indicates heterogeneous variances. Hence, the Brown and Forsythe F-test [20] was applied. The results (p-value = 2.9 $\cdot$ 10^{-11} for $\mu_{emp,max}$ and p-value = 5.0 $\cdot$ 10^{-27} for $q_{titer,emp,max}$) show that statistically significant differences (on a 5%-level) exist in both cases. The results of the post hoc tests, to identify differences between individual groups, are listed in Table 4.

Table 4. Results of the test regarding differences in means: differences concerning growth rate ($d_{\mu_{emp},diff}$ and corresponding p-value) and concerning production rate ($d_{q_{titer,emp},diff}$ and corresponding p-value).

Clones	$d_{\mu_{emp},diff}$ [h^{-1}]	p-Value [-]	Clones	$d_{q_{titer,emp},diff}$ [$1 \cdot 10^{-10}$ mg cell^{-1} h^{-1}]	p-Value [-]
5 vs. 3	0.0041	$8 \cdot 10^{-6}$	2 vs. 3	9.3	$2 \cdot 10^{-15}$
5 vs. 2	0.0039	$1 \cdot 10^{-6}$	2 vs. 4	7.5	$5 \cdot 10^{-14}$
4 vs. 3	0.0030	$3 \cdot 10^{-4}$	**2 vs. 5**	7.4	$6 \cdot 10^{-12}$
4 vs. 2	0.0028	$3 \cdot 10^{-5}$	2 vs. 1	6.4	$1 \cdot 10^{-11}$
5 vs. 1	0.0023	$5 \cdot 10^{-4}$	1 vs. 3	3.0	$7 \cdot 10^{-6}$
1 vs. 3	0.0018	$3 \cdot 10^{-2}$	5 vs. 3	1.9	$10 \cdot 10^{-5}$
1 vs. 2	0.0015	$1 \cdot 10^{-2}$	4 vs. 3	1.8	$2 \cdot 10^{-4}$
4 vs. 1	0.0012	$3 \cdot 10^{-2}$	1 vs. 4	1.1	$4 \cdot 10^{-2}$
5 vs. 4	0.0011	0.19	**1 vs. 5**	1.0	$4 \cdot 10^{-2}$
2 vs. 3	0.0003	1	5 vs. 4	0.1	1

It can be seen that most populations show statistically significant differences between each other (p-values $\ll 0.05$), except clone 2 and clone 3 concerning growth rate (p-value = 1), and clone 4 candompared to clone 5 concerning both, growth rate (p-value = 0.19) and production rate (p-value = 1). The biggest difference in terms of growth rate has been found between clone 5 and clone 3, with a difference of 0.0041 h^{-1} (see Table 4, row 1, columns 1–3). A positive value in column 2 means that the left clone in column 1 has a higher $\mu_{emp,max}$ than the right clone in column 1.

Clone 2 has a significantly higher specific productivity than any of the other clones (see Table 4, rows 1 to 4, columns 4–6). All differences between clone 2 and the compared clone are positive and statistically significant (p-values < 0.05). The following is cell line 1 with significantly higher production rates than clones 3, 4 and 5 (see Table 4, rows 5, 8 and 9 in columns 4–6).

To investigate whether a theoretical clonal cell population showing high growth rates is more favorable than cell lines with higher production rates, clonal populations (here referred to in a generalist way as clone A and clone B) are considered which are inversely related to each other. This means that the following criteria are fulfilled:

- The averaged empirical growth rate of clone A, $\overline{\mu}_{emp,A}$, is statistically significantly higher than the averaged empirical growth rate of clone B, $\overline{\mu}_{emp,B}$, i.e., $\overline{\mu}_{emp,A} > \overline{\mu}_{emp,B}$.
- The averaged empirical production rate of clone A, $\overline{q}_{titer,emp,A}$, is statistically significantly lower than averaged empirical production rate of clone B, $\overline{q}_{titer,emp,B}$, i.e., $\overline{q}_{titer,emp,A} < \overline{q}_{titer,emp,B}$.

This holds for the comparisons "clone 1 vs. clone 5" and "clone 2 vs. clone 5" of Section 3.1. Therefore, the differences between these clones in terms of growth rate and production rate (highlighted in bold font in Table 4) are considered in the following.

It should be noted that the averaged growth and production rates differ from the model parameters maximum growth rate μ_{max} and maximum production rate $q_{titer,max}$, used within a cell culture model. For this reason, the presented findings regarding differences between clonal populations have also been calculated on a percentage basis, to keep the same ratios within the simulation-based investigations. The empirical growth rate of clone 1 was approximately 7.6% higher than that of clone 5 and 10% higher than that of clone 2. The empirical production rate was 10.5% lower than that of clone 5 and 74% higher than that of clone 2.

In order to know how these clones would behave in a typical cell expansion process (from vial to production vessel) and at the final production phase, a representation was created which is explained in the following section. Growth rates and production rates are assumed to remain the same at the larger scales of operation.

3.2. Model Adaption of a Mechanistic Cell Culture Model for Prediction Using Bayesian Parameter Estimation

To display the cell growth behavior of a cell line, a growth model has to be applied and adapted based on experimental data. Since only clone 5 was available for further experiments, cell expansion processes from 5 mL and 10 mL in parallel to 500 mL have been performed at ExcellGene SA for this clone. At 5 and 10 mL scales, viable cell density, viability and volumetric titer have been measured. Calculated growth rates and production rates have been used to define the prior distributions of μ_{max} and $q_{titer,max}$ in the following.

At 500 mL scale, cells have been cultivated over a period of 8 days and substrates (glucose and glutamine) and metabolites (lactate and ammonia) were measured in addition to viable cell density, viability and volumetric titer (to also adapt parameters characterizing substrate uptake and death rate). Based on these experiments, a growth model (see Section 2.2), which had been already applied to other CHO cell lines, was used here while applying Bayesian parameter estimations. This approach consists of the following steps:

In a first step, the prior knowledge about model parameters had to be quantified. In the second step, experimental data were added, and a Markov chain Monte Carlo algorithm is used to find the posterior probability distributions of the model parameters to be estimated. The obtained posterior distributions contained information from prior knowledge and new experimental data.

3.2.1. Prior Knowledge

To quantify the prior probability distributions of model parameters, data from the stability study and data from additional experiments at 5 mL and 10 mL with the same clone, clone 5, have been used in the following way:

The maximum growth rate of clone 5 over all subcultivations of the stability study was $\mu_{emp,max} = 0.035\,\mathrm{h}^{-1}$. Additional experiments at 5 and 10 mL-scales revealed growth rates of 0.046 and $\mu_{emp,max} = 0.048\,\mathrm{h}^{-1}$, respectively. The additional experiments provide one measurement per day, allowing the computation of the growth rate per day. The stability study provides data at the beginning and at the end of each subcultivation (with a duration of 3 or 4 days each). Consequently, the maximum growth rate cannot be approximated as precisely as using daily measurements. Nevertheless, it is considered for determination of the prior distribution but with less weight (1/3) than the approximations of further experiments (2/3).

The maximum production rate of clone 5 over all subcultivations of the stability study revealed $q_{titer,emp,max} = 10.7 \cdot 10^{-10}\,\mathrm{mg\,cell}^{-1}\,\mathrm{h}^{-1}$. The maximum production rates of clone 5, based on additional experiments at 5 and 10 mL-scales, were $q_{titer,emp,max} = 5.9 \cdot 10^{-10}\,\mathrm{mg\,cell}^{-1}\,\mathrm{h}^{-1}$ and $q_{titer,emp,max} = 6.8 \cdot 10^{-10}\,\mathrm{mg\,cell}^{-1}\,\mathrm{h}^{-1}$, respectively. The reason for the variation of these values is unknown, but the variation (uncertainty) itself is information also included in the prior probability study. (A higher uncertainty signifies less weight for the prior mean within the parameter estimation process).

Based on this information, mean and variance have been calculated to characterize the prior probability distribution of maximum growth rate μ_{max} and maximum production rate $q_{titer,max}$ according to Equation (A1). These are listed in Table 5, including the corresponding coefficient of variation (cv).

Table 5. Prior parameter values for maximum growth rate μ_{max} (mean, variance and coefficient of variation (cv)) and empirical production rate $q_{titer,max}$ (mean, variance and coefficient of variation (cv)).

Parameter	Mean	Variance	cv
μ_{max}	$0.0428\,\mathrm{h}^{-1}$	$5.36 \cdot 10^{-5}\,\mathrm{h}^{-1}$	17%
$q_{titer,max}$	$7.8 \cdot 10^{-10}\,\mathrm{mg\,cell}^{-1}\,\mathrm{h}^{-1}$	$6.56 \cdot 10^{-20}\,\mathrm{mg\,cell}^{-1}\,\mathrm{h}^{-1}$	33%

3.2.2. Posterior Distributions

Bayesian parameter estimation has been performed using a Markov chain Monte Carlo (MCMC) algorithm considering cultivation data (at 500 mL over 8 days) and prior distributions as described in Section 2.7. Measured and simulated time course data are presented in Figure 4. It can be seen that reasonable agreement between measured and simulated data can be achieved by the set of model parameters used, although more experimental data between day 4 and day 8 could have helped to define more precisely when the cells entered into the stationary phase. Prior (before parameter estimation) and posterior (after parameter estimation) distributions are shown in Figure 5. Posterior means of estimated model parameters and values of the fixed model parameters are presented in Table A1. It can be concluded from Figure 4 together with Figure 5 that parameters μ_{max} and $q_{titer,max}$ represent rather well the measured data: Posterior distributions (red solid lines) are much narrower than the prior distributions (blue dashed lines), thereby reducing uncertainty for these model parameters. This means that uncertainty has been reduced for these model parameters. Furthermore, the means moved slightly to the right in the case of maximum growth rate μ_{max} and strongly to the left in case of $q_{titer,max}$. Posterior distributions of the remaining model parameters do not differ much from their prior distributions.

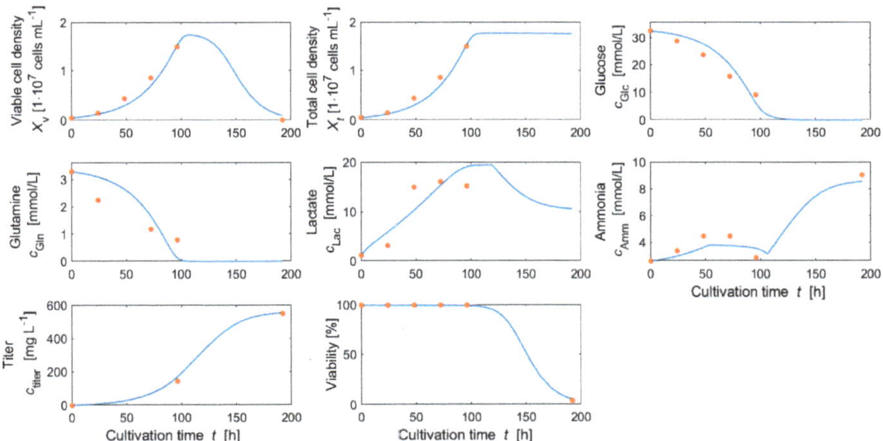

Figure 4. Measurements (red dots) and simulated time profiles (blue solid lines) of viable and total cell density, glucose concentration, glutamine concentration, lactate concentration, ammonia concentration, volumetric titer and viability after Bayesian parameter estimation (based on the maximum a posteriori estimate (MAP)) at 500 mL scale (TubeSpin bioreactor 600™).

3.3. Uncertainty-Based Upstream Process Simulation—Comparison of Three Clonal Populations with Different Growth and Production Rates

In this section, we describe the application of the adapted model perform upstream simulations for three different theoretical cell lines, named A, B and C, under consideration of variabilities observed in Section 3.1. The reference clone A is characterized by the model parameter distributions obtained in the previous section (parameter estimation for clone 5). The two other clones B and C are defined as showing lower growth rates than clone A, but higher production rates than clone A, as listed in Table 6. In order to choose realistic values concerning the differences between clones A, B and C, the differences obtained in Section 3.1 concerning growth rate and production rate have been applied. Empirical growth rates for experimentally analyzed cell lines 5 and 1 showed an averaged difference of 7.6% and for 5 and 2 an averaged difference of 10%. Therefore, model parameter μ_{max} of clone B was chosen to be 7.6% lower than μ_{max} of clone A and μ_{max} of clone C was chosen to be 10% lower than μ_{max} of clone A.

Empirical production rates $q_{\text{titer,max}}$ for cell lines 5 and 1 showed averaged differences of 10.5% and 74%, respectively. Therefore, model parameter $q_{\text{titer,max}}$ of clone B was chosen to be 10.5% higher than $q_{\text{titer,max}}$ of clone A, and $q_{\text{titer,max}}$ of clone C was chosen to be 74% higher than $q_{\text{titer,max}}$ of clone A. A suitable inoculum train protocol was defined for each clonal cell line. Furthermore, these simulations were used to investigate and illustrate the impact of differences in growth and production rates between all three clones regarding duration of the inoculum train, volumetric titer in production and overall process productivity for a batch process.

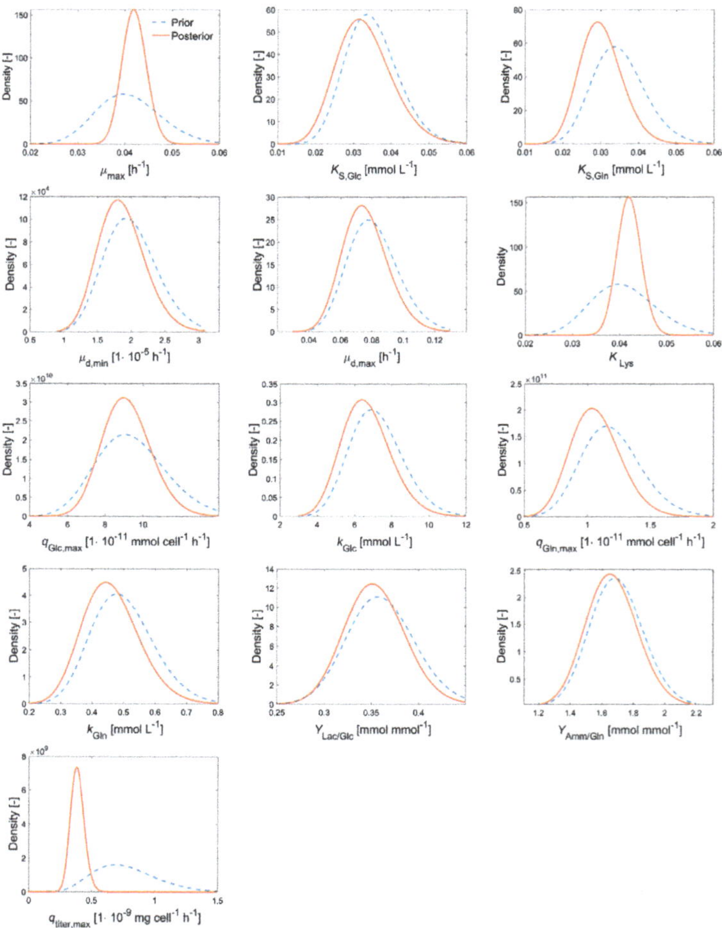

Figure 5. Earlier derived (a priori) parameter distributions (before parameter estimation, blue dashed lines) and posterior parameter distributions (after parameter estimation, red solid lines) of estimated model parameters of the process model presented in Table 2 based on experimental data from cultivation in 500 mL (TubeSpin 600™ bioreactor) with a cultivation time of 8 days.

Table 6. Model parameters maximum growth rate μ_{max} and maximum production rate $q_{titer,max}$ of the three clones used for the simulation.

Clone	Remark	μ_{max} $[\mathrm{h}^{-1}]$	$q_{titer,max}$ $[1 \cdot 10^{-10}$ mg cell^{-1}h$^{-1}]$
A	reference clone	0.042	3.90
B	7.6% lower μ_{max}, 10.5% higher $q_{titer,max}$	0.039	4.31
C	10% lower μ_{max}, 74% higher $q_{titer,max}$	0.038	6.90

To digitally display an upstream process, the following inputs have been defined:

Volumes: The simulated upstream process consisted of six scales of operation with the following volumes: 10 mL → 120 mL → 1.5 L → 15 L → 100 L → 1000 L (production). This setup enabled the use of a highly similar type of cultivation approach (orbital shaking) at all scales. Note: ExcellGene has considerable experience with scale-up cultures in both orbital shaken and standard stirred systems to have sufficient confidence in the matching impacts of critical parameters in both approaches (not published). Passaging strategy: Cells were passaged, i.e., subcultivated, as soon as a required cell density for transfer was reached, using the predicted viable cell density. The required cell biomass was based on the optimal cell density for inoculation at $5 \cdot 10^5$ cells mL^{-1}. Initial concentrations: viable cell density, $X_{v,0} = 5.3 \cdot 10^5$ cells mL^{-1}; viability $= 100\%$; glucose, $c_{Glc,0} = 32.6$ mmol L^{-1}; glutamine $c_{Gln,0} = 3.3$ mmol L^{-1}; lactate, $c_{Lac,0} = 0.001$ mmol L^{-1}; ammonia, $c_{Amm,0} = 2.6$ mmol L^{-1}, titer, $c_{titer,0} = 0$ mg L^{-1}; and volume, $V = 0.01$ L. Furthermore, a limiting substrate was assumed to have initial value $c_{LS,0} = 2$ mmol L^{-1}.

The corresponding simulated time profiles for trends in viable cell density and titer are presented in Figure 6.

It turned out that the designed inoculum trains seemed suitable for cell expansion of all three clonal populations. When the inoculum cell densities were fulfilled, cells did not enter into the stationary phase during the inoculum train, and transfer cell densities were within an acceptable range (maximum cell density below $1 \cdot 10^7$ cells mL^{-1}).

The durations of the inoculum train cultures ranged from 298 to 333 h. Obviously, lower growth rates cause longer cultivation times. Clone B needed 24 h and clone C 35 h more than clone A. Clone A and clone B, concerning the predicted volumetric titers in the production vessel, differed by a 10.5% higher production rate of clone B, resulting in a 13% higher volumetric titer during the first hours of the production phase. However this difference shrunk over time: After 25 h clone A reached 13 mg L^{-1} and clone B 15 mg L^{-1}. Between 50 and 100 h in the production vessel, clone A compensated for the disadvantage through a 7.6% higher growth rate. After 100 h, clone A presentd a titer of 222 mg L^{-1}, 7% more than clone B with 207.0 mg L^{-1}. Nevertheless, after 168 h (7 days) clone B reached a higher volumetric titer (558 mg L^{-1}) than clone A (539 mg L^{-1}). This was due to the fact that the higher growth rate of clone A led to an earlier beginning of the death phase (here in batch mode) compared to clone B. Putting the volumetric titer in relation to the overall cultivation time and accepting an overall error of about 10%, both clones led to a similar overall process productivity (1.16 mg L^{-1}h^{-1} for clone A and 1.14 mg L^{-1}h^{-1} for clone B).

A clearer impact was observed for clone C, having a 10% lower maximum growth rate combined with a 74% higher production rate as compared to clone A. Already, after the first 25 h in the production vessel, clone C reached 25 mg L^{-1} on average (clone A and B only 13 and 15 mg L^{-1}, respectively), and after 7 days (168 h) clone C reached 876 mg L^{-1} (clone A and B only 539 and 558 mg L^{-1}, respectively). This is an increase of 337 mg L^{-1} (63.5% of the volumetric titer generated with clone A).

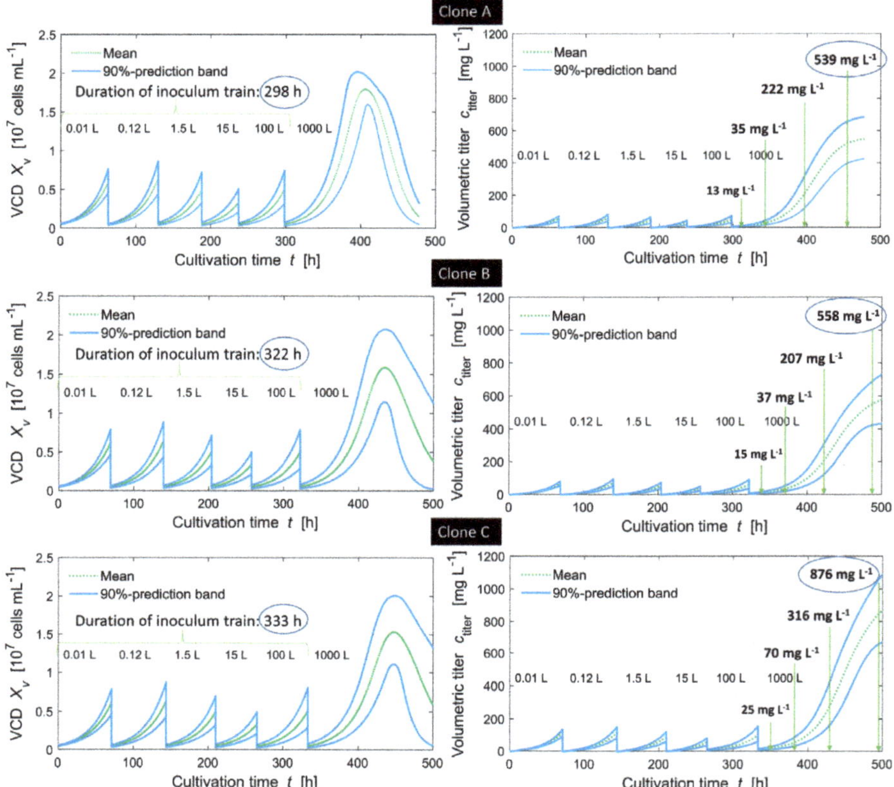

Figure 6. Simulated viable cell density (VCD) and volumetric titer over inoculum train cultures (5 scales) and production scale (1000 L) for clones A (above), B (middle) and C (below). The maximum growth rate of clone B was 7.6% lower than that of clone A, and the maximum cell-specific production rate of clone B was 10.5% higher than that of clone A. The maximum growth rate of clone C was 10.5% lower than that of clone A and the maximum cell-specific production rate of clone C was 74% higher than that of clone A.

It should be noted that the presented model-based method can be further extended to fed-batch processes which are most frequently applied in industrial large scale manufacturing or to perfusion mode. However, the batch process has been considered in this study because the focus was not to find an optimal operating mode for the production bioreactor, but rather to consider how phenotypic differences effect cell growth in the inoculum train, which contributes significantly to the manufacturing time and overall process productivity, yet is rarely considered in literature [26].

A comparison of different CHO host cell lines for batch, fed-batch and perfusion modes was recently reported in [6]. They found that differences in phenotypic properties affect cell growth and productivity regardless of process mode (batch, fed-batch or perfusion) or cell culture media.

For a better illustration, Figure 7 shows how variabilities in model parameters $\mu_{\max}$ and $q_{\mathrm{titer,max}}$ propagate onto the output uncertainty in form of probability distributions (histograms) at each interesting point in time.

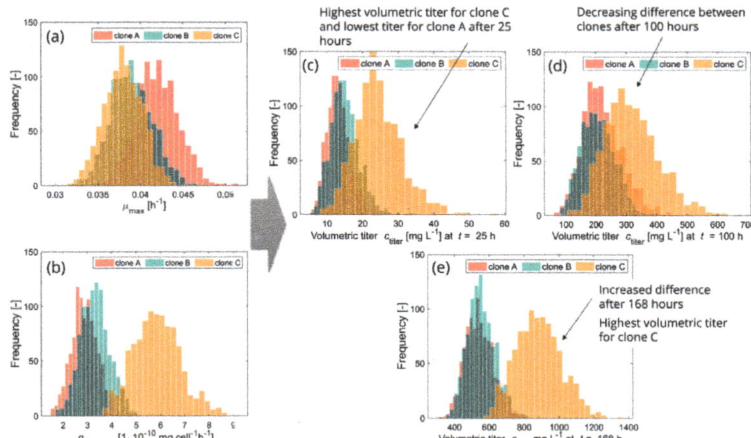

Figure 7. Histograms of inputs and outputs for clone A in the background (red), clone B in the middle (turquoise) and clone C in the front (orange). Inputs: Probability distributions of maximum growth rate (**a**) and maximum production rate (**b**) for three different clones. Probabilities are defined by the given means μ_{max} and $q_{titer,max}$ and the corresponding coefficients of variation (cv). Outputs: Probability distributions of volumetric titer after 25 (**c**), 100 (**d**) and 168 (**e**) hours for the three clones.

It thus becomes visible how the output distribution changes over time. In accordance with the results presented in Figure 6, there is not a huge difference concerning volumetric titer between the distributions of clone A and clone B: not after 25 h (Figure 7c), 100 h (Figure 7d) or 168 h (Figure 7e) of production. The distributions are almost overlapping, although after 100 h clone A shows a higher mean than clone B, as described above. The distribution of clone C instead differs clearly from those of clone A and clone B (small overlap) after 25 h in production. After 100 h of production there are larger overlapping areas between all three clones, indicating a decline in differences between them. However, after 168 h (7 days) of production, a clear difference (smaller overlap) is visible between clone C and clones A and B, whereas clone A and B are almost totally overlapping. In this case and under the assumptions of equal stability and quality, clone C would be the recommended clone for moving forward.

Nevertheless, it may be the case that two or more clonal populations differ in a different proportion to each other in terms of phenotypic characteristics than the here discussed three. The following section tries to address such.

3.4. Impacts of Differences in Growth and Production Rates on Inoculum Train and Titer at Production Scale—General Considerations and a Decision Criterion

To judge the effects of growth rate and specific productivity in numerous clonal cell populations, one needs to know the resulting overall process productivities (=Space-Time-Yield: volumetric titer in production/overall cultivation time, including inoculum train).

We determined these effects for a realistic cell expansion setup and based on model parameter ranges derived from the previous sections (μ_{max} = 0.397 h^{-1} ±10%, $q_{titer,max}$ = 5 · 10^{-10} mg cell^{-1}h^{-1} ±10%). For each parameter combination, the two response numbers (volumetric titer in production and overall process productivity) were obtained by upstream simulations as before. These results were then adapted to corresponding response surfaces (see Figures A4–A6), which visualize the effects of maximum growth rate and maximum production rate on each response quantity.

Multiple linear regression has been performed for the responses after 50, 100 and 168 h in the production vessel. Due to their different orders of magnitude, all variables have been scaled (transformed) to the range of [0, 1]. The results are presented in Table 7.

Table 7. Results of a multiple linear regression (R^2, β-coefficients, standard error (SE) and p-value) in two variables, maximum growth rate μ_{max} (coefficient β_1) and maximum production rate $q_{titer,max}$ (coefficient β_2) at 50, 100 and 168 h of production. Response values are volumetric titer in production and overall process productivity (=Space-Time-Yield: volumetric titer in production/overall cultivation time, including inoculum train).

Response Variable	Factor	R^2	β−Coefficients (β_1, β_2)	Standard Error (SE)	p-Value
Volumetric titer (50 h)		0.989			
	μ_{max}		0.42	0.016	$< 1 \cdot 10^{-20}$
	$q_{titer,max}$		0.57	0.012	$< 1 \cdot 10^{-20}$
Overall process productivity (50 h)		0.991			
	μ_{max}		0.61	0.013	$< 1 \cdot 10^{-20}$
	$q_{titer,max}$		0.38	0.010	$< 1 \cdot 10^{-20}$
Volumetric titer (100 h)		0.992			
	μ_{max}		0.69	0.013	$< 1 \cdot 10^{-20}$
	$q_{titer,max}$		0.31	0.010	$< 1 \cdot 10^{-20}$
Overall process productivity (100 h)		0.991			
	μ_{max}		0.75	0.014	$< 1 \cdot 10^{-20}$
	$q_{titer,max}$		0.25	0.011	$< 1 \cdot 10^{-20}$
Volumetric titer (168 h)		0.998			
	μ_{max}		0.33	0.008	$< 1 \cdot 10^{-20}$
	$q_{titer,max}$		0.67	0.006	$< 1 \cdot 10^{-20}$
Overall process productivity (168 h)		0.996			
	μ_{max}		0.54	0.009	$< 1 \cdot 10^{-30}$
	$q_{titer,max}$		0.46	0.007	$< 1 \cdot 10^{-30}$

All regressions have an R^2-value very close to one, meaning that the applied model is suitable to present the correlation between factors and response variables. All determined β-coefficients, which describe the correlation of μ_{max} (β_1) and $q_{titer,max}$ (β_2) for the investigated response variable, show p-values less than 0.05 (meaning that they are statistically significant to a 5%-level). Due to the scaling of both factors, the β-coefficients stayed within the range of 0 and 1. It is interesting to see that the impact and the relation between both factors, μ_{max} and $q_{titer,max}$, varies depending on which point in time in production is considered.

Regarding volumetric titer as a response variable, it can be observed that after 50 h in the production vessel, the impact of $q_{titer,max}$ ($\beta_2 = 0.57$) was 1.4 times higher than the impact of μ_{max} ($\beta_1 = 0.42$). After 100 h in production, this changed. Then, the impact of μ_{max} ($\beta_1 = 0.69$) was 2.2 times higher than the impact of $q_{titer,max}$ ($\beta_2 = 0.31$). However, after 168 h (7 days) in production, $q_{titer,max}$ ($\beta_2 = 0.67$) was again higher than μ_{max} (two times $\beta_1 = 0.33$). The decreasing impact of μ_{max} after 168 h can be explained because cells probably entered in the stationary/death phase (here batch-mode is assumed) while cells were still producing titer.

Considering overall process productivity, μ_{max} has a higher impact than $q_{titer,max}$, regardless of the considered point in time (see β-coefficients for the overall process productivity in Table 7). It was 1.6 times higher after 50 h, three times higher after 100 h and 1.17 times higher after 168 h cultivation time in the production vessel compared to $q_{titer,max}$. Obviously, therefore, growth rates have a higher impact on the output per year.

This regression analysis was performed within a range of 0.397 h^{-1} ± 10% for μ_{max} and $5 \cdot 10^{-10}$ mg cell^{-1}h^{-1} ± 10% for $q_{titer,max}$); it should be noted, however, that the results of the stability study showed a higher variation of the production rate than that of the growth rate. Therefore, $q_{titer,max}$ has been varied ± 50%, and response surfaces for both response variables, volumetric titer and overall process productivity, have been estimated as shown in Figure 8.

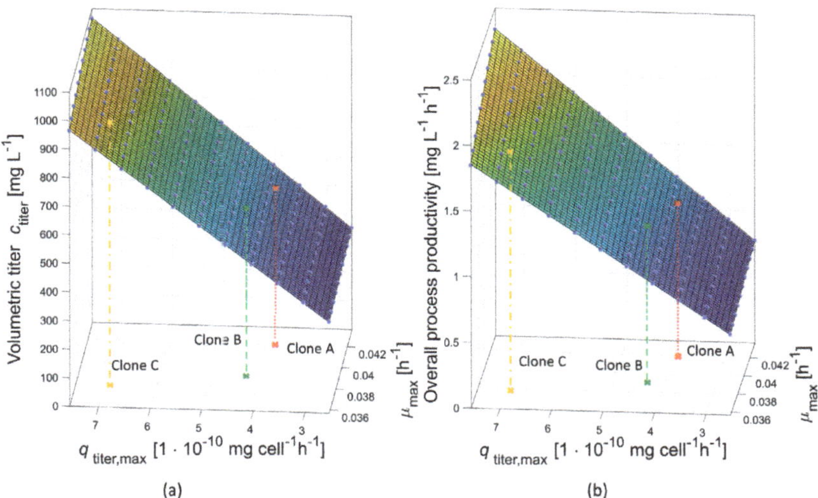

Figure 8. Response surface for volumetric titer after 168 h (7 days) of production over maximum production rate and maximum growth rate (**a**). Overall process productivity (=Space-Time-Yield: volumetric titer after 168 h (7 days) in production/overall cultivation time, including inoculum train) over maximum production rate and maximum growth rate (**b**). The reference clone A (red solid line) and two compared clones, clone B (green dashed line, 7.6% lower μ_{max} and 10.5% higher $q_{titer,max}$) and clone C (orange dotted line, 10% lower μ_{max} and 74% higher $q_{titer,max}$) are placed in the graphs.

Table 8. Results, volumetric titer and overall process productivity (=Space-Time-Yield: volumetric titer after 168 h (7 days) in production/overall cultivation time, including inoculum train) for reference clone A, and two clones to be compared, clone B and clone C. These clones differ in terms of maximum growth rates and maximum production rates, as listed in the corresponding rows.

Clone	Remark	μ_{max} [h^{-1}]	$q_{titer,max}$ [$1 \cdot 10^{-10}$ mg cell^{-1}h^{-1}]	Volumetric Titer [mg L^{-1}]	Overall Process Productivity [mg L^{-1} h^{-1}]
A	reference clone	0.042	3.9	539	1.16
B	7.6% lower μ_{max}, 10.5% higher $q_{titer,max}$	0.039	4.31	558	1.14
C	10% lower μ_{max}, 74% higher $q_{titer,max}$	0.0378	6.9	876	1.75

Choosing a clonal population showing a lower growth rate but a higher production rate will only be favorable if the productivity is high enough. As stated before, and summarized in Table 8, clones A and B delivered very similar process productivities (1.16 mg L^{-1} h^{-1} for clone A and 1.14 mg L^{-1} h^{-1} for clone B). The final titers for clone A and B were 539 and 558 mg L^{-1}, respectively—a negligible difference. When taking 74% higher productivity for C, then a more significant difference is obtained with 876 mg L^{-1} after 168 h, and an 1.75 mg L^{-1} h^{-1} overall process productivity enhancement over A and B is seen.

The response surface models can be used, therefore, to approximate volumetric titer and overall process productivity for a realistic combination of growth rate and production rate, and help with the decision processes. These simulations can be used to determine to what extent growth rate or production rate must differ to cause a difference of at least 5% in the response variables.

4. Conclusions

A model-based approach in combination with statistical methods was applied to study the impacts of the cell-specific growth rate ("growth rate") and cell-specific production rate ("production rate" or "specific productivity") on overall product yield using parameters such as time needed for the inoculum train cultures, the volumetric titer during final production phase and the overall process productivity (=Space-Time-Yield: volumetric titer in production/overall cultivation time, including inoculum train). For three theoretical clonal populations an inoculum train protocol was defined, suitable for all of them: Cell line A showed a 7.6% higher maximum growth rate, cell line B a 10.5% higher production rate than cell line A and cell line C a 10% lower maximum growth rate and a 74% higher production rate than cell line A. For all three cell lines, a prediction model of an inoculum train, including predictive uncertainty arising from model parametric uncertainty (due to biological variabilities), has been utilized. For cell line A (higher μ_{max}) the inoculum train would take 298 h until inoculation of cells into the production bioreactor (1000 L), for B (higher $q_{titer,max}$) it would take 322 h and for C (higher $q_{titer,max}$) 333 h. Cell line A would generate a volumetric titer of approximately 539 mg L^{-1} after 168 h in the final production vessel, B would result in 558 mg L^{-1} and C would result in 876 mg L^{-1}, assuming a batch process.

Moreover, response surface modeling was applied to quantify the effects of both parameters on volumetric titer and overall process productivity at specific points in time in production. Based on the results of a simulation using mathematical process models in combination with statistical methods, decision criteria can be provided that can help to evaluate different clonal cell lines for future manufacturing purposes. This can be seen as a support tool in addition to the characterization of biochemical, biophysical and functionality properties to asses the quality of the final product. Assuming little or no quality differences in the products obtained in cell culture, the growth rate of a clonal cell population has the higher impact (up to three times) on the overall process productivity, and thus, on the output per year, and clones with higher production rates have the potential to generate significantly more volumetric titer in production.

It has not escaped our attention that modern processes in large scale manufacturing are most frequently fed-batch processes with production run times exceeding the herein-discussed 7-day batch processes. These fed-batch processes can increase volumetric titers quite dramatically. Nevertheless, the batch process evaluation is a good first step to obtain quick results. They can be further expanded once a smaller number of clonally derived cell lines have been chosen to also involve fed-batch processes. Moreover, shorter batch processes have certain advantages for some products—for example, reducing the negative impacts of certain losses in production campaigns, such as contaminations or disruptions from instrument failures. Thus, the preferred mode for most efficient use of a given manufacturing facility would be to shorten overall production time phases (in the largest bioreactor) while maximizing growth in inoculum train cultures and to achieve the highest maximal density in the so-called N-1 cultures (i.e., the culture preceding the production vessel). In spite of this, the authors of this article hope to having provided a useful discussion on the complex relationships between different phenotypes of CHO cells, particularly those that have major impacts on overall productivity in manufacturing.

Author Contributions: Conceptualization, T.H.R. and B.F.; methodology, T.H.R. and B.F.; software, T.H.R.; validation, T.H.R.; formal analysis, T.H.R.; investigation, T.H.R., S.M., F.M.W. and B.F.; resources, S.M. and F.M.W.; data curation, T.H.R. and S.M.; writing—original draft preparation, T.H.R.; writing—review and editing, F.M.W. (major), B.F., R.P. and T.H.R.; visualization, T.H.R. and B.F.; supervision, B.F. and F.M.W.; project administration, T.H.R. and B.F.; funding acquisition, B.F. and T.H.R. All authors have read and agreed to the published version of the manuscript.

Funding: This research received no external funding. The article processing charge (APC) was funded partially by Ostwestfalen-Lippe University of Applied Sciences and Arts.

Institutional Review Board Statement: Not applicable.

Informed Consent Statement: Not applicable.

Data Availability Statement: Not applicable.

Acknowledgments: We acknowledge support for the Open Access fees by Ostwestfalen-Lippe University of Applied Sciences and Arts (TH OWL) through the funding program Open Access Publishing.

Conflicts of Interest: The authors declare no conflict of interest.

Abbreviations

The following abbreviations and symbols are used in this manuscript:

α	Shape parameter of a gamma distribution (-)
$\beta = (\beta_1, \beta_2)$	Regression coefficients (impact)
λ	Scale parameter of a gamma distribution (-)
μ	Growth rate (h^{-1})
μ_d	Death rate (h^{-1})
$\mu_{d,max}$	Maximum death rate (h^{-1})
$\mu_{d,min}$	Minimum death rate (h^{-1})
μ_{emp} ($\mu_{emp,max}$)	(Maximum) empirical cell-specific growth rate (h^{-1})
μ_{max}	Maximum cell-specific growth rate (h^{-1})
$\bar{\mu}_{emp,A}$ ($\bar{\mu}_{emp,B}$)	Averaged empirical growth rate of clone A (B) (h^{-1})
c_{Amm} ($c_{Amm,0}$)	(Initial) ammonia concentration (mmol L^{-1})
c_{Glc} ($c_{Glc,0}$)	(Initial) glucose concentration (mmol L^{-1})
c_{Gln} ($c_{Gln,0}$)	(Initial) glutamine concentration (mmol L^{-1})
c_{Lac} ($c_{Lac,0}$)	(Initial) lactate concentration (mmol L^{-1})
c_{LS} ($c_{LS,0}$)	(Initial) limiting substrate concentration (mmol L^{-1})
c_{titer} ($c_{titer,0}$)	(Initial) volumetric titer (product concentration) (mg L^{-1})
$c_{titer,i}$	Volumetric titer (product concentration) at point in time t_i (mg L^{-1})
CHO	Chinese Hamster Ovary
CI	Confidence interval
cv	coefficient of variation
$d_{\mu_{emp},diff}$	Difference in terms of growth rate μ (h^{-1})
$d_{q_{titer,emp},diff}$	Difference in terms of production rate q_{titer} (mg $cell^{-1}$ h^{-1})
F_{sample}	Change of volume due to sampling [L h^{-1}]
i	Running index (-)
j	Running index (-)
K_{Amm}	Correction factor for ammonia uptake (-)
K_{Lys}	Cell lysis constant (h^{-1})
$K_{S,Glc}$	Monod kinetic constant for glucose (mmol L^{-1})
$K_{S,Gln}$	Monod kinetic constant for glutamine (mmol L^{-1})
$K_{S,LS}$	Monod kinetic constant for limiting substrate (mmol L^{-1})
k_{Glc}	Monod kinetic constant for glucose uptake (mmol L^{-1})
k_{Gln}	Monod kinetic constant for glutamine uptake (mmol L^{-1})
k_{LS}	Monod kinetic constant for uptake of limiting substrate (mmol L^{-1})
MAP	Maximum a posteriori
max	Maximum value
MCMC	Markov Chain Monte Carlo
q_{Amm} ($q_{Amm,uptake,max}$)	(Maximum) cell-specific ammonia uptake rate (mmol $cell^{-1}$ h^{-1})
q_{Glc} ($q_{Glc,max}$)	(Maximum) cell-specific glucose uptake rate (mmol $cell^{-1}$ h^{-1})
q_{Gln} ($q_{Gln,max}$)	(Maximum) cell-specific glutamine uptake rate (mmol $cell^{-1}$ h^{-1})
q_{Lac} ($q_{Lac,uptake,max}$)	(Maximum) cell-specific lactate uptake rate (mmol $cell^{-1}$ h^{-1})
q_{LS} ($q_{LS,max}$)	(Max.) cell-specific uptake rate of limiting substrate (mmol $cell^{-1}$ h^{-1})
q_{titer} ($q_{titer,max}$)	(Maximum) cell-specific production rate (mg $cell^{-1}$ h^{-1})
$q_{titer,emp}$ ($q_{titer,emp,max}$)	(Maximum) empirical cell-specific production rate (mg $cell^{-1}$ h^{-1})
$\bar{q}_{titer,clone,ref}$, $\bar{q}_{titer,clone,compared}$	Average empirical production rate of reference or compared clone

R^2	Coefficient of determination
RSM	Response surface models
sd	Standard deviation
SE	Standard error
t	Time (h)
t_i	Point in time with index i (h)
V	Volume (L)
Via	Viability (%)
X_t	Total cell density (cells L^{-1})
X_v	Viable cell density (cells L^{-1})
$X_{v,i}$	Viable cell density at point in time with index i (cells L^{-1})
Y	Arbitrary random variable (-)
$Y_{Amm/Gln}$	Kinetic production constant for ammonia (mmol $mmol^{-1}$)
$Y_{Lac/Glc}$	Kinetic production constant for lactate (mmol $mmol^{-1}$)

Appendix A

Appendix A.1. Choice of the Prior Distribution

To perform Bayesian parameter estimation, prior distributions have to be quantified to form probability distributions. An appropriate type of distribution has to be chosen in accordance with the available knowledge. In this contribution, a gamma distribution has been chosen to describe the probability distribution of model parameters. This assumption is based on the fact that the considered random variables can only adopt positive values, and furthermore, the gamma distribution is well suited for representing the realistic range based on the available prior knowledge. It is defined by the parameters α(shape) and λ(rate).

To characterize the individual distribution of a variable Y (here μ_{max} or $q_{titer,max}$), the corresponding mean ($E(Y)$) and variance ($V(Y)$) are used to compute the distribution parameters rate (α) and shape (λ), according to:

$$\alpha = \frac{E(Y)^2}{Var(Y)} \quad \text{and} \quad \lambda = \frac{Var(Y)}{E(Y)}. \tag{A1}$$

Appendix B

Appendix B.1. Supplementary Figures

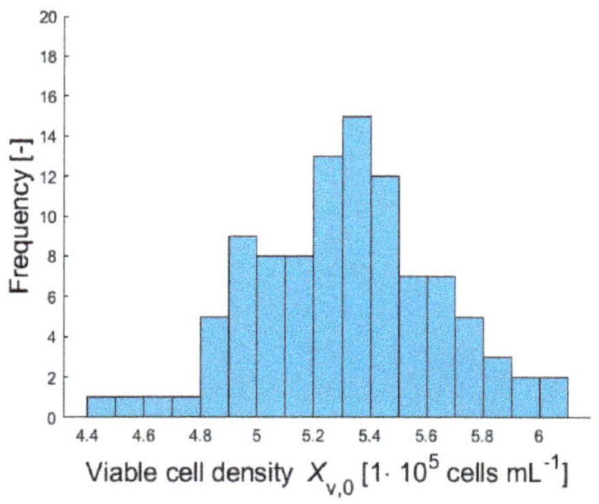
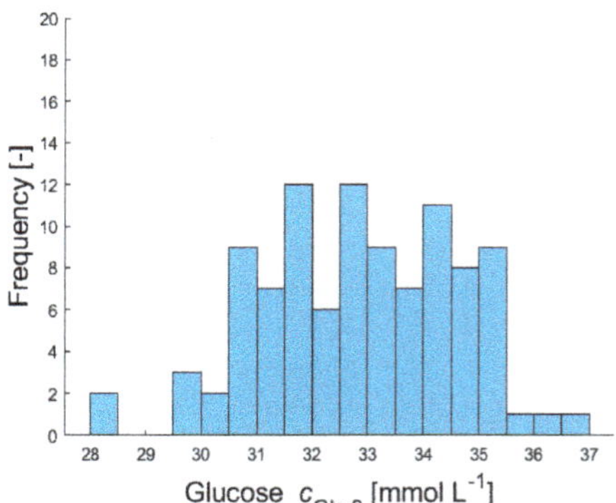

Figure A1. *Cont.*

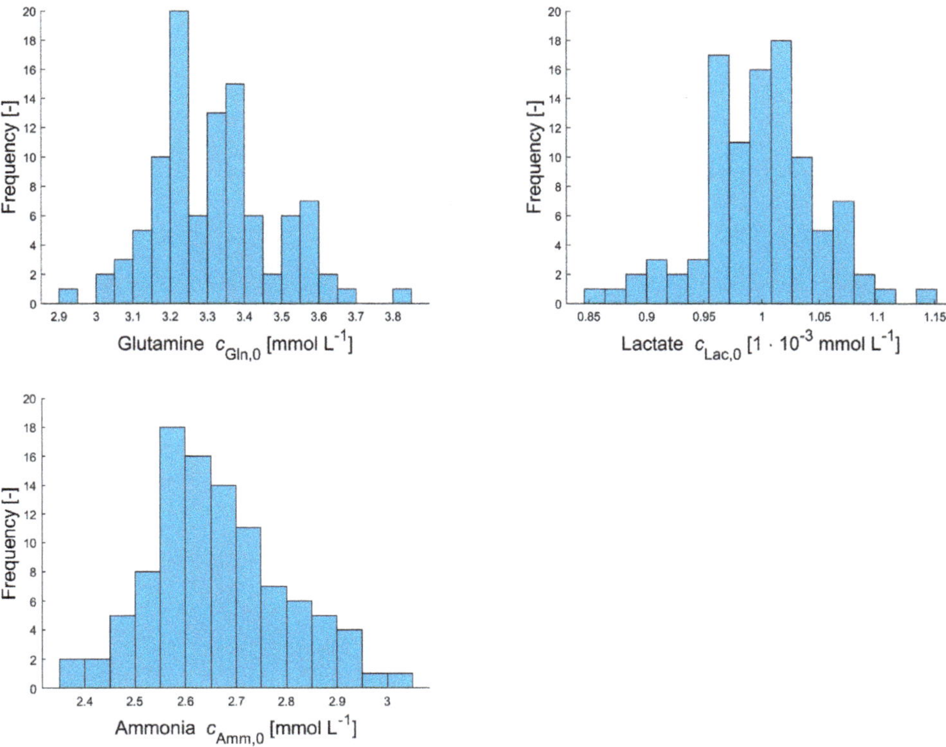

Figure A1. Histograms of sampled initial concentrations of viable cells X_{v0}, glucose c_{Glc}, glutamine c_{Gln}, lactate c_{Lac} and ammonia c_{Amm} in the first cultivation vessel for upstream simulations.

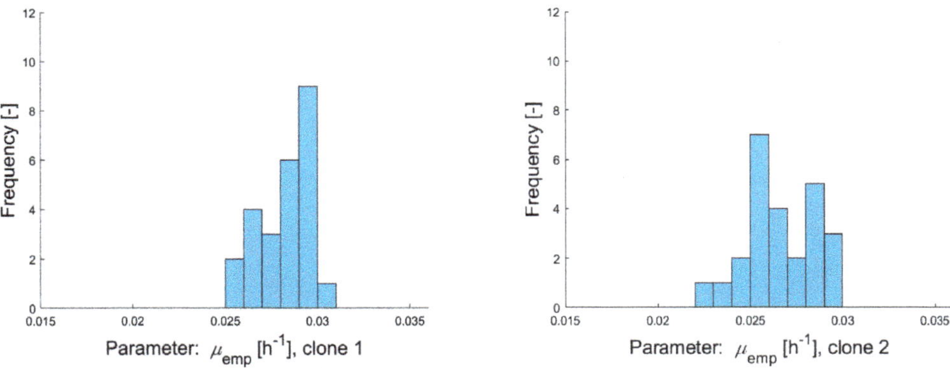

Figure A2. *Cont.*

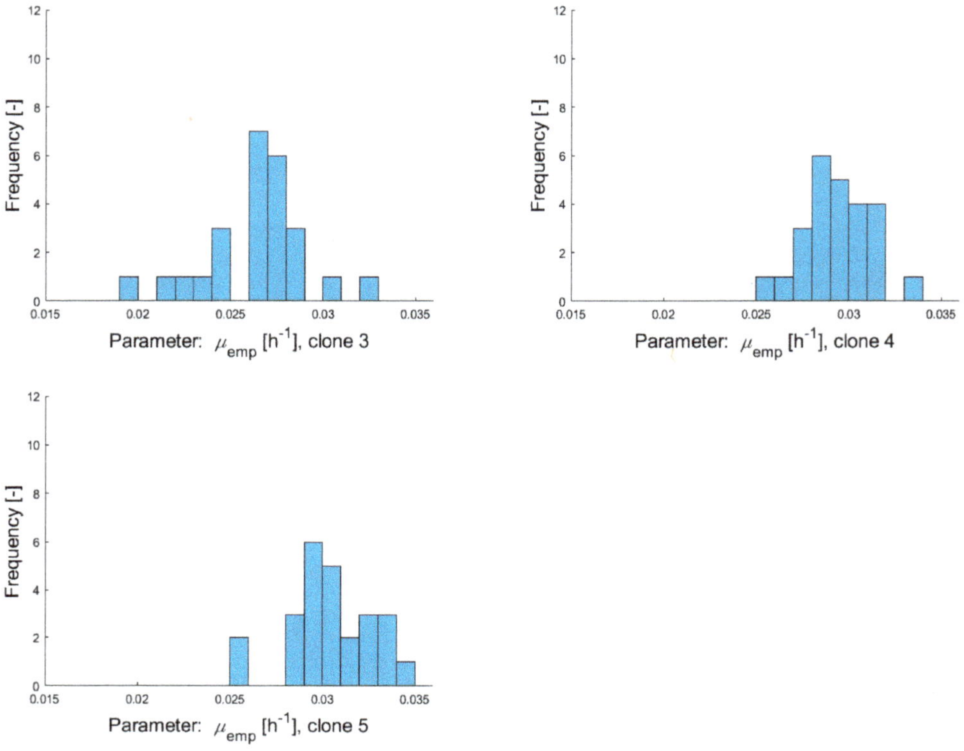

Figure A2. Histograms of empirical growth rates for clonal populations clone 1–clone 5 calculated over several subcultivation steps with culture volumes of 5 mL (during a period of 13 weeks).

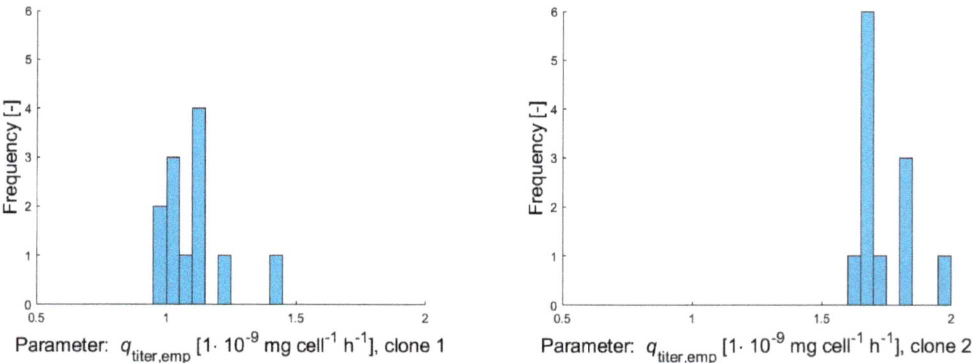

Figure A3. *Cont.*

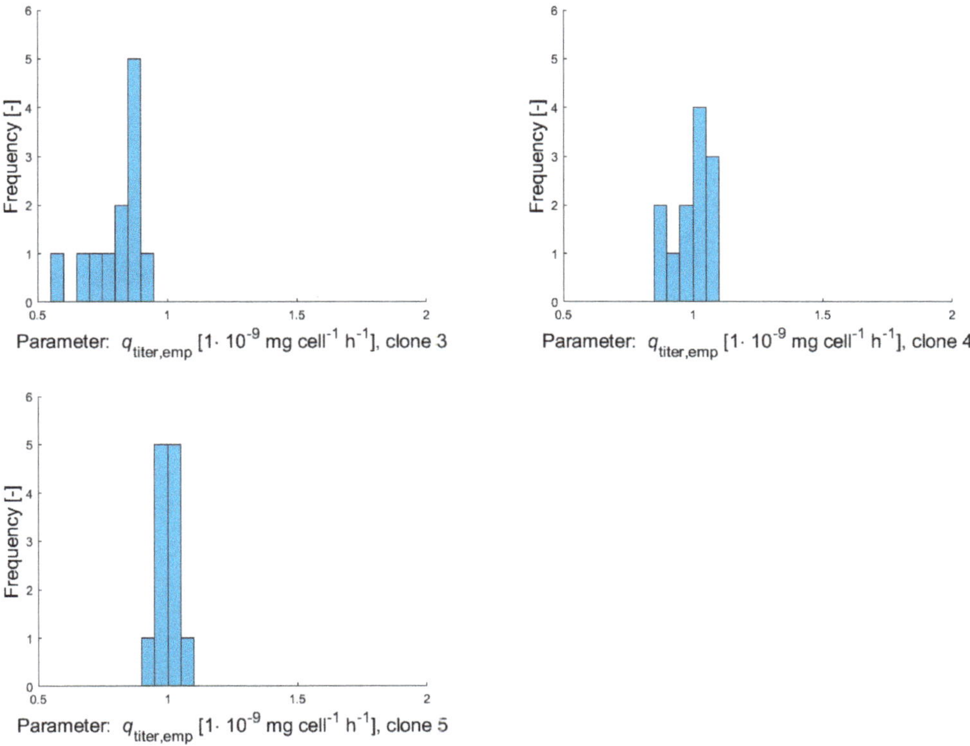

Figure A3. Histograms of empirical production rates for clonal populations clone 1–clone 5 calculated over several subcultivation steps with culture volumes of 5 mL (during a period of 13 weeks).

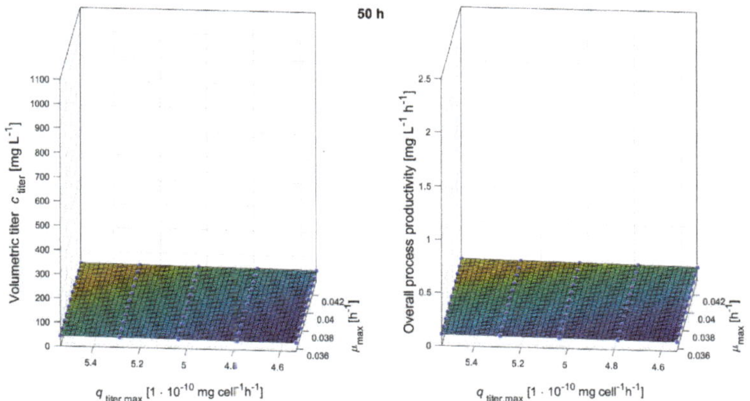

Figure A4. Response surfaces showing the impacts of maximum growth rate and maximum production rate on volumetric titer and overall process productivity after 50 h of production.

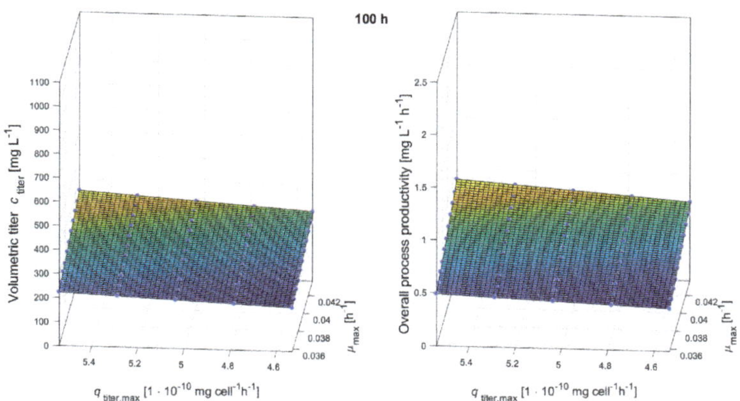

Figure A5. Response surfaces showing the impacts of maximum growth rate and maximum production rate on volumetric titer and overall process productivity after 100 h of production.

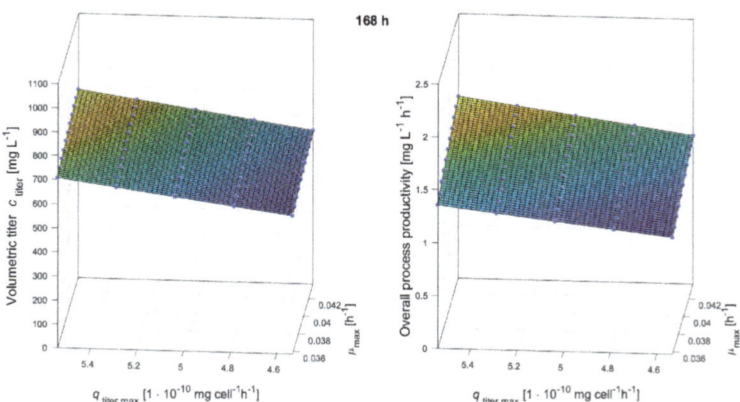

Figure A6. Response surfaces showing the impacts of maximum growth rate and maximum production rate on volumetric titer and overall process productivity after 168 h (7 days) of production.

Appendix B.2. Supplementary Tables

Table A1. Modeled variables and model parameters included in the underlying model (symbols, units and descriptions).

Variable/Parameter	Unit	Initial Value	Description
X_t	cells L^{-1}	$5.3 \cdot 10^5$	Total cell density
X_v	cells L^{-1}	$5.3 \cdot 10^5$	Viable cell density
c_{Glc}	mmol L^{-1}	5.9	Glucose concentration
c_{Gln}	mmol L^{-1}	3.3	Glutamine concentration
c_{Lac}	mmol L^{-1}	0.01	Lactate concentration
c_{Amm}	mmol L^{-1}	2.6	Ammonia concentration
c_{titer}	mg L^{-1}	0	Volumetric product (antibody) titer
V	L	0.01	Volume

Table A1. *Cont.*

Model Parameter	Unit	Posterior Estimate/Fixed Value	Description
μ_{max}	h^{-1}	0.042	Maximum cell-specific growth rate
$K_{S,Glc}$	$mmol\ L^{-1}$	0.03	Monod kinetic constant for glucose
$K_{S,Gln}$	$mmol\ L^{-1}$	0.03	Monod kinetic constant for glutamine
$K_{S,LS}$	$mmol\ L^{-1}$	0.16 (fixed)	Monod kinetic constant for limiting substrate
$\mu_{d,min}$	h^{-1}	$1.0 \cdot 10^{-5}$	Minimum cell-specific death rate
$\mu_{d,max}$	h^{-1}	0.08	Maximum cell-specific death rate
K_{Lys}	h^{-1}	$1.2 \cdot 10^{-4}$	Cell lysis constant
$q_{Glc,max}$	$mmol\ cell^{-1}\ h^{-1}$	$9.7 \cdot 10^{-11}$	Maximum cell-specific glucose uptake rate
k_{Glc}	$mmol\ L^{-1}$	6.2	Monod kinetic constant for glucose uptake
$q_{Gln,max}$	$mmol\ cell^{-1}\ h^{-1}$	$1.1 \cdot 10^{-11}$	Maximum cell-specific glutamine uptake rate
k_{Gln}	$mmol\ L^{-1}$	0.5	Monod kinetic constant for glutamine uptake
$q_{LS,max}$	$mmol\ cell^{-1}\ h^{-1}$	$1.1 \cdot 10^{-11}$ (fixed)	Max. cell-specific uptake rate of limiting substrate
k_{LS}	$mmol\ L^{-1}$	8.15 (fixed)	Monod kinetic constant for uptake of limiting substrate
$Y_{Lac/Glc}$	$mmol\ mmol^{-1}$	0.35	Kinetic production constant for lactate
$q_{Lac,uptake,max}$	$mmol\ cell^{-1}\ h^{-1}$	$1.0 \cdot 10^{-11}$ (fixed)	Cell-specific maximum lactate uptake rate
$Y_{Amm/Gln}$	$mmol\ mmol^{-1}$	1.67	Kinetic production constant for ammonia
$q_{Amm,uptake,max}$	$mmol\ cell^{-1}\ h^{-1}$	$4.5 \cdot 10^{-10}$ (fixed)	Cell-specific maximum ammonia uptake rate
K_{Amm}	-	1.9 (fixed)	Correction factor for ammonia uptake
$q_{titer,max}$	$mg\ cell^{-1}\ h^{-1}$	$3.9 \cdot 10^{-10}$	Cell-specific maximum production rate

References

1. Walsh, G. Biopharmaceutical benchmarks 2018. *Nat. Biotechnol.* **2018**, *36*, 1136–1145. [CrossRef] [PubMed]
2. Wurm, F. CHO Quasispecies—Implications for Manufacturing Processes. *Processes* **2013**, *1*, 296. [CrossRef]
3. Wurm, F.; Wurm, M. Cloning of CHO Cells, Productivity and Genetic Stability—A Discussion. *Processes* **2017**, *5*, 20. [CrossRef]
4. Browne, S.; Al-Rubeai, M. Selection Methods for High-Producing Mammalian Cell Lines. *Trends Biotechnol.* **2007**, *25*, 425–432. [CrossRef]
5. Lakshmanan, M.; Kok, Y.J.; Lee, A.P.; Kyriakopoulos, S.; Lim, H.L.; Teo, G.; Poh, S.L.; Tang, W.Q.; Hong, J.; Tan, A.H.M.; et al. Multi–omics profiling of CHO parental hosts reveals cell line–specific variations in bioprocessing traits. *Biotechnol. Bioeng.* **2019**, *116*, 2117–2129. [CrossRef]
6. Reinhart, D.; Damjanovic, L.; Kaisermayer, C.; Sommeregger, W.; Gili, A.; Gasselhuber, B.; Castan, A.; Mayrhofer, P.; Grünwald-Gruber, C.; Kunert, R. Bioprocessing of Recombinant CHO-K1, CHO-DG44, and CHO-S: CHO Expression Hosts Favor Either mAb Production or Biomass Synthesis. *Biotechnol. J.* **2019**, *14*, e1700686. [CrossRef]
7. Beketova, E.V.; Ibneeva, L.R.; Abdulina, Y.A.; Dergousova, E.A.; Filatov, V.L.; Kozlovsky, S.V.; Shilov, E.S.; Datskevich, P.N.; Rozov, F.N. Optimized dual assay for the transgenes selection and screening in CHO cell line development for recombinant protein production. *Biotechnol. Lett.* **2019**, *41*, 929–939. [CrossRef]
8. Porter, A.J.; Racher, A.J.; Preziosi, R.; Dickson, A.J. Strategies for selecting recombinant CHO cell lines for cGMP manufacturing: Improving the efficiency of cell line generation. *Biotechnol. Prog.* **2010**, *26*, 1455–1464. [CrossRef]
9. Wang, B.; Albanetti, T.; Miro-Quesada, G.; Flack, L.; Li, L.; Klover, J.; Burson, K.; Evans, K.; Ivory, W.; Bowen, M.; et al. High-throughput screening of antibody-expressing CHO clones using an automated shaken deep-well system. *Biotechnol. Prog.* **2018**, *34*, 1460–1471. [CrossRef]
10. Narayanan, H.; Luna, M.F.; von Stosch, M.; Bournazou, M.N.C.; Polotti, G.; Morbidelli, M.; Butté, A.; Sokolov, M. Bioprocessing in the Digital Age—The Role of Process Models. *Biotechnol. J.* **2019**. [CrossRef]
11. Xie, X.; Schenkendorf, R. Robust Process Design in Pharmaceutical Manufacturing under Batch-to-Batch Variation. *Processes* **2019**, *7*, 509. [CrossRef]
12. Frahm, B. Seed Train Optimization for Cell Culture. In *Animal Cell Biotechnology*; Pörtner, R., Ed.; Springer protocols, Humana Press: New York, NY, USA, 2014; pp. 355–367.
13. Hernández Rodríguez, T.; Posch, C.; Schmutzhard, J.; Stettner, J.; Weihs, C.; Pörtner, R.; Frahm, B. Predicting industrial-scale cell culture seed trains-A Bayesian framework for model fitting and parameter estimation, dealing with uncertainty in measurements and model parameters, applied to a nonlinear kinetic cell culture model, using an MCMC method. *Biotechnol. Bioeng.* **2019**, *116*, 2944–2959. [CrossRef]
14. Hernández Rodríguez, T.; Frahm, B. Design, Optimization, and Adaptive Control of Cell Culture Seed Trains. *Methods Mol. Biol.* **2019**, *2095*, 251–267. [CrossRef]
15. Hernández Rodríguez, T.; Frahm, B. Digital Seed Train Twins and Statistical Methods. *Adv. Biochem. Eng. Biotechnol.* **2020**. [CrossRef]

16. Deppe, S.; Frahm, B.; Hass, V.C.; Hernández Rodríguez, T.; Kuchemüller, K.B.; Möller, J.; Pörtner, R. Estimation of Process Model Parameters. *Methods Mol. Biol.* **2019**, *2095*, 213–234. [CrossRef]

17. Zhang, X.; Stettler, M.; de Sanctis, D.; Perrone, M.; Parolini, N.; Discacciati, M.; de Jesus, M.; Hacker, D.; Quarteroni, A.; Wurm, F. Use of orbital shaken disposable bioreactors for mammalian cell cultures from the milliliter-scale to the 1,000-liter scale. *Adv. Biochem. Eng. Biotechnol.* **2009**, *115*, 33–53. [CrossRef]

18. R Core Team. *R: A Language and Environment for Statistical Computing*; R Foundation for Statistical Computing: Vienna, Austria, 2020.

19. Bartlett, M.S. Properties of sufficiency and statistical tests. *Proc. R. Soc. Lond. Ser. A Math. Phys. Sci.* **1937**, *160*, 268–282. [CrossRef]

20. Brown, M.B.; Forsythe, A.B. The Small Sample Behavior of Some Statistics Which Test the Equality of Several Means. *Technometrics* **1974**, *16*, 129–132. [CrossRef]

21. Benjamini, Y.; Hochberg, Y. Controlling the False Discovery Rate: A Practical and Powerful Approach to Multiple Testing. *J. R. Stat. Soc. Ser. B Methodol.* **1995**, *57*, 289–300. [CrossRef]

22. Kern, S.; Platas-Barradas, O.; Pörtner, R.; Frahm, B. Model-based strategy for cell culture seed train layout verified at lab scale. *Cytotechnology* **2016**, *68*, 1019–1032. [CrossRef]

23. Möller, J.; Hernández Rodríguez, T.; Müller, J.; Arndt, L.; Kuchemüller, K.B.; Frahm, B.; Eibl, R.; Eibl, D.; Pörtner, R. Model uncertainty-based evaluation of process strategies during scale-up of biopharmaceutical processes. *Comput. Chem. Eng.* **2019**, 106693. [CrossRef]

24. MATLAB. *Version 9.9.0 (R2020b)*; The MathWorks Inc.: Natick, MA, USA, 2010.

25. Kim, Y.J.; Han, S.K.; Yoon, S.; Kim, C.W. Rich production media as a platform for CHO cell line development. *AMB Express* **2020**, *10*, 93. [CrossRef] [PubMed]

26. Böhl, O.J.; Schellenberg, J.; Bahnemann, J.; Hitzmann, B.; Scheper, T.; Solle, D. Implementation of QbD strategies in the inoculum expansion of a mAb production process. *Eng. Life Sci.* **2020**, *27*, 9. [CrossRef]

Article

New Insights from Locally Resolved Hydrodynamics in Stirred Cell Culture Reactor

Fabian Freiberger [1], Jens Budde [1], Eda Ateş [1], Michael Schlüter [2], Ralf Pörtner [1,*] and Johannes Möller [1]

[1] Institute of Bioprocess and Biosystems Engineering, Hamburg University of Technology, Denickestr. 15, 21073 Hamburg, Germany; fabian.freiberger@tuhh.de (F.F.); jens.budde@tuhh.de (J.B.); eda.ates@tuhh.de (E.A.); johannes.moeller@tuhh.de (J.M.)

[2] Institute of Multiphase Flows, Hamburg University of Technology, Eißendorfer Str. 38, 21073 Hamburg, Germany; michael.schlueter@tuhh.de

* Correspondence: poertner@tuhh.de; Tel.: +49-40-42878-2886

Abstract: The link between hydrodynamics and biological process behavior of antibody-producing mammalian cell cultures is still not fully understood. Common methods to describe dependencies refer mostly to averaged hydrodynamic parameters obtained for individual cultivation systems. In this study, cellular effects and locally resolved hydrodynamics were investigated for impellers with different spatial hydrodynamics. Therefore, the hydrodynamics, mainly flow velocity, shear rate and power input, in a single- and a three-impeller bioreactor setup were analyzed by means of CFD simulations, and cultivation experiments with antibody-producing Chinese hamster ovary (CHO) cells were performed at various agitation rates in both reactor setups. Within the three-impeller bioreactor setup, cells could be cultivated successfully at much higher agitation rates as in the single-impeller bioreactor, probably due to a more uniform flow pattern. It could be shown that this different behavior cannot be linked to parameters commonly used to describe shear effects on cells such as the mean energy dissipation rate or the Kolmogorov length scale, even if this concept is extended by locally resolved hydrodynamic parameters. Alternatively, the hydrodynamic heterogeneity was statistically quantified by means of variance coefficients of the hydrodynamic parameters fluid velocity, shear rate, and energy dissipation rate. The calculated variance coefficients of all hydrodynamic parameters were higher in the setup with three impellers than in the single impeller setup, which might explain the rather stable process behavior in multiple impeller systems due to the reduced hydrodynamic heterogeneity. Such comprehensive insights lead to a deeper understanding of the bioprocess.

Keywords: CHO DP-12; computational fluid dynamics; bioreactor characterization; hydrodynamic gradients; process development; critical shear stress; Kolmogorov length scale; operational space

Citation: Freiberger, F.; Budde, J.; Ateş, E.; Schlüter, M.; Pörtner, R.; Möller, J. New Insights from Locally Resolved Hydrodynamics in Stirred Cell Culture Reactors. *Processes* **2022**, *10*, 107. https://doi.org/10.3390/pr10010107

Academic Editor: Francesca Raganati

Received: 11 November 2021
Accepted: 22 December 2021
Published: 5 January 2022

1. Introduction

Mammalian cell culture processes are state-of-the-art for the production of therapeutic antibodies. However, the influence of the bioreactor hydrodynamics on the cell culture process are still not fully understood [1–6]. Therefore, bioreactor design and scale-up in today's biopharma industry rely mostly on empirical correlations, experience, and engineering heuristics. Common methods for scaling-up mammalian cell-based production processes aim to keep the reactor geometry and certain process parameters such as the average volumetric power input constant [7–11].

These "rules of thumb" methods are limited as they can hardly be used to describe or estimate an appropriate operation range with respect to shear effects on cellular behavior. Platas et al. [12] observed an approx. constant cell specific growth rate, μ_{max}, over a broader stirrer operation range in reactor systems with multiple impellers compared to reactor systems with only one impeller. The cell growth rate in reactors with multiple impellers decreased much slower with increasing agitation rates and corresponding mean power inputs than in reactors with less impellers. Therefore, shear effects seem to be less pronounced in the case of multiple impellers.

An approach frequently discussed in the literature to predict critical conditions with respect to shear effects to be expected in stirred tank bioreactors is the Kolmogorov eddy length scale [12,13], for which cell harm is predicted if the eddies are in the order of magnitude as the cell diameter. This approach should predict an acceptable power input—and thus an adequate agitation rate—to be estimated without extensive experiments. Conventionally, the system averaged energy dissipation rate is used to calculate the Kolmogorov length scale, which does not consider local gradients, even if the local power input can differ by several orders of magnitude within the reactor. Furthermore, this length scale is usually compared considering an average cell diameter, even though the cell diameter is widely distributed. However, while this hypothesis seems to be proven for microcarrier cultures, it is still under discussion for suspension cells [14–19]. A broad variety of studies on lethal and sub-lethal responses of mammalian cell cultures to shear stress has been published where a detailed overview can be found in Chalmers et al. The investigated magnitude of the average volumetric power input ranges from 10^1 to almost 10^9 W m^{-3} [3]. However, the vast majority of these studies only considered the estimated or averaged hydrodynamics.

Taken together, more insights into locally resolved hydrodynamics are needed. Since local maxima can reach much higher values than the averaged, locally resolved hydrodynamics could provide more detailed information on the hydrodynamics. These can be identified and quantified by the use of computational fluid dynamics (CFD) methods, which have become an established tool for cell culture process development [5,10,20–22]. CFD can be used to predict the locally resolved hydrodynamic behavior of cell cultivation systems and has already been successfully applied for the design of stem cell expansion processes in stirred tank reactors and wave bags with microcarriers [23–25].

In this study, we investigated to what extent the cellular and hydrodynamic effects change with different spatial hydrodynamics of different stirrers. Furthermore, it was discussed whether locally resolved hydrodynamics could help to explain the cellular effects. Therefore, the hydrodynamics, mainly flow velocity, shear rate, and power input, in a single- and a three-impeller bioreactor setup were analyzed by means of CFD simulations. Then, cultivation experiments with antibody-producing Chinese hamster ovary (CHO) cells were performed at various agitation rates in both reactor setups. Finally, conventional process design parameters such as the average volumetric power input and the Kolmogorov length scale were evaluated by means of the obtained data. As the model system, a reactor setup was chosen that could be equipped with various numbers of impellers and operated without baffles at constant culture volumes. Because surface aeration is sufficient for a wide range of operation conditions, the impact of bubble aeration was neglected in these investigations.

2. Materials and Methods

2.1. Cultivation Procedures

The cell line used in all cultivation experiments was CHO DP-12, which was kindly provided by Prof. Dr. Thomas Noll (University of Bielefeld, Bielefeld, Germany). All cultivation experiments were performed in TC-42 medium (Xell AG, Bielefeld, Germany), supplemented with 8 mmol L^{-1} L-glutamine and 200 nmol L^{-1} methotrexate.

2.1.1. Pre-Culture

Pre-cultivations were carried out in 125 mL shake flasks with 40 mL culture volume for the first culture after thawing and 250 mL shake flasks with 80 mL culture volume for expanding the cells [26]. Culture conditions were adjusted to 37 °C, 5% CO$_2$, 85% relative humidity, and 200 rpm. From pre-cultures, 15×10^7 cells in total were harvested, centrifuged for 10 min at $300 \times g$, and resuspended in 10 mL fresh medium for inoculating the stirred tank bioreactor with an initial cell concentration of 1×10^6 cells mL^{-1}.

2.1.2. Main Culture

For bioreactor cultivations, the stirred tank bioreactor Vario 1000 (MDX Biotech GmbH, Nörten-Hardenberg, Germany) was used with 150 mL culture volume and headspace aeration by gas mixtures of air and CO_2. Only at the end of the cultivations at cell densities of around 10×10^6 cells mL^{-1}, the reactor was additionally bubble-aerated with pure oxygen. Thus, effects of gas bubbles and additional turbulences, caused by baffles could be neglected and observed effects could be referred to the impellers. The dissolved oxygen tension (DO) was controlled at 40% of air saturation in all cultivations. pH was controlled at 7.1 by the addition of CO_2 via headspace or by adding 0.5 M Na_2CO_3. The temperature was set to 37 °C. To investigate the effect of hydrodynamics caused by stirring on the cell culture process, cultivation experiments were performed at different agitation rates for a reactor setup with one pitched blade impeller (MDX Biotech GmbH, Nörten-Hardenberg, Germany) (single impeller system, SIS) and a setup with two additional six blade impellers (self-made at Hamburg University of Technology, Nörten-Hardenberg, Germany) (triple impeller system, TIS) (see Appendix A Figure A1). The agitation rates chosen for the SIS were 200 to 1400 rpm in 200 rpm steps, while TIS cultivations at agitation rates of 770, 930, 1080, 1200 and 1400 rpm were run. In all experiments, the pitched blade impeller was operated in the down flow direction. Agitation rates for the TIS were chosen while following a discarded working hypothesis.

2.2. Analytical Methods

Cell densities were determined with the Z2 Particle Counter (Beckman Coulter, Brea, CA, USA). All particle distributions were recorded in triplicate. The cell viability was measured using the flow cytometer CytoFLEX (Beckman Coulter, Brea, CA, USA) after staining the cells with 1 μg mL^{-1} DAPI. Samples above 2×10^6 cells mL^{-1} were diluted ten-fold before staining. Antibody concentrations were determined with Protein-A binding sensors in the OCTET according to the manufacturer's protocol (Pall Corporation, Port Washington, NY, USA). For a comparison of the processes, in addition to growth curves, the space time yield (STY) related to the antibody concentration was calculated as the quotient of the maximum antibody concentration and the cultivation time until the maximum concentration was reached.

2.3. CFD Simulations

All CFD models were developed in COMSOL Multiphysics 5.3 to 5.5 (COMSOL AB, Stockholm, Sweden). The reactor geometry was fully implemented in COMSOL, except for the six blade impeller, which was imported from a CAD file. Reactor vessel and impellers were modeled as accurately as possible and necessary. Additionally, all reactor inserts, namely the impeller shaft, temperature, pH and DO probes as well as sampling tubes and aeration tubes were implemented. As the physics module, the mixture module was chosen with turbulent flow conditions. The module contains the Reynolds Averaged Navier–Stokes equation and continuity equation as governing equations. Standard parameters were used to model the turbulence with the k-ε-model. The geometry was meshed to 1.69 million mesh elements for the SIS and 1.78 million mesh elements for the TIS. To solve the stationary studies, the PARDISO solver was set up as a direct block structured solver. For further, detailed information about the CFD models, see the model reports attached as Supplementary Materials (Files S1 and S2). Due to numerical reasons, agitation rates of up to 800 rpm for the reactor setup with a single impeller and 600 rpm for the reactor setup with three impellers were simulated. Velocities and shear rates above the given agitation rates were extrapolated linearly while the energy dissipation rates were extrapolated quadratically, according to the simplified dependency given by Hu et al. [1]. The computed values from the CFD simulations and extrapolations can be found in Appendix A (Tables A1 and A2). A mesh refinement study was performed for all shown models (not shown). No dependencies on the degree of meshing were observed.

To highlight local distributions of hydrodynamic parameters, the computed velocities, shear rates, and energy dissipation rates were averaged over horizontal cut planes along the reactor height in steps of one millimeter. The obtained mean values were then plotted in comparison to the reactor height.

2.4. Calculation of the Critical Energy Dissipation Rate

The Kolmogorov length scale represents the size of the smallest turbulence eddies in a fluid flow and can be calculated from the energy dissipation rate ε, the kinematic viscosity ν, and the density ρ of the fluid.

$$\lambda = \left(\frac{\nu^3 \rho}{\varepsilon} \right)^{\frac{1}{4}} \tag{1}$$

According to the hypothesis, significant cell damage occurs when the smallest turbulence eddies are within the same order of magnitude as the cell diameter d_c. Thus, the equation for the Kolmogorov length was solved for the critical energy dissipation rate ε_{krit} where the cell diameter d_c is the cell diameter, which is in the same order as the Kolmogorov length λ, ν is the kinematic viscosity, and ρ is the density of the fluid.

$$\varepsilon_{krit} = \frac{\nu^3 \rho}{d_c^4} \tag{2}$$

Distributions of the cell diameter d_c were determined alongside the cell number in the Z2 Particle Counter. More details on the relation of turbulence eddies and cell damage according to the Kolmogorov length scale hypothesis can be found in Section 3.3.

3. Results

In the following, the results from the CFD study as well as from the cell cultivation experiments are shown for the single-impeller setup (SIS) and the three-impeller setup (TIS). Cultivation results are presented for both reactor setups to demonstrate the biological responses to different hydrodynamic conditions in the different setups. Finally, it was examined to what extent the observed effects can be explained on the basis of the hydrodynamic parameters determined via CFD simulations, where mainly the fluid velocity, power input, or energy dissipation rate were considered. It was investigated whether locally resolved parameters and parameter distributions had an advantage in this respect compared to the averaged values. This is further discussed in Section 4.

3.1. Characterization of Hydrodynamics of Single- and Multiple-Impeller Setups with CFD Methods

For both bioreactor setups (SIS and TIS, respectively), the hydrodynamic parameters fluid velocity u, shear rate γ, and energy dissipation rate ε were evaluated in the CFD study. Locally resolved values of these parameters were determined along the reactor height.

For example, the hydrodynamic parameters u and ε for 400 rpm are shown in Figures 1 and 2a (SIS) and Figure 2b (TIS), plotted in longitudinal section (for shear rate γ, see Figure A2). To gain a deeper numerical insight into locally resolved values, the CFD computed values for velocity, shear rate, and energy dissipation rate were averaged for horizontal slices along the reactor height. The mean values were then plotted over these slices to quantitatively visualize locally resolved values for all simulated agitation rates (slice plots in Figures 1 and 2c,d).

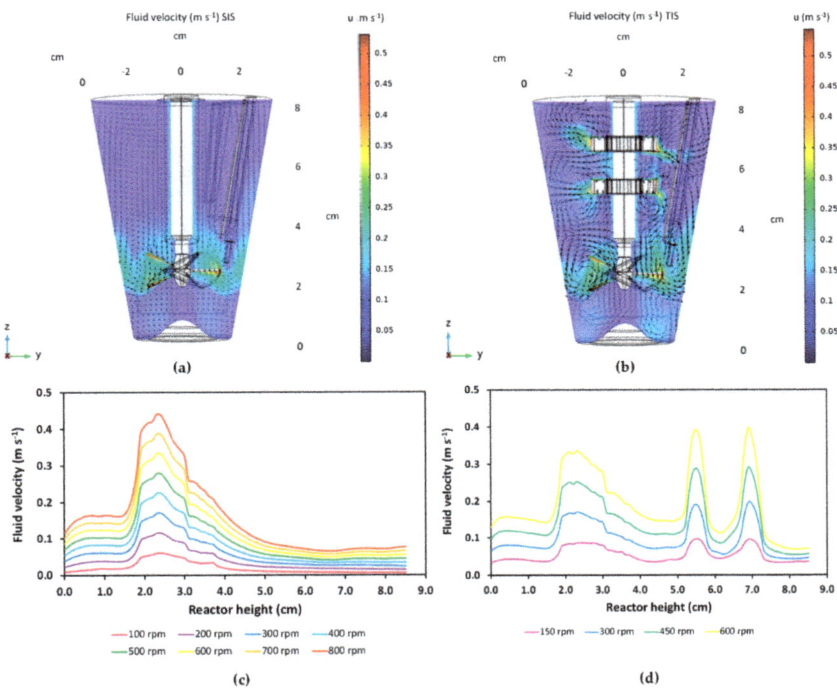

Figure 1. Fluid velocity u in the SIS and TIS. (**a**) Fluid velocity in the SIS; (**b**) Fluid velocity in the TIS. The shown data correspond to an agitation rate of 400 rpm. (**c**) Mean fluid velocity along the reactor height (slice plot) in the SIS for different agitation rates; (**d**) mean fluid velocity along the reactor height (slice plot) in the TIS for different agitation rates. In (**c,d**), the flow velocity was averaged across the diameter in steps of one millimeter along the reactor height.

From the plots of the fluid velocity u (see Figure 1a,b), the flow pattern in the respective reactor setups can be derived. Both setups were simulated and operated in down flow mode. In the SIS (Figure 1a), the fluid is pressed downward and sideways by the impeller at an approximately 2.5 cm reactor height. Then, it rises to the surface at the vessel walls and comes down again close to the stirrer shaft, yielding a typical axial flow pattern. The fluid velocity u reaches maximal values near the impeller blades up to 0.55 m s^{-1}. With increasing distance from the impeller, the fluid velocity u decreases to approx. 0.05 m s^{-1}. In the TIS (Figure 1b), the fluid in the upper parts of the reactor is pushed to the side by the six blade impellers, known as radial flow pattern at approx. 5.5 cm and 7 cm reactor height. At the vessel walls, it flows upward and back downward to the impeller close to the stirrer shaft, resulting in a typical radial flow pattern. The system averaged fluid velocity u_{av} was about twice as high as in the SIS with averaged 0.11 m s^{-1}.

The plots in Figure 1c,d show general high local values as well as high maxima in narrow areas close to the impeller at about a 2.5 cm reactor height, while large parts of the reactor stayed at values up to five-fold lower. In the TIS, additional peaks could be observed at 5.5 cm and 7 cm reactor height due to the added impellers. It can be seen that the velocity in both setups reached the highest values in regions close to the impellers. Nevertheless, due to the additional impellers, the velocity profile in the TIS was more homogenized than in the SIS.

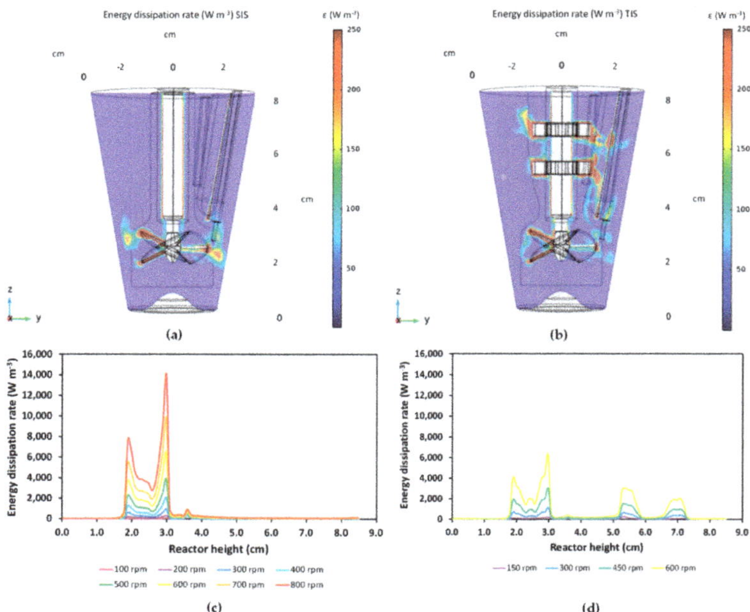

Figure 2. Energy dissipation rate ε in the SIS and TIS: (**a**) energy dissipation rate in the SIS; (**b**) energy dissipation rate in the TIS. The shown data correspond to an agitation rate of 400 rpm. (**d**) Mean energy dissipation rate along the reactor height (slice plot) in the SIS for different agitation rates; (**d**) mean energy dissipation rate along the reactor height (slice plot) in the TIS for different agitation rates. In (**c,d**), the energy dissipation rate was averaged across the diameter in steps of one millimeter along the reactor height.

The local energy dissipation rates are shown in Figure 2. In the SIS, the energy dissipation rate $\varepsilon_{\max}$ reached values up to 139 kW m^{-3} directly at the impeller (see Figure 2a). The slice plots (see Figure 2c,d) revealed that locally resolved values for the energy dissipation rate showed quite sharp and narrow peaks (also compare Figure A2 for the shear rate γ). At 800 rpm, the peaks of the energy dissipation rate close to the impeller in 2 cm to 3 cm reactor height are up to 280 times higher compared to the rest of the cultivation system. In this area, the mean energy dissipation rate ε across the diameter showed a maximum $\varepsilon_{\max,SP}$, with extent values of over 14 kW m^{-3}, despite it staying between 50 and 100 W m^{-3} in the largest parts of the reactor. The system average volumetric energy dissipation rate ε_{av}, which corresponds to the averaged volumetric power input P/V, did not exceed 110 W m^{-3} for 800 rpm. In the TIS, additional peaks could again be found in the height of the six blade impellers at 5.5 cm and 7.0 cm reactor height. At 600 rpm, the slice plot maximum $\varepsilon_{\max,SP}$ of the pitched blade impeller at 3 cm reactor height was over 6 kW m^{-3}, and the maximum of the six blade impellers was only slightly over 3 kW m^{-3}.

The slice plots (averaged values across the diameter) were obtained for different agitation rates to evaluate the spatial distribution of the investigated hydrodynamic parameters within both reactor systems (see Figures 1 and 2c,d). They all showed peaks for all parameters in the area of the impellers, which were magnitudes higher than in the remaining parts of the system. From the plots, it can be concluded that the characteristic profiles of the six blade impellers were added to the profile of the pitched blade impeller. Pre-investigations to this study showed that these profiles were moved but not altered with a change in the impeller position. In the TIS, maxima from slice plots $\varepsilon_{\max,SP}$ for the pitched blade impeller tended to have the same magnitudes as in the SIS for corresponding agitation rates. However, the characteristic peaks of the six blade impellers were added to the peak of the pitched blade impeller, which led to a higher energy dissipation rate ε_{av} averaged for the whole system, but lower peaks for the same system averaged energy dissipation rate ε_{av}. Hence, setups with multiple impellers yielded flatter energy dissipation rate ε distributions for the same averaged power inputs P/V in the same reactor vessels. Further values for agitation rates, averaged dissipation rates ε_{av}, and maximum energy dissipation rates ε_{max} for both systems can be found in Tables A1 and A2 in Appendix A.

3.2. Cultivation Results

Cultivation experiments with CHO DP-12 cells were run in both reactor systems at different agitation rates. For all samples of the cultivation experiments, cell densities and antibody concentrations were quantified. Mean exemplary data containing viable cell densities and viabilities for cultivations at 400, 800, and 1200 rpm in the SIS are shown in Figure 3.

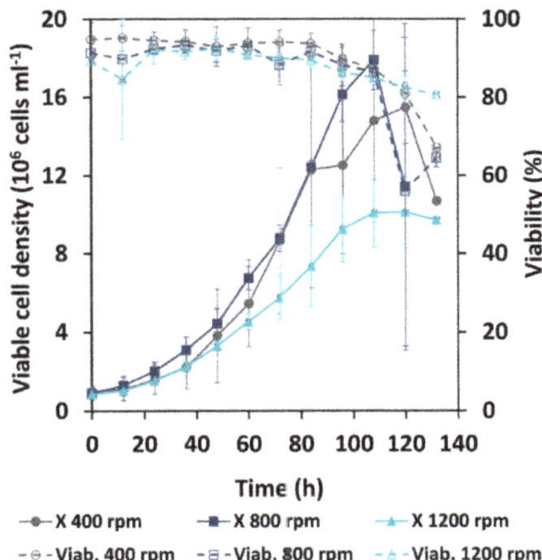

Figure 3. Mean growth curves (viable cell density X and viability) for the cultivations in the SIS at different agitation rates and respective averaged power inputs: 400 rpm (16 W m^{-3}); 800 rpm (110 W m^{-3}); 1200 rpm (356 W m^{-3}). Data were averaged from three cultivations each.

Figures 4 and 5 show the results of the cultivation experiments performed in the TIS. Surprisingly, the effect of the agitation rate on maximum cell density, viability, or growth rate seemed to be less pronounced as that for the SIS, even at 1400 rpm. Obviously, the stable operational range was broader when using multiple impellers.

From Figures 3 and 5 (The maximum cell densities are summarized in Figure 5), it can be concluded that the maximal cell density and the cell specific growth rate (not shown) were more or less unaffected by the agitation rate up to 800 rpm. For 1000 rpm and higher agitation rates, a decrease in maximum cell densities and cell specific growth rate was observed. Furthermore, the viability had already started to decrease in processes at high agitation rates above 1000 rpm before reaching the maximum cell density.

With respect to the antibody production behavior of CHO DP-12 cells, the space time yield (STY) was calculated for all performed processes (Figure 6). For the SIS experiments, the highest STYs with up to 2.9 mg L^{-1} h^{-1} are observed between agitation rates of 400 rpm and 800 rpm; for higher agitation rates, the STY declined and at 1400 rpm, the SIS of the antibody concentration in all samples was even below the detection limit of 0.15 g L^{-1}. For the TIS, the STYs were approx. constant at around 1.5 mg L^{-1} h^{-1} for agitation rates between 770 rpm and 1400 rpm. These results also support the idea of a broader stable operational range when using multiple impellers.

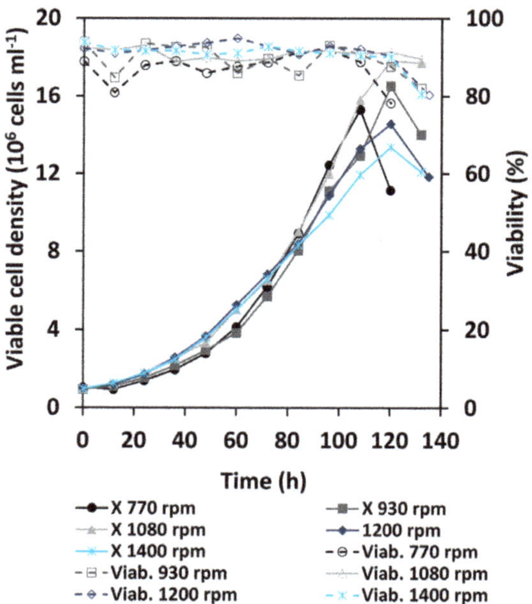

Figure 4. Growth curves (viable cell density X and viability) for the cultivations in the TIS at different agitation rates and respective averaged power inputs. The agitation rate settings range from 770 rpm or 454 W m^{-3} to 1400 rpm or 4742 W m^{-3}. One experiment was performed at each agitation rate.

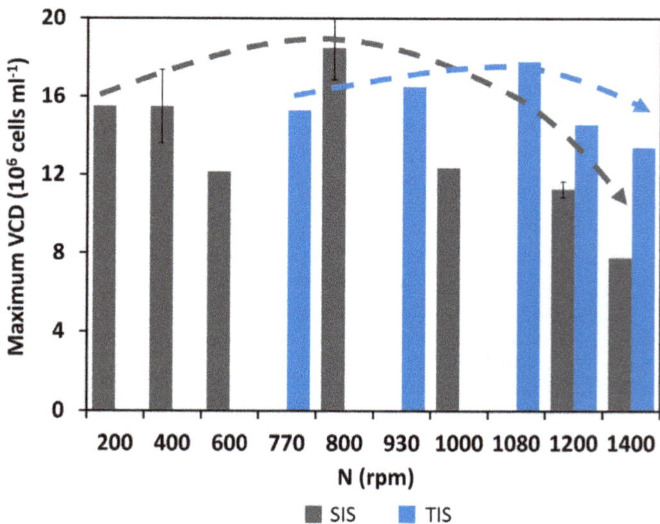

Figure 5. Maximum viable cell densities (VCD) compared for experiments in both reactor setups: cultivation results from the SIS (grey bars); cultivation results from the TIS (blue bars). Data for 400, 800, and 1200 rpm in the SIS were averaged from three cultivations each. Dashed arrows indicate qualitative trends for each reactor setup.

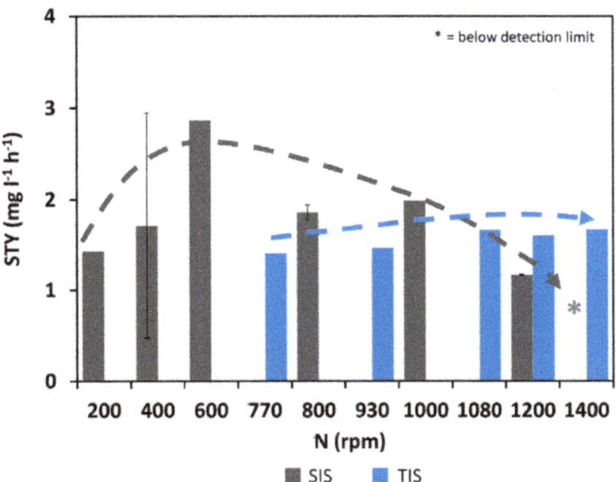

Figure 6. Antibody space time yields (STY) compared for experiments in both reactor setups: cultivation results from the SIS (grey bars); cultivation results from the TIS (blue bars). Data for 400, 800, and 1200 rpm in the SIS were averaged from three cultivations each. Dashed arrows indicate expected trends for each reactor setup.

3.3. Comparison of Cellular Behavior and Hydrodynamic Parameters

In the following, the presented experimental findings from cell cultures and hydrodynamic parameters are compared. In particular, it must be noted that the cells in the multi-stage reactors can be cultivated stably over a much broader operation range. This comparison will be made on the basis of various definitions for the energy dissipation rate including averaged and locally resolved ones. On one hand, parameters frequently used for process design, average power input P/V, and Kolmogorov length scale λ based on average cell diameters $d_{c,av}$ are discussed. On the other hand, locally resolved parameters such as energy dissipation rate ε distributions from CFD simulations, critical energy dissipation rate ε_{krit} distributions calculated from cell size d_c distributions, maximum energy dissipation rates ε_{max}, maximum energy dissipation rates from slice plots $\varepsilon_{max,SP}$ and variance coefficients of hydrodynamic parameters u, γ, and ε will be considered.

The Kolmogorov length scale λ [12,13] is a commonly used tool to evaluate the impact of hydrodynamics on mammalian cell cultures. With respect to flow induced cell damage, it is assumed that significant cell damage has to be expected when the size of the smallest turbulence eddy length is in the order of magnitude of the cell size. Therefore, for a known cell diameter d_c, the respective critical energy dissipation rate ε_{krit} and hence, an appropriate agitation rate, can be estimated. Usually, the system averaged energy dissipation rate ε_{av} is used to calculate the Kolmogorov length scale λ and compared to a mean cell diameter d_c. The concept seems to work for microcarrier cultures, but is under discussion for suspension cells [14–19]

As an example, the average cell size d_c of CHO DP-12 cells with about 10 μm would result in a critical energy dissipation rate ε_{krit} of about 32 kW m^{-3}, which would be far above the calculated average energy dissipation rates ε_{av} even at 1400 rpm, where a strong decrease in maximum viable cell density and antibody productivity were observed. Vice versa, an agitation rate calculated for process design with this approach would be way too high and might lead to undesired process behavior. Therefore, in order to evaluate a possible relationship between the cell size d_c distribution and the Kolmogorov length scale λ, critical energy dissipation rate ε_{krit} distributions (see Figures 7 and 8) were calculated from actual cell size d_c distributions. Exemplary cell size d_c distributions obtained from cultivation data are shown in the Appendix A (Figure A3). It was observed that cell size distributions neither expressively differed with changing agitation rates nor were dependent on the cultivation system.

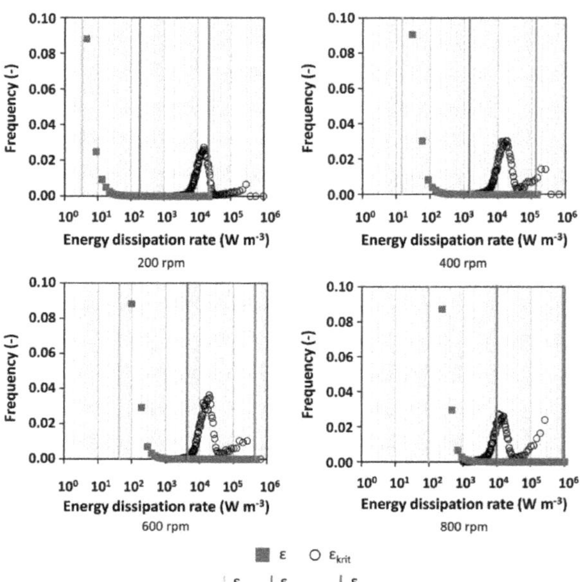

Figure 7. Critical energy dissipation rate ε_{krit} distributions (black circles) calculated from the cell size distributions in the SIS compared to the energy dissipation rate distribution in the reactor system (grey squares) obtained from CFD data. The averaged energy dissipation rates from CFD simulations are indicated in blue, their maxima in red, maximum energy dissipation rates calculated from slice plots are indicated in purple.

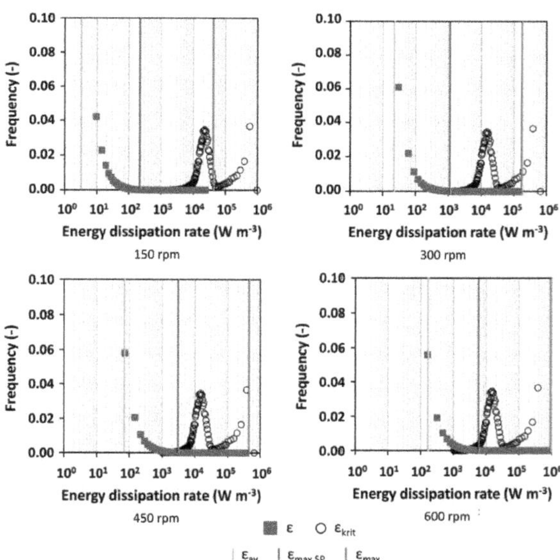

Figure 8. Critical energy dissipation rate distributions (black circles) in the TIS compared to the energy dissipation rate distribution in the reactor system (grey squares). The averaged energy dissipation rates from CFD simulations are indicated in blue, their maxima in red, maximum energy dissipation rates calculated from slice plots are indicated in purple.

The distribution of the normalized frequency of the energy dissipation rate ε from CFD data for different agitation rates is shown in Figures 7 and 8. Furthermore, the averaged and maximum energy dissipation rates ε_{av} and ε_{max} from CFD simulations as well as the maxima of the slice plots are indicated in the diagrams.

In the cultivation experiments (see Section 3.2) with the SIS, no decrease in cell growth or antibody productivity was observed up to 800 rpm. The averaged energy dissipation rate ε stayed a magnitude below the critical energy dissipation rate ε_{krit} distribution curve, calculated from the cell size distributions. The maximum energy dissipation rates ε_{max} from the CFD simulations were already above the main cell peak of the critical energy dissipation rate ε_{krit} distribution at the lowest agitation rate. At 800 rpm, the maximum value was right below the peak that represents the main cell population. With increasing agitation rates, the maximum energy dissipation rate ε_{max} also increases and, according to the Kolmogorov length scale hypothesis, for half of the cell population, shear related effects would have to be expected.

With CFD tools, it is possible to obtain normalized volumetric distributions of hydrodynamic parameters for the simulated systems [21,27]. Since the so far compared parameters might not provide satisfying information on the impact of shear forces, the actual normalized energy dissipation rate ε distribution in the bioreactor was compared to the critical energy dissipation rate ε_{krit} distribution, calculated from the cell size distributions (see Figures 7 and 8).

It was found that the energy dissipation rate in the largest part of the vessel volume stayed below the critical energy dissipation rate distribution curve. Only in a very small fraction did the energy dissipation rate in the reactor reach the area of the critical energy dissipation rate distribution. Although agitation rates above 800 rpm cannot be simulated, it can be concluded that the energy dissipation rate distribution curve of the reactor will move further into the critical energy dissipation rate distribution curve, according to extrapolations of the average and maximum values.

Additionally, for the TIS, the critical energy dissipation rate distributions were calculated and compared to the actual energy dissipation rate distribution in the cultivation system as well as the maximum energy dissipation rate from the slice plots. The results are shown in Figure 8.

For comparison of the different reactor systems, the plots for 600 rpm and 800 rpm in the SIS and 450 rpm in the TIS are shown in Figure 9. Their average volumetric power inputs were within the same order of magnitude (47 W m^{-3} to 107 W m^{-3}) and clear differences are recognized.

In all cultivations in the TIS at 770 rpm to 1400 rpm, no noteworthy changes in cell growth or antibody productivity were present. From extrapolations of the slice plot maxima, which resulted in 64 kW m^{-3} at 1400 rpm, which was just at the end of the main cell population peak (modal value at 10 kW m^{-3} to 20 kW m^{-3}), it can be derived that the maximum from slice plots might not be a suitable parameter to make a connection between hydrodynamics and biology. Additionally, for this setup, the energy dissipation rate stayed below the critical energy dissipation rate distribution curve in the vast majority of the reactor volume. To have an optical impression of critical volume fractions, the areas in the cultivation system with energy dissipation rates above 1000 W m^{-3} were identified. Detailed pictures can be found in the Appendix A (Figure A4).

For further analysis of both reactor systems, the hydrodynamic homogeneity was evaluated. Therefore, variance coefficients were calculated as the standard deviation from slice plot data divided by their mean value, which are shown in Figure 10.

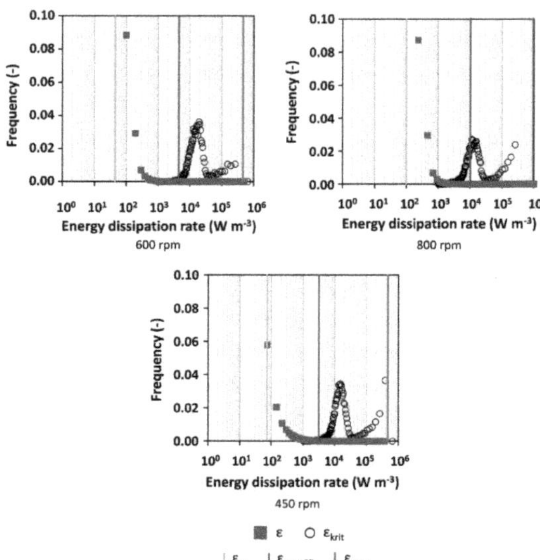

Figure 9. Critical energy dissipation rate distributions (black circles) compared to the energy dissipation rate distribution in the SIS (600 rpm and 800 rpm) and in the TIS (450 rpm) (grey squares). The averaged energy dissipation rates from CFD simulations are indicated in blue, their maxima in red. Maximum energy dissipation rates calculated from slice plots are indicated in purple.

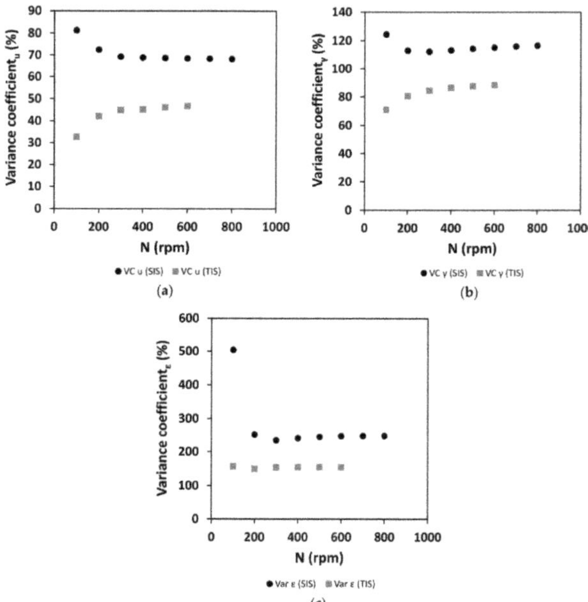

Figure 10. Variance coefficients for hydrodynamic parameters: (**a**) fluid velocity u, (**b**) shear rate γ, and (**c**) energy dissipation rate ε in both cultivation systems. The coefficients were calculated from standard deviations from slice plot data divided by their mean value. They represent a measure of hydrodynamic homogeneity of the respective reactor system.

It can be seen that the variance coefficient of all three hydrodynamic parameters (fluid velocity u, shear rate γ, and energy dissipation rate ε) was higher in the SIS. This means that the SIS is a more heterogeneous system than the TIS. This might possibly be an explanation for the more stable process behavior of the TIS with respect to a changing agitation rate.

4. Discussion

The previously shown results from this study are discussed in the following under different aspects. The process stability that depends on hydrodynamics is debated and the application of the Kolmogorov length scale hypothesis with modern CFD methods is highlighted. Finally, consequences for the scale-up and scale-down of the bioreactors are drawn.

4.1. Hydrodynamic Differences for Single and Multi-Stage Impellers

Certain differences in flow pattern, velocity, and other hydrodynamic parameters have been observed for multiple impellers in the same stirred reactor system. It was found from CFD studies that the system averaged fluid velocity, shear rate, and energy dissipation increased with the addition of multiple impellers. The maximum shear rate and energy dissipation rate also increased in the reactor setup with multiple impellers. Only the maximum fluid velocity remained almost constant as it depends on the agitation rate solely. Overall, it can be concluded that the use of multiple impellers leads to a more uniform fluid flow and therefore to a lower variance coefficient of the hydrodynamic parameters.

4.2. Stability of Cell Culture Processes Related to Locally Resolved Hydrodynamics

Cultivation results for both reactor setups showed clear differences in the process behavior with respect to changing agitation rates. While the maximum cell density decreased up to 60% in the SIS, and antibody productivity decreased up to 100% above 1000 rpm (averaged 203 W m^{-3}), no changes were recorded for agitation rates up to 1400 rpm (averaged 4742 W m^{-3}) in the TIS. Platas et al. [28] reported a rather stable process behavior at higher agitation rates when using a bioreactor with multiple impellers instead of a single impeller. We identified that local hydrodynamic maxima and hydrodynamic heterogeneities, represented by variant coefficients were less pronounced in the TIS at the same averaged energy dissipation rate, resulting in a more homogeneous fluid flow. From the cultivation data, it could be concluded that this leads to a more robust process behavior.

4.3. Evaluation of Concepts for Estimation of Shear Related Parameters

From the results of the cultivation experiments, it can be concluded that the average volumetric power input or energy dissipation rate is clearly not suitable as a universal shear-related parameter for bioprocess design. The two examined reactor systems showed a different process behavior for the same averaged energy dissipation rates. Even if the average dissipation rate increases way faster in the TIS with increasing agitation rate, the process is still more stable with respect to the agitation rate in the TIS. Maximum energy dissipation rates are also not suitable.

None of the parameters calculated from CFD data corresponded to the cultivation process behavior with changing agitation rate for both cultivation systems. Hence, it can be assumed that these parameters are not sufficient as indicators for the impact of hydrodynamics on the cells. However, solely, the maximum energy dissipation rates from the slice plots $\varepsilon_{\text{max,SP}}$ could possibly show a connection to the cellular effects.

The Kolmogorov length scale is commonly used to describe the size of smallest eddies that might lead to cell damage [14–19]. In this study, it was calculated in the form of distributions instead of one fixed length for the whole bioreactor due to strong hydrodynamic heterogeneities in the cultivation systems. In addition, cell size distributions instead of a mean cell size were considered. The resulting critical energy dissipation rate distribution, calculated from cell size distributions, was compared to the averaged and maximum energy dissipation rate as well as to the maxima from slice plots and volumetric distributions of the energy dissipation rate, all from CFD simulations.

It was found that the Kolmogorov length scale is not suitable to describe a link between hydrodynamics and cell damage, if the averaged or maximum energy dissipation rate is used since they do not correlate with any biological observations (Figures 3–6). Furthermore, critical volume fractions of the cultivation vessels do not deliver the desired insights, despite the fact that the energy dissipation rate distribution curve moves into the critical region with increasing agitation rate. Nevertheless, extrapolations of the maximum energy dissipation rates from slice plots showed a slightly faster increase for the TIS. Since cells grew with roughly the same growth rate up to the highest agitation rate in this setup, this parameter does not seem to be suitable to describe a link between cell growth and hydrodynamics, but still seems to be a closer estimation than the other parameters.

The hydrodynamic heterogeneity of a bioreactor, which was quantified in this study with variance coefficients of the hydrodynamic parameters fluid velocity, shear rate, and energy dissipation rate, might be a useful parameter to estimate the suitability of a cultivation system. The calculated variance coefficients of all hydrodynamic parameters were higher in the TIS than in the SIS, which might explain the rather stable process behavior in multiple impeller systems due to the improved hydrodynamic homogeneity.

4.4. Learning for Reactor Scale-Up and Scale-Down

When performing scale-up of bioreactors, common criteria that are kept constant are the impeller tip speed or the averaged power input [7–10]. In this study, the averaged power input increased much faster with increasing agitation rate in the TIS than in the SIS. However, no prominent differences in process behavior were found in cultivations at different agitation rates in the TIS, while in the SIS, the maximum cell densities and antibody productivities were reduced at high agitation rates. Consequently, the cultivation data showed that the biological process behavior does not solely depend on single process parameters such as the averaged power input, which are commonly used in process design, but rather on their homogeneity. Since maxima of hydrodynamic parameters also increased faster in the TIS and no relevant changes in process behavior were recorded, it is likely that the whole reactor system and locally resolved hydrodynamics need to be considered. Therefore, locally resolved hydrodynamic parameters were carved out of the CFD data as slice-plots and as volume fraction specific distributions. It was observed that strong hydrodynamic heterogeneities were present in stirred tank reactor systems, which became steeper with increasing agitation rate. A multi-parametrical approach considering locally resolved hydrodynamics resulting from geometrical characteristics could be the consequence for process scale-up. In turn, the investigation of local hydrodynamic effects has been the focus of late. Local distributions of the shear stress and the Kolmogorov length scale have been evaluated for spinner flasks, but have not been compared to cell size distributions thus far [27]. Furthermore, the calculation of cellular residence times in differently mixed reactor areas could provide deeper insights on the actual impact of hydrodynamic phenomena on cellular behavior [22,29]. Another recent study showed that it is worth considering the local distributions of conventional process design parameters such as $k_L a$ [10]. In addition, the local determination of the power input shows promising results for the design of cultivation processes for human mesenchymal stem cells. These approaches could be combined with novel uncertainty-based model evaluation methods [6,30]. It is noteworthy that the six blade impellers, specifically designed for the TIS in this study, worked immediately for the cultivation of CHO DP-12 cells.

5. Conclusions

In this study, parameters obtained from CFD simulations were linked to cell culture cultivation experiments to investigate the influence of hydrodynamic indifferent reactor setups on cell growth and antibody productivity. It was shown that hydrodynamically more uniform conditions and the resulting flatter hydrodynamic profiles might be the reason for the broader stable operational space regarding the agitation rate of a stirred tank with multiple impellers, which was previously observed by Platas et al. [28]. The evaluation of the CFD results clearly showed stronger pronounced hydrodynamic heterogeneities for the same power input in single impeller setups. Hence, the use of conventional, mostly averaged process design parameters, needs to be questioned and rather, local gradients should be considered, especially for scale-up [22,29]. Furthermore, it was found that the Kolmogorov length scale hypothesis might not be appropriate to describe the influence of hydrodynamics on mammalian cell culture processes without a hydrodynamic characterization of the full cultivation system including all hydrodynamic gradients. Overall, a deeper insight into local hydrodynamic gradients in cell culture reactors was obtained, but the reliable prediction of design parameters prior to an experimental evaluation still remains difficult.

Supplementary Materials: The following supporting information can be downloaded at: https://www.mdpi.com/article/10.3390/pr10010107/s1, File S1: Supplementary File SIS Model Report.pdf, File S2: Supplementary File TIS Model Report.pdf.

Author Contributions: Conceptualization, F.F., J.M., R.P. and M.S.; Methodology, F.F.; Software, F.F. and J.B.; Validation, F.F. and J.M.; Formal analysis, F.F. and E.A.; Investigation, F.F., E.A. and J.B.; Resources, R.P.; Data curation, F.F., E.A. and J.B.; Writing—original draft preparation, F.F., J.M., and R.P.; Writing—review and editing, M.S.; Project administration, J.M. and R.P. All authors have read and agreed to the published version of the manuscript.

Funding: This research received no external funding.

Institutional Review Board Statement: Not applicable.

Informed Consent Statement: Not applicable.

Conflicts of Interest: The authors declare no conflict of interest.

Appendix A

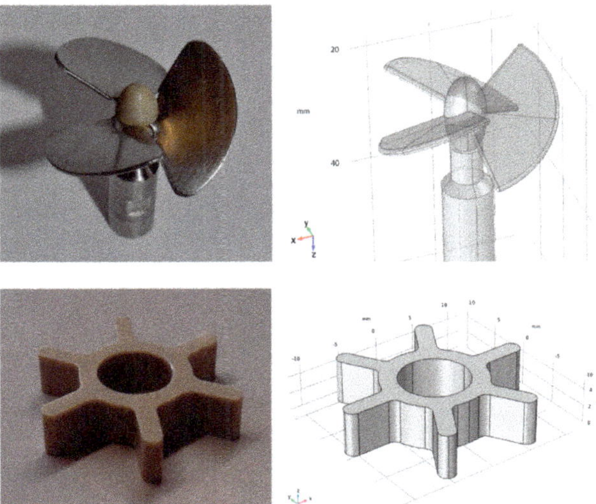

Figure A1. Pitched blade impeller (**above**) and six blade impeller (**below**) with corresponding CAD models. For the pitched blade impeller, a Newton number Ne of 0.35 was determined with an empiric correlation [7], while for the six blade impeller, 3.95 was considered, following another empiric correlation [31].

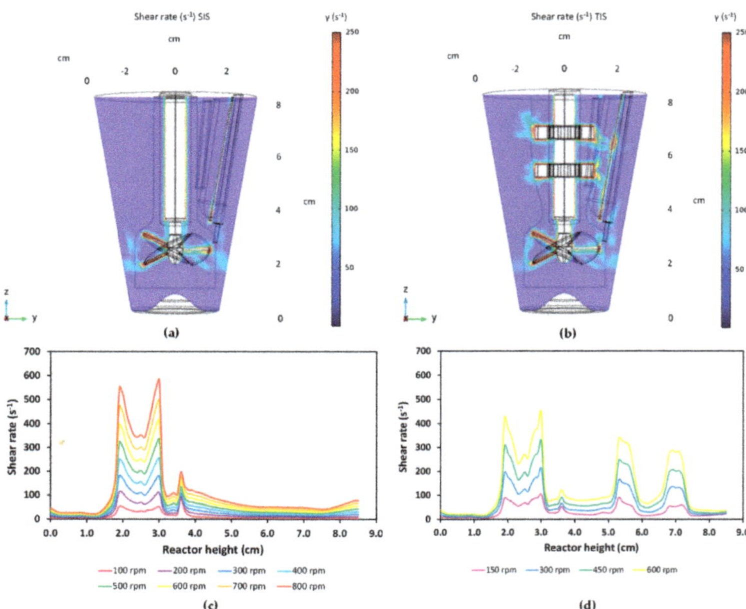

Figure A2. Shear rate γ in the SIS and TIS: (**a**) shear rate in the SIS; (**b**) shear rate in the TIS. The shown data correspond to an agitation rate of 400 rpm. (**d**) Mean shear rate along the reactor height (slice plot) in the SIS for different agitation rates; (**d**) mean shear rate along the reactor height (slice plot) in the TIS for different agitation rates. In (**c**,**d**), the shear rate was averaged across the diameter in steps of one millimeter along the reactor height.

In both cases, for most parts of the reactor systems, a quite uniform distribution of the shear rate could be observed. The mean shear rate did not exceed 100 s^{-1}, but reached almost 600 s^{-1} at the height of the stirrer in the SIS, and the shear rate showed values of up to 9167 s^{-1} at 400 rpm. Only small parts of the reactor reached values above 1000 s^{-1}, namely at the stirrer. In the largest part along the reactor height, the mean shear rate did not exceed 100 s^{-1}, but reached almost 600 s^{-1} at the height of the stirrer. The maximum values in the TIS were similar for equal agitation rates with 9383 s^{-1} at 400 rpm, but again, two additional peaks could be found in the slice plots, representing the added impellers. However, the values between the stirrers were still comparably low, below 100 s^{-1}.

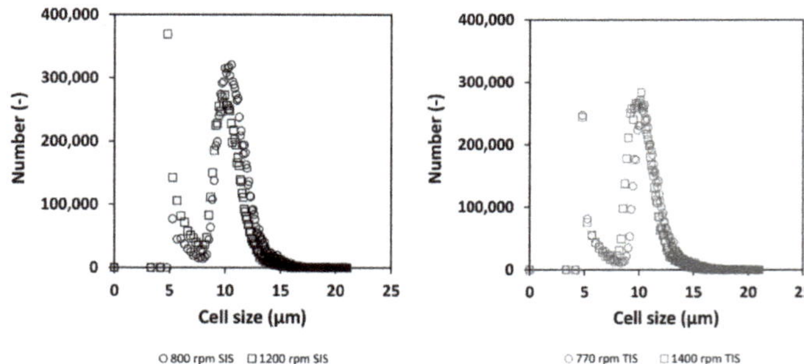

Figure A3. Exemplary cell size distributions from cultivation data in the SIS (800 rpm and 1200 rpm) and in the TIS (770 rpm and 1400 rpm). Distributions for all cultivations were taken from samples at 72 h cultivation time.

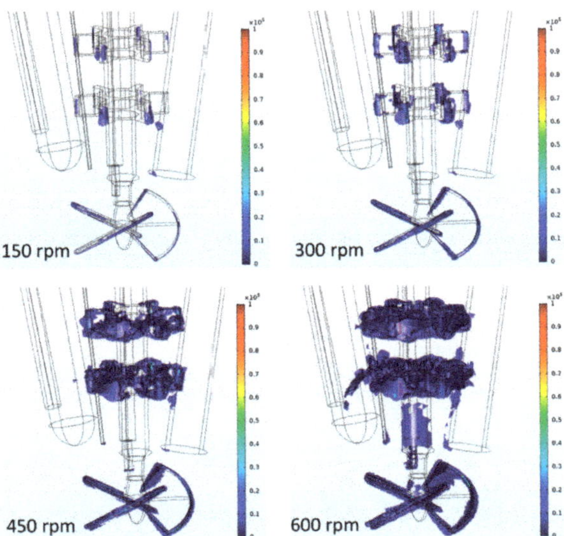

Figure A4. Areas in the TIS with energy dissipation rates above 1000 W m^{-3}.

The regions above 1000 W m^{-3} are all located in thin layers on the pitched blade impellers, small areas around the six blade impellers, and very small parts at the edges of probes and other inserts.

With a defined threshold for the critical energy dissipation rate, the critical volume fractions of the cultivation system can be calculated from CFD data. They are linearly dependent on the power input or energy dissipation rate. Due to the respective number of impellers, the different reactor setups result in different critical reactor volume fractions for the same power input. The critical areas were smaller for equal averaged power inputs in the TIS due to flatter gradients. Cultivation experiments in the SIS showed a decrease in cell growth and antibody productivity above 1000 rpm. In the TIS, no changes were recorded. Since the critical reactor volume for 1400 rpm was about five-fold higher in the TIS (0.45% at 4.7 kW m^{-3}) than in the SIS (0.07% at 465 W m^{-3}), it can be concluded that the critical volume fraction might not be a suitable parameter to also evaluate the influence of hydrodynamics on the process.

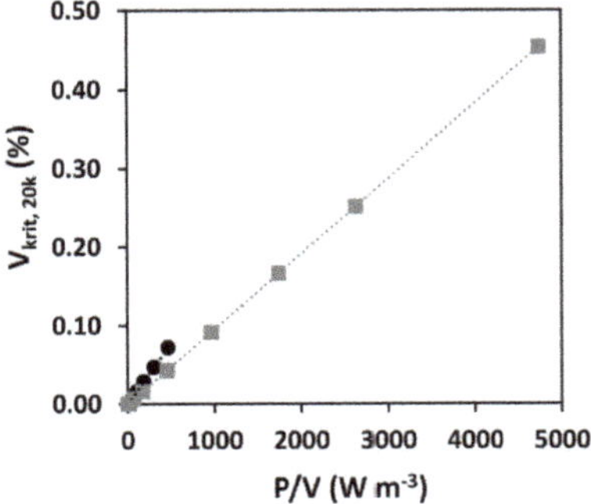

Figure A5. Critical reactor volume fractions with energy dissipation rates above 20 kW m^{-3} plotted over the corresponding averaged power input. Black circles represent values for the SIS, grey squares those of the TIS.

Table A1. Hydrodynamic parameters in the SIS based on CFD simulations. All values above 800 rpm were extrapolated linear (velocity and shear rate) or quadratic, respectively (energy dissipation rate).

N (rpm)	u_{av} (m s^{-1})	γ_{av} (s^{-1})	ε_{av} (W m^{-3})	ε_{max} (W m^{-3})	$\varepsilon_{max,SP}$ (W m^{-3})
200	0.032	10.06	2.39	22,169	214
400	0.068	21.23	15.35	146,510	1575
600	0.103	32.22	47.35	491,767	4959
800	0.138	43.15	107.12	1,159,628	11,001
1000	0.171	53.56	203.56	2,265,066	20,239
1200	0.205	64.27	345.05	3,932,425	33,194
1400	0.240	74.99	540.50	6,286,050	50,385

Table A2. Hydrodynamic parameters in the TIS based on CFD simulations. All values above 600 rpm were extrapolated linearly (velocity and shear rate) or quadratically, respectively (energy dissipation rate).

N (rpm)	u_{av} (m s^{-1})	γ_{av} (s^{-1})	ε_{av} (W m^{-3})	ε_{max} (W m^{-3})	$\varepsilon_{max,SP}$ (W m^{-3})
150	0.039	13.24	3.63	45,738	221
300	0.081	28.62	23.86	194,000	1118
450	0.125	43.93	72.90	479,000	3014
600	0.159	55.53	176.93	1,010,000	6247
770	0.208	72.98	454.35	2,007,732	12,103
930	0.252	88.15	964.25	3,401,191	20,258
1080	0.292	102.36	1744.79	5,172,104	30,686
1200	0.325	113.74	2633.29	6,952,684	41,252
1400	0.379	132.69	4742.49	10,725,752	63,867

References

1. Hu, W.; Berdugo, C.; Chalmers, J.J. The Potential of Hydrodynamic Damage to Animal Cells of Industrial Relevance: Current Understanding. *Cytotechnology* **2011**, *63*, 445–460. [CrossRef] [PubMed]
2. Godoy-Silva, R.; Chalmers, J.J.; Casnocha, S.A.; Bass, L.A.; Ma, N. Physiological responses of CHO cells to repetitive hydrodynamic stress. *Biotechnol. Bioeng.* **2009**, *103*, 1103–1117. [CrossRef] [PubMed]
3. Chalmers, J.J.; Ma, N. Hydrodynamic Damage to Animal Cells. In *Animal Cell Culture*; Al-Rubeai, M., Ed.; Springer International Publishing: Basel, Switzerland, 2015; pp. 169–183.
4. Nienow, A.W.; Scott, W.H.; Hewitt, C.J.; Thomas, C.R.; Lewis, G.; Amanullah, A.; Kiss, R.; Meier, S.J. Scale-down studies for assessing the impact of different stress parameters on growth and product quality during animal cell culture. *Chem. Eng. Res. Des.* **2013**, *91*, 2265–2274. [CrossRef]
5. Jossen, V.; Eibl, D.; Eibl, R. Numerical Methods for the Design and Description of In Vitro Expansion Processes of Human Mesenchymal Stem Cells. *Adv. Biochem. Eng. Biotechnol.* **2020**, *177*, 185–228. [CrossRef]
6. Arndt, L.; Wiegmann, V.; Kuchemüller, K.B.; Baganz, F.; Pörtner, R.; Möller, J. Model-based workflow for scale-up of process strategies developed in miniaturized bioreactor systems. *Biotechnol. Prog.* **2021**, *37*, e3122. [CrossRef]
7. Doran, P.M. *Bioprocess Engineering Principles*, 2nd ed.; Academic Press: Cambridge, MA, USA, 2013.
8. Storhas, W. *Bioverfahrensentwicklung*; John Wiley & Sons: Hoboken, NJ, USA, 2003; ISBN 3-527-28866-X.
9. Xu, S.; Hoshan, L.; Jiang, R.; Gupta, B.; Brodean, E.; O'Neill, K.; Seamans, T.C.; Bowers, J.; Chen, H. A practical approach in bioreactor scale-up and process transfer using a combination of constant P/V and vvm as the criterion. *Biotechnol. Prog.* **2017**, *33*, 1146–1159. [CrossRef]
10. Scully, J.; Considine, L.B.; Smith, M.T.; McAlea, E.; Jones, N.; O'Connell, E.; Madsen, E.; Power, M.; Mellors, P.; Crowley, J.; et al. Beyond heuristics: CFD-based novel multiparameter scale-up for geometrically disparate bioreactors demonstrated at industrial 2kL–10kL scales. *Biotechnol. Bioeng.* **2020**, *117*, 1710–1723. [CrossRef]
11. Rosseburg, A.; Fitschen, J.; Wutz, J.; Wucherpfennig, T.; Schlüter, M. Hydrodynamic inhomogeneities in large scale stirred tanks—Influence on mixing time. *Chem. Eng. Sci.* **2018**, *188*, 208–220. [CrossRef]
12. Kolmogorov, A. Dissipation of Energy in the Locally Isotropic Turbulence. *Dokl. Akad. Nauk SSSR* **1941**, *32*, 538–541.
13. Kolmogorov, A. The Local Structure of Turbulence in Incompressible Viscous Fluid for Very Large Reynolds Numbers. *Dokl. Akad. Nauk SSSR* **1941**, *30*, 301–305.
14. Cherry, R.S.; Papoutsakis, E.T. Hydrodynamic Effects on Cells in Agitated Tissue Culture Reactors. *Bioprocess Eng.* **1986**, *1*, 29–41. [CrossRef]
15. Cherry, R.S.; Papoutsakis, E.T. Physical mechanisms of cell damage in microcarrier cell culture bioreactors. *Biotechnol. Bioeng.* **1988**, *32*, 1001–1014. [CrossRef]
16. Cherry, R.S.; Papoutsakis, E.T. Growth and death rates of bovine embryonic kidney cells in turbulent microcarrier bioreactors. *Bioprocess Eng.* **1989**, *4*, 81–89. [CrossRef]

17. Hua, J.; Erickson, L.E.; Yiin, T.Y.; Glasgow, L.A. A review of the effects of shear and interfacial phenomena on cell viability. *Crit. Rev. Biotechnol.* **1993**, *13*, 305–328. [CrossRef]
18. Cherry, R.S. Animal cells in turbulent fluids: Details of the physical stimulus and the biological response. *Biotechnol. Adv.* **1993**, *11*, 279–299. [CrossRef]
19. Croughan, M.S.; Hamel, J.F.; Wang, D.I. Hydrodynamic Effects on Animal Cells Grown in Microcarrier Cultures. *Biotechnol. Bioeng.* **1987**, *29*, 130–141. [CrossRef]
20. Wang, G.; Haringa, C.; Tang, W.; Noorman, H.; Chu, J.; Zhuang, Y.; Zhang, S. Coupled metabolic-hydrodynamic modeling enabling rational scale-up of industrial bioprocesses. *Biotechnol. Bioeng.* **2020**, *117*, 844–867. [CrossRef]
21. Ueki, M.; Tansho, N.; Sato, M.; Kanamori, H.; Ito, Y.; Kato, Y. Improved cultivation of CHO cells in bioreactor with reciprocal mixing. *Authorea* **2020**. [CrossRef]
22. Kuschel, M.; Fitschen, J.; Hoffmann, M.; von Kameke, A.; Schlüter, M.; Wucherpfennig, T. Validation of Novel Lattice Boltzmann Large Eddy Simulations (LB LES) for Equipment Characterization in Biopharma. *Processes* **2021**, *9*, 950. [CrossRef]
23. Jossen, V.; Schirmer, C.; Mostafa Sindi, D.; Eibl, R.; Kraume, M.; Pörtner, R.; Eibl, D. Theoretical and practical issues that are relevant when scaling up hMSC microcarrier production porocesses. *Stem Cells Int.* **2016**, *2016*, 4760414. [CrossRef]
24. Isu, G.; Morbiducci, U.; Nisco, G.d.; Kropp, C.; Marsano, A.; Deriu, M.A.; Zweigerdt, R.; Audenino, A.; Massai, D. Modeling methodology for defining a priori the hydrodynamics of a dynamic suspension bioreactor. Application to human induced pluripotent stem cell culture. *J. Biomech.* **2019**, *94*, 99–106. [CrossRef]
25. Kaiser, S.; Jossen, V.; Schirmaier, C.; Eibl, D.; Brill, S.; van den Bos, C.; Eibl, R. Fluid Flow and Cell Proliferation of Mesenchymal Adipose-Derived Stem Cells in Small-Scale, Stirred, Single-Use Bioreactors. *Chem. Ing. Tech.* **2013**, *85*, 95–102. [CrossRef]
26. Möller, J.; Kuchemüller, K.B.; Steinmetz, T.; Koopmann, K.S.; Pörtner, R. Model-assisted Design of Experiments as a concept for knowledge-based bioprocess development. *Bioprocess Eng.* **2019**, *42*, 867–882. [CrossRef]
27. Ghasemian, M.; Layton, C.; Nampe, D.; Zur Nieden, N.I.; Tsutsui, H.; Princevac, M. Hydrodynamic characterization within a spinner flask and a rotary wall vessel for stem cell culture. *Biochem. Eng. J.* **2020**, *157*, 107533. [CrossRef]
28. Platas Barradas, O.; Jandt, U.; Minh Phan, L.D.; Villanueva, M.E.; Schaletzky, M.; Rath, A.; Freund, S.; Reichl, U.; Skerhutt, E.; Scholz, S.; et al. Evaluation of Criteria for Bioreactor Comparison and Operation Standardization for Mammalian Cell Culture. *Eng. Life Sci.* **2012**, *12*, 518–528. [CrossRef]
29. Fitschen, J.; Hofmann, S.; Wutz, J.; Hoffmann, M.; Wucherpfennig, T.; Schlüter, M. Novel evaluation method to determine the local mixing time distribution in stirred tank reactors. *Chem. Eng. Sci. X* **2021**, *10*, 100098. [CrossRef]
30. Möller, J.; Hernández Rodríguez, T.; Müller, J.; Arndt, L.; Kuchemüller, K.B.; Frahm, B.; Eibl, R.; Eibl, D.; Pörtner, R. Model uncertainty-based evaluation of process strategies during scale-up of biopharmaceutical processes. *Comput. Chem. Eng.* **2020**, *134*, 106693. [CrossRef]
31. Bates, R.L.; Fondy, P.L.; Corpstein, R.R. Examination of Some Geometric Parameters of Impeller Power. *Ind. Eng. Chem. Proc. Des. Dev.* **1963**, *2*, 310–314. [CrossRef]

Article

Seed Train Intensification Using an Ultra-High Cell Density Cell Banking Process

Jan Müller *, Vivian Ott , Dieter Eibl and Regine Eibl

Institute of Chemistry and Biotechnology, School of Life Sciences and Facility Management, ZHAW Zurich University of Applied Sciences, 8820 Wädenswil, Switzerland; vivian.ott@zhaw.ch (V.O.); dieter.eibl@zhaw.ch (D.E.); regine.eibl@zhaw.ch (R.E.)
* Correspondence: jan.mueller@zhaw.ch

Abstract: A current focus of biopharmaceutical research and production is seed train process intensification. This allows for intermediate cultivation steps to be avoided or even for the direct inoculation of a production bioreactor with cells from cryovials or cryobags. Based on preliminary investigations regarding the suitability of high cell densities for cryopreservation and the suitability of cells from perfusion cultivations as inoculum for further cultivations, an ultra-high cell density working cell bank (UHCD-WCB) was established for an immunoglobulin G (IgG)-producing Chinese hamster ovary (CHO) cell line. The cells were previously expanded in a wave-mixed bioreactor with internal filter-based perfusion and a 1 L working volume. This procedure allows for cryovial freezing at 260×10^6 cells mL^{-1} for the first time. The cryovials are suitable for the direct inoculation of N−1 bioreactors in the perfusion mode. These in turn can be used to inoculate subsequent IgG productions in the fed-batch mode (low-seed fed-batch or high-seed fed-batch) or the continuous mode. A comparison with the standard approach shows that cell growth and antibody production are comparable, but time savings of greater than 35% are possible for inoculum production.

Keywords: Chinese hamster ovary cells; cryopreservation; monoclonal antibodies; N−1 perfusion; process intensification; upstream processing

Citation: Müller, J.; Ott, V.; Eibl, D.; Eibl, R. Seed Train Intensification Using an Ultra-High Cell Density Cell Banking Process. *Processes* **2022**, *10*, 911. https://doi.org/10.3390/pr10050911

Academic Editor: Raja Ghosh

Received: 7 April 2022
Accepted: 1 May 2022
Published: 5 May 2022

Publisher's Note: MDPI stays neutral with regard to jurisdictional claims in published maps and institutional affiliations.

1. Introduction

The prevalence of biopharmaceuticals in the pharmaceutical market has been steadily increasing in terms of approvals and sales, reaching USD 270 billion in global sales in 2020 [1]. Biopharmaceuticals include, among others, enzymes, hormones, blood-clotting factors, and vaccines. The largest and best-selling group, mAbs, is mainly produced in CHO cells [2]. As a result of continuous improvements to production cell lines and cell culture media in recent decades, product titers of up to 5 g L^{-1} in standard fed-batch processes have become state of the art in biopharmaceutical production, and maximum titers in the range of 10 g L^{-1} have already been achieved [3–5].

The focus of biopharmaceutical research and production has now shifted from improving cell lines and media to intensifying production processes to achieve time and cost savings. A frequently used approach is seed train intensification, especially through N−1 perfusion, in which perfusion cultivation is done as the final step of inoculum production to generate UHCDs exceeding 100×10^6 cells mL^{-1}. These cells can subsequently be used to inoculate a production bioreactor. On the one hand, the inoculum production steps can be reduced, and on the other hand a continuous process or a high-seed fed-batch process can be directly implemented with these cells instead of the otherwise usual low-seed fed-batch (standard fed-batch) process. High-seed fed-batch is defined as fed-batch processes, in which the inoculation cell density of the production process is increased to $4–10 \times 10^6$ cells mL^{-1}, compared to a maximum of 0.5×10^6 cells mL^{-1} in standard processes, thus avoiding unproductive phases [6]. Successful N−1 perfusions have already been achieved. Schulze et al. demonstrated that cell densities of up to 100×10^6 cells mL^{-1} can be achieved in a wave-mixed perfusion bioreactor bag with an internal filter for an IgG-producing CHO DG44 cell line and that these cells are suitable as inoculum for batch experiments in stirred tank bioreactors (STRs) with a 15 mL working volume. Comparable maximum viable cell densities (VCDs) and slightly increased productivities were observed for cells from N−1 perfusion [7]. Stepper et al. used tangential flow filtration (TFF) in the cultivation of IgG-producing CHO DG44 cells in STRs with a cell density of up to 45×10^6 cells mL^{-1} (3.1 L and 16 L working volume), with the aim of subsequently starting a high-seed fed-batch experiment [8]. In addition, Xu et al. used STRs with alternating TFF (ATF) up to a volume of 200 L and reached about 10×10^6 cells mL^{-1} (mAb-producing CHO K1 cell lines) before inoculating the production bioreactor (STR, up to 1000 L) [9]. Wright et al. tested the growth performance of enzyme-producing CHO cells by inoculating their spin tube bioreactors from an N−1 perfusion STR (10 L working volume) with ATF at cell densities up to 100×10^6 cells mL^{-1} [10]. The fact that cell densities >200×10^6 cells mL^{-1} are also possible in the perfusion mode has already been shown by Clincke et al. (IgG-producing CHO cell line) in a wave-mixed bag with TFF [11] and by Müller et al. (IgG-producing CHO DP-12 cell line) in a wave-mixed perfusion bag with an internal filter [12]. In these two studies, however, no further bioreactors were inoculated with the cells from the perfusion process.

Besides N−1 perfusion, another starting point for the seed train intensification is the cell banking process. There are two approaches, both based on increasing the number of frozen cells compared to the standard cell bank, and both with a much longer tradition than N−1 perfusion: (1) the freezing of high volumes (large volume cell banks) and (2) the freezing of high cell densities (high cell density cell banks). More than thirty years ago, Ninomiya et al. first described an approach to freezing human–human and mouse–mouse hybridomas with cell densities of up to 150×10^6 cells mL^{-1} [13]. However, a serum-containing medium was used; now chemically defined media are state of the art, and dimethyl sulphoxide (DMSO) is almost exclusively used as the cryoprotective additive. In 2002, Heidemann et al. published an approach to freezing 50–100 mL cell suspension with 20–40×10^6 cells mL^{-1} in cryobags to directly inoculate 2 L stirred bioreactors [14]. Subsequently, further approaches were developed, in which bioreactors in the perfusion mode were used for cell bank production. Here, cryovials or cryobags with up to 110×10^6 cells mL^{-1} were frozen [10,15–17]. Similar approaches are also currently outlined by bioreactor and media manufacturers [18,19]. Both variants, high cell density as well as high volume, have the advantage of avoiding cell propagation in shake flasks and directly inoculating a larger bioreactor instead. Besides the resulting time and labor savings, another advantage is that fewer manual operations and fewer culture vessels are required, which reduces the risk of contamination. Furthermore, process steps in which pH and dissolved oxygen are not actively controlled are eliminated, and the cells are provided with more consistent conditions. When using cryobags, it must be noted that they have disadvantages compared to cryovials. The flexible material is less robust and special equipment is required for controlled freezing [20]. In fact, not every cryobag available on the market is suitable for freezing cells at −196 °C [21,22], and cryobags are much more expensive than cryovials.

Robust production cell lines and progressive improvements in commercially available media have helped to simplify process development in recent years. However, further intensification of cell banking by freezing higher cell densities as well as inoculation at densities above 100×10^6 cells mL^{-1} as a result of perfusion have not yet been published. In fact, most processes continue to be based on standard cell banks and upstream processing through several passages in shake flasks. Therefore, this work aimed to determine the suitability of cells, which were produced through perfusion with a wave-mixed bioreactor, achieving cell densities of over 150×10^6 cells mL^{-1}, as inoculum for subsequent batch experiments. In addition, an approach in which CHO cells were frozen in the UHCD range over 200×10^6 cells mL^{-1} was investigated for the first time.

2. Materials and Methods

2.1. Cell Line and Medium

For all experiments, an IgG-producing ExpiCHO-S cell line (Gibco, Waltham, MA, USA) was used. As a basal medium, 0.66× concentrated High-Intensity Perfusion CHO medium (Gibco) was used for inoculum production, batch experiments in the shake flasks, and as a starting medium in the perfusion processes. With the switch to the perfusion mode, High-Intensity Perfusion CHO Medium 1× concentrated was used. The basal and perfusion medium were supplemented with 4 mmol L^{-1} L-glutamine and 0.1% Anti-Clumping Agent (Gibco). To maintain selection pressure, 400 nmol L^{-1} methotrexate (Sigma-Aldrich, St. Louis, MO, USA) was added to all passages of inoculum production except the first passage after thawing and the last passage before production trials.

2.2. Experimental Design

Figure 1 outlines the experimental design of the study. In the first stage of the study, a comparison between batch experiments with standard inoculum production and direct inoculation from cryovials was performed. In addition, cryovials with cell densities between 90 and 250×10^6 cells mL^{-1} were frozen, and the growth and production behavior of these frozen cells was compared to standard cryovials with 15–40×10^6 cells mL^{-1}. The second stage consisted of establishing perfusion experiments, checking cell growth and production performance by inoculating batch experiments over the course of a perfusion process, and finally freezing the UHCD-WCB. Lastly, the growth and production behavior of the UHCD-WCB was tested by directly inoculating batch experiments and a perfusion bioreactor.

2.3. Standard Inoculum Production in Shake Flasks

Standard inoculum production took place over a period of 7 d in disposable shake flasks (Corning, Corning, NY, USA). Cryovials (2 mL, Brand, Wertheim, Deutschland) with a VCD of 15×10^6 cells mL^{-1} were thawed and transferred to a 125 mL shake flask with a 40 mL working volume. The cells were passaged every second or third day with a VCD of 0.3–0.5×10^6 cells mL^{-1}, and 250 mL and 500 mL shake flasks were used with 80 mL and 160 mL working volumes, respectively. Shake flasks were incubated in a shaking incubator with a 120 rpm shaking speed at an amplitude of 25 mm, 37 °C, 8% CO_2, and 80% relative humidity.

2.4. Direct Inoculation with Cells from Cryovials

For direct inoculation of batch experiments (B-CV-15, B-90–B-250, B-UHCD) in shake flasks and the perfusion experiment P05 with cells from cryovials, either vials from WCBs with standard freezing cell densities of $\leq 40 \times 10^6$ cells mL^{-1} or cryovials from the freezing experiments and the UHCD-WCB with $\geq 90 \times 10^6$ cells mL^{-1} were used. For this purpose, the vials were removed from the cryotank and thawed for 1–2 min at 37 °C in a water bath, and then the cell suspension was transferred directly into the shake flask (2 mL cryovial) or the wave-mixed bioreactor (5 mL cryovial). B-CV-15 and B-UHCD were performed as triplicates, with batches B-90–B-250 as duplicates.

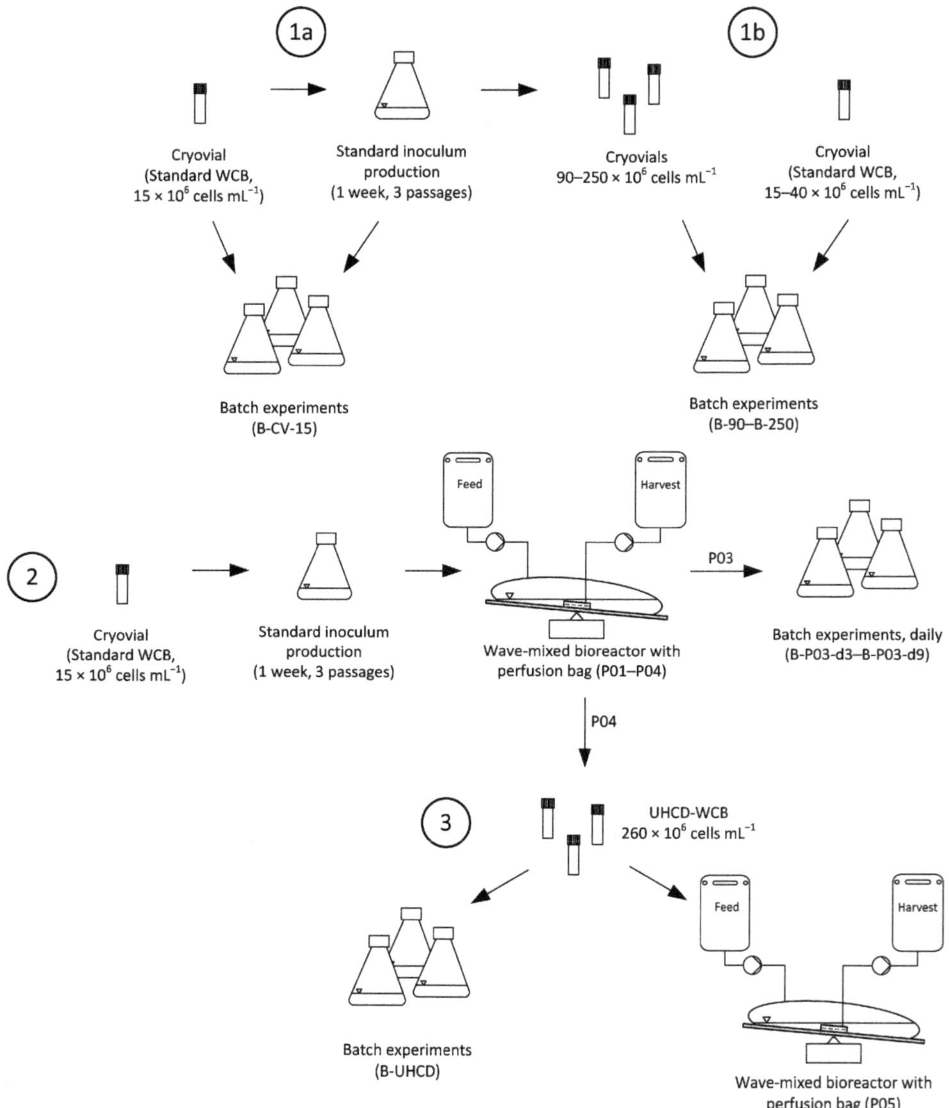

Figure 1. Schematic representation of the experiment design: (**1a**) Direct inoculation out of cryovials (B-CV-15) vs. standard inoculum production, freezing of different VCDs; (**1b**) Direct inoculation from standard cryovials vs. cryovials with higher cell densities (B-90–B250; 90–250 × 10⁶ cells mL⁻¹); (**2**) Perfusion processes (P01–P04) with standard inoculum production, daily inoculation of batch experiments (B-P03-d3–B-P03-d9), and freezing of the UHCD-WCB; (**3**) Performance test of the UHCD-WCB in batch (B-UHCD) and perfusion (P05) experiments.

2.5. Batch Cultivation in Shake Flasks

All batch experiments were performed in 125 mL disposable shake flasks (Corning) with a 40 mL working volume. Shake flasks were incubated in a shaking incubator at a 120 rpm shaking speed, an amplitude of 25 mm, 37 °C, 8% CO_2, and 80% relative humidity.

A starting VCD of 1×10^6 cells mL^{-1} was targeted. Samples for atline analyses (Section 2.10) were taken daily. Batch experiments were terminated when viability fell below 40%.

2.6. Perfusion Cultivations in Wave-Mixed Bioreactors

All cultivations were carried out with the Ready to Process Wave 25 control unit from GE Healthcare (now Cytiva, Marlborough, MA, USA). 2 L wave-mixed bioreactors with integrated filter membrane (Flexsafe RM 2 L perfusion pro 1.2 μm; for P01–P04) from Sartorius (Göttingen, Germany) and Cytiva (Cellbag, 2 L, BC10, pHOPT, DOOPT II and Perfusion; for P05) were used. The inoculum was produced with the standard method (Section 2.3) for P01–PC4; P05 was inoculated directly from a 5 mL UHCD cryovial. The processes were inoculated with a VCD of 0.6–1.2×10^6 cells mL^{-1}. Cultivations were performed at 1 L working volume, 37 °C, with overlay aeration of 0.2 vvm, pH ≤ 7.2 by addition of CO_2, and dissolved oxygen (DO) controlled to $\geq 40\%$ by addition of O_2. The rocking rate (20–40 rpm) and rocking angle (6–12°) were manually adjusted with the growth of the cells and the accompanying oxygen demand. Samples were taken daily during the batch phase and twice daily during the perfusion phase. Perfusion was started on day two of the cultivation, between 3 and 6×10^6 cells mL^{-1}. Depending on the current VCD, the specific growth rate, and the time until the next sampling, the perfusion rate D was adjusted to ensure a minimum cell-specific perfusion rate (CSPR) of 55 pL cell^{-1} d^{-1} (Equation (1)).

$$D = \mathrm{CSPR}_{min} \times \mathrm{VCD}_{\text{next sample}} \tag{1}$$

For P01 and P02, 10 L perfusion medium was prepared, and D was limited to maximal 3.1 vvd, for P05 to 7 vvd with 15 L perfusion medium. P03 (15 L perfusion medium) and P04 (10 L perfusion medium) had no perfusion rate limit since the cells were used for further experiments. In the bioreactors, a constant glucose concentration of 3 g L^{-1} was targeted by the continuous addition of a 200 g L^{-1} glucose solution. Perfusion experiments P01–P03 and P05 were terminated when the perfusion medium was depleted, and P04 ended with the freezing of the UHCD-WCB.

2.7. Cell Growth and Production Performance after $N-1$ Perfusion

In order to characterize the growth and production behavior of the cells grown in perfusion mode, a sterile cell suspension was taken daily from perfusion experiment P03 between day 3, at 7×10^6 cells mL^{-1}, and day 9, at 170×10^6 cells mL^{-1}. Batch experiments (B-P02-d3–B-P03-d9) in shake flasks were performed as duplicates according to Section 2.5.

2.8. Freezing Cells from Shake Flask Cultivations

To determine the maximum possible freezing cell density, the cells were expanded according to the standard inoculum production described in Section 2.3. Afterward, the cell suspension was concentrated by centrifugation at 500 g for 5 min and resuspended in ice-cold $0.66\times$ concentrated supplemented High-Intensity Perfusion CHO medium with 10% DMSO. The following freezing cell densities were chosen: 90×10^6 cells mL^{-1}, 115×10^6 cells mL^{-1}, 150×10^6 cells mL^{-1}, 180×10^6 cells mL^{-1}, and 250×10^6 cells mL^{-1}. Immediately after resuspension and aliquotation in 2 mL cryovials (Brand), the cells were stored in freezing containers (Mr. Frosty, Nalgene, Waltham, MA, USA) for 24 h in a -80 °C freezer and subsequently transferred to the liquid nitrogen cryotank.

2.9. Freezing Cells from Perfusion Cultivation

The inoculum for the perfusion experiment (P04) was produced with the standard method (Section 2.3) with MTX addition in passages 2 and 3 as well as in the wave-mixed bag. The perfusion cultivation was performed as described in Section 2.6. The cells were grown for 6 d. At the start of the freezing process, the cell suspension in the wave-mixed bag was cooled to 10 °C on a water-cooled rocking platform (Sartorius), and the cell suspension was transferred to centrifuge tubes (175 mL, Falcon, Corning, NY, USA) afterward. From here on out, the cell suspension was kept on ice if possible. The entire freezing procedure is shown in Figure 2. The suspension was centrifuged at 500 g for 5 min. The largest part of supernatant (75% of the centrifugation volume) was discarded; 25% of the centrifugation volume of ice-cold fresh medium containing 14% DMSO was added; and the cell pellet was resuspended. The cell suspension was centrifuged again; the complete supernatant was removed; and the cell pellet was resuspended with a volume of 1:1 of ice-cold medium containing 14% DMSO. The resulting DMSO concentration was 10.5% DMSO (v/v), and the VCD was 260×10^6 cells mL^{-1}.

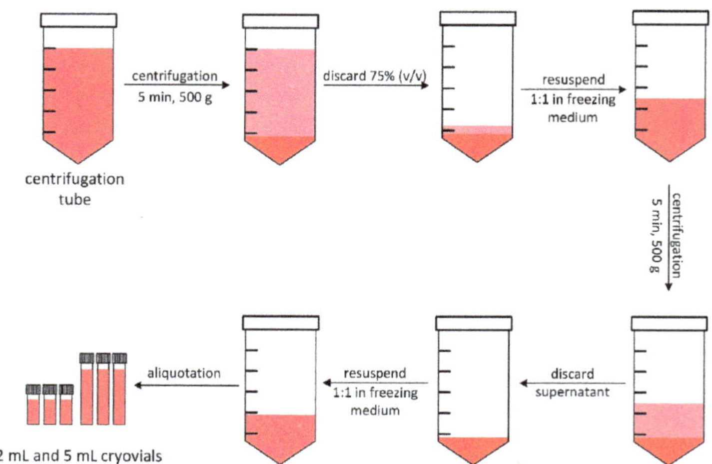

Figure 2. Scheme of the preparation of the cell suspension for freezing of the UHCD-WCB.

The cell suspension was transferred to 2 mL and 5 mL cryovials (Brand) containing 1.5 mL and 4.5 mL cell suspensions and stored in the freezing container (Mr. Frosty, Nalgene) for 24 h at $-80\,°C$. The UHCD-WCB was transferred to the cryotank at $-196\,°C$ for long-term storage.

2.10. Analytical Methods

During cultivation, VCD, total cell density, viability, cell diameter, compactness, and the aggregation rate were determined using a Cedex HiRes analyzer (Roche Diagnostics, Basel, Switzerland). For the determination of the concentration of glucose, glutamine, ammonium, lactate, and IgG, the Cedex Bio analyzer (Roche Diagnostics) was used.

2.11. Statistical Evaluation of Experiments

In diagrams showing a positive control in multiple determination, the arithmetic mean value is shown with a tolerance interval with a coverage p of 90% and a confidence α of 90%, calculated according to Howe [23] and Guenther [24].

3. Results

3.1. Direct Inoculation of Batch Experiments with Cells from Cryovials

A preliminary study was performed to compare the growth and production performance of ExpiCHO-S cells in shake flasks inoculated from both a standard inoculum production and directly from cryovials with 15×10^6 cells mL^{-1}. The courses of VCD, viability, and glucose and IgG concentration are shown in Figure 3. As expected, there was a lag phase after thawing, so the growth rate of 0.0344 ± 0.0028 h^{-1} in the experiments inoculated from cryovials (B-CV-15_1, B-CV-15_2, B-CV-15_3) was 16.0% lower on the first day than that of experiments inoculated from shake flasks using standard inoculum production, therefore being inoculated in the exponential growth phase (0.0409 ± 0.0013 h^{-1}, control). On day 2, the growth rate was still 8.4% lower. Due to the higher growth rates on days 3 and

4, an almost identical peak VCD was achieved after 4 d with $14.2 \pm 0.2 \times 10^6$ cells mL^{-1} compared to $14.6 \pm 0.4 \times 10^6$ cells mL^{-1} for the control group. Due to the initial lag phase, however, the growth curve is slightly offset in time, so the death phase also occurred later. The viability was still 35–88% on day 5, whereas it was already $\leq$12% for classical inoculum production. Along with the later death phase, glucose as the main carbon source was depleted later.

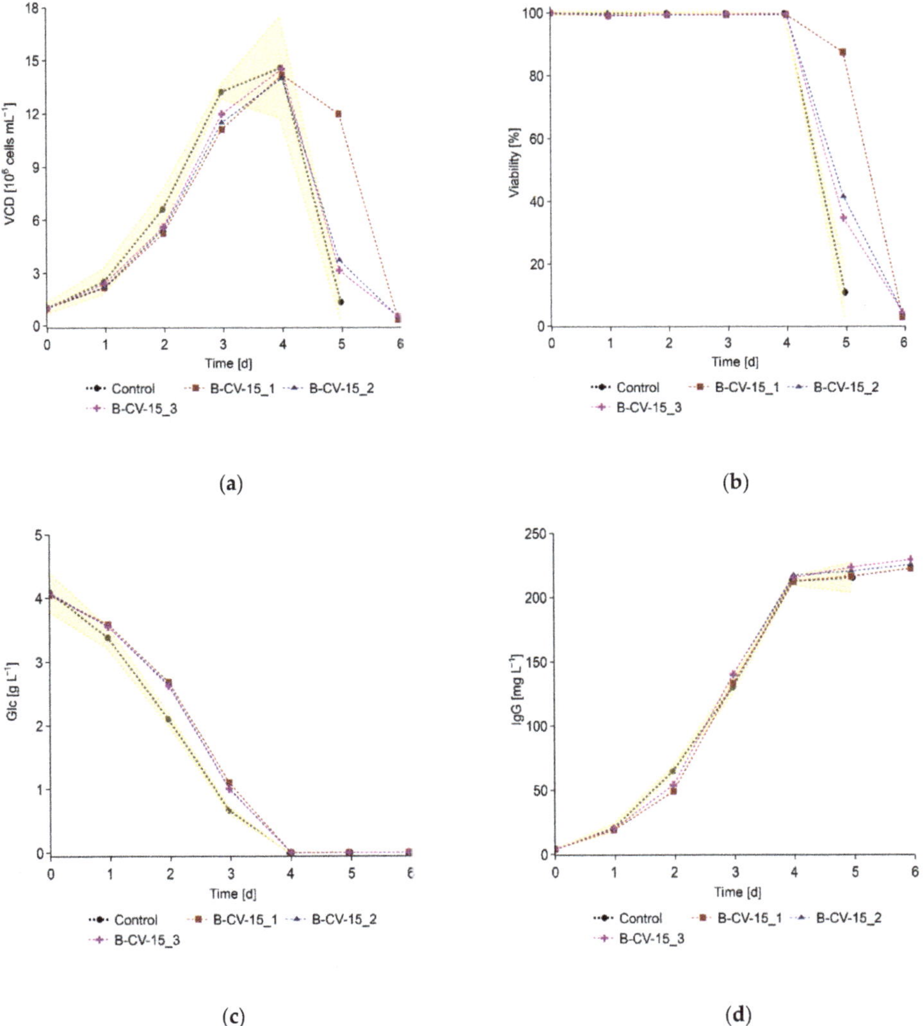

Figure 3. Course of (**a**) VCD, (**b**) viability, (**c**) glucose (Glc), and (**d**) IgG concentration during the batch experiments inoculated directly from cryovials (nomenclature: B = Batch, CV = inoculated out of cryovial, 15 = VCD in cryovial ($\times 10^6$ cells mL^{-1})). Control: batch experiments with standard inoculum production ($n = 3$).

IgG production was similar. The titers on day 5 were comparable for both approaches, with 220 ± 3 mg L^{-1} for direct inoculation from cryovials and 215 ± 2 mg L^{-1} for standard inoculum production. The courses of the glucose consumption rate and the IgG production rate are shown in the Appendix A (Figure A1).

Since comparable VCDs and IgG titers were achieved for batch experiments with direct inoculation from cryovials compared to standard inoculum production after five days of batch cultivation, cryovials with different VCDs were frozen (Section 2.8) and used again for directly inoculating shake flasks. The growth curves and viability courses as well as the concentration courses of glucose and IgG are shown in Figure 4.

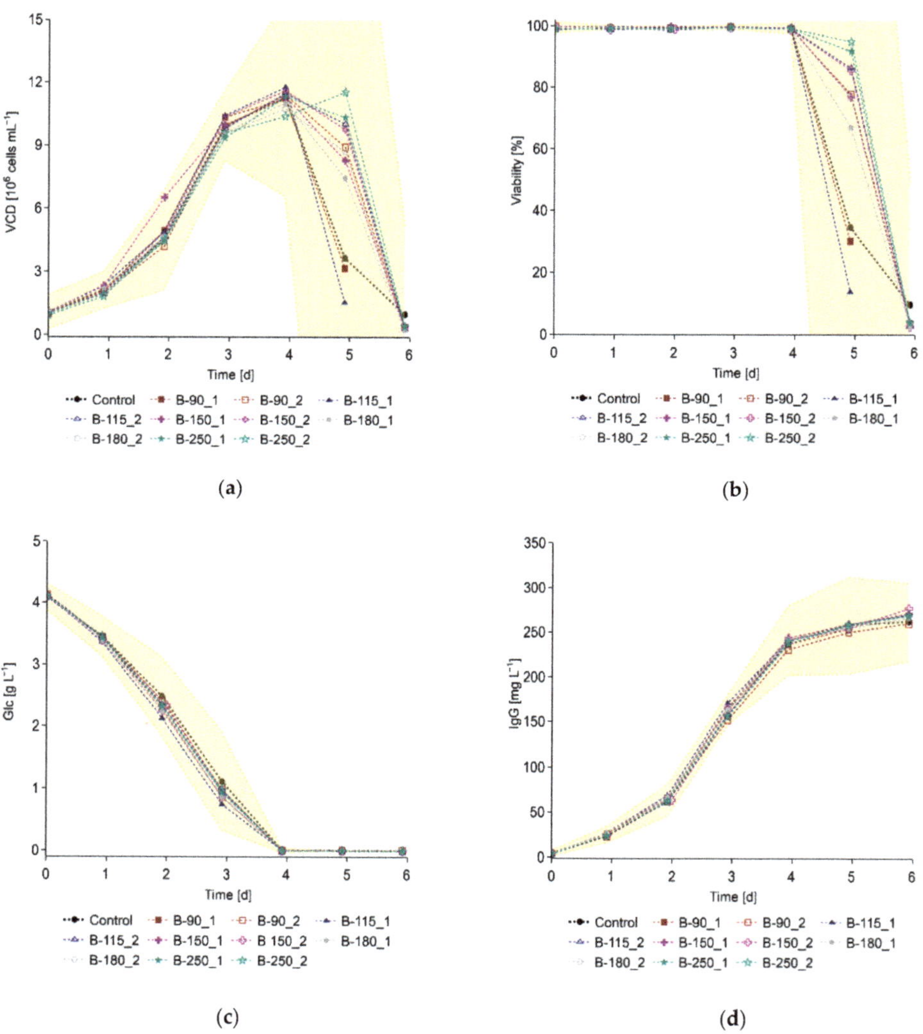

Figure 4. Course of (**a**) VCD, (**b**) viability, (**c**) glucose (Glc), and (**d**) IgG concentration during the batch experiments inoculated directly from cryovials containing high VCDs from 90 to 250×10^6 cells mL^{-1} (nomenclature: B = Batch, 90/115/150/180/250 = VCD in cryovial ($\times 10^6$ cells mL^{-1})). Control: batch experiments inoculated directly from cryovials containing lower VCDs (15–40 $\times 10^6$ cells mL^{-1}, $n = 3$).

Growth was independent of the freezing VCD until day 4. Growth rates from day 0 to day 3 ranged from 0.0310 ± 0.0027 h^{-1} for the batch experiments inoculated from cryovials containing 180×10^6 cells mL^{-1} to 0.0325 ± 0.0023 h^{-1} for the batch experiments inoculated from cryovials containing 90×10^6 cells mL^{-1}. Viability remained >98% in all shake flasks until day 4. Maximum VCDs were comparable and ranged from 10.7×10^6 cells mL^{-1} (B-180_2) to 11.8×10^6 cells mL^{-1} (B-115_1). Differences could be seen based on the differently progressed death phases on day 5. The shake flasks with the highest freezing VCD were the most viable (92% and 95% for the duplicate), and the control group had the lowest viability ($35 \pm 31\%$). No change was detected in glucose consumption and IgG production. On day 4, glucose was depleted in all experiments. On day 5, when the first batch experiments were stopped due to low viability, IgG concentrations in all shake flasks were between 250 mg L^{-1} and 269 mg L^{-1}. The courses of the glucose consumption rate and the IgG production rate can be found in the Appendix A (Figure A2).

Although these results show that a higher freezing cell density minimally delays growth and the death phase occurs a few hours later, high viability is obtained after thawing regardless of the freezing cell density. The peak VCDs are comparable, and IgG production also shows no differences. Since it could be shown that the establishment of a UHCD-WCB with $>200 \times 10^6$ cells mL^{-1} is possible with regard to cell growth and IgG production, the expansion of a sufficiently large number of cells was considered in the next step. Therefore, perfusion experiments were performed.

3.2. UHCD Perfusion Cultivations and Inoculation of Batch Experiments out of a Perfusion Process

The second part of the study aimed to check whether the growth and production behavior of the cell suspension changes with increasing VCD during a perfusion process with complete cell retention. For this purpose, a perfusion process was established, whereby the first two cultivations P01 and P02 were not used for further experiments, whereas daily batch experiments were inoculated from P03. The growth and viability curve as well as the course of the perfusion rate and the CSPR of the perfusion experiment are shown in Figure 5.

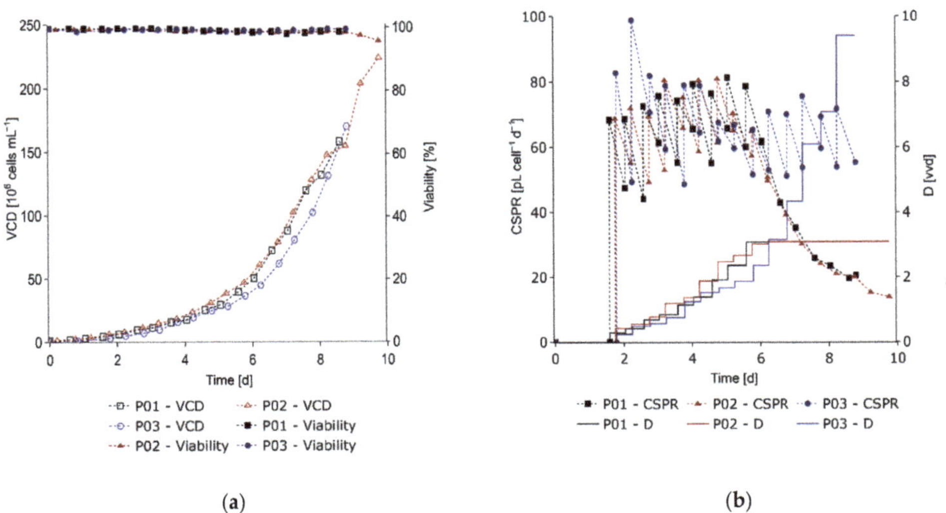

(a)   (b)

Figure 5. (a) Course of the VCD and viability of P01, P02, and P03, and (b) course of perfusion rate D and the CSPR of P01, P02, and P03.

The wave-mixed bioreactors were inoculated with 1.0 (P01), 1.2 (P02), and 0.6×10^6 cells mL^{-1} (P03). Perfusion was started after 1.6–1.8 d, and the perfusion rate was gradually increased. For the first two perfusion processes P01 and P02, the maximum perfusion rate was limited to 3.1 vvd, causing the CSPR and thus the growth rate μ to decrease after about 6 d. For P03, D was increased to a maximum of 9.4 vvd between 8.2 d and 8.8 d to provide the cells with the best possible conditions, as P03 was used for daily inoculation of batch experiments. The maximum cell density of P01 was 158×10^6 cells mL^{-1} with 98.6% viability after 8.6 d, of P02 224×10^6 cells mL^{-1} with 95.7% viability after 9.8 d, and of P08 170×10^6 cells mL^{-1} with 99.4% viability after 8.8 d. The growth rate from the start of cultivation to the limit of the perfusion rate was comparable for all three cultivations: 0.0267 ± 0.0100 h^{-1} for P01 (until 6.0 d), 0.0295 ± 0.0089 h^{-1} for P02 (until 5.8 d), and 0.0259 ± 0.0096 h^{-1} for P03 (until 8.8 d).

The cell growth and product formation of batch experiments inoculated from the perfusion cultivation P03 were characterized. The growth and viability curves as well as the courses of the glucose and IgG concentration of the experiments are shown in Figure 6. The experiments that were started on days 3 and 4 were considered as the positive control. It was found that the growth rate on the first day of the batch experiments increased during the perfusion cultivation until day 7, at 62×10^6 cells mL^{-1}. The growth rate on the first day was 0.0301 h^{-1} and 0.0289 h^{-1} for the shake flask duplicate inoculated on day 7. The control group, on the other hand, had a growth rate of 0.0254 ± 0.0034 h^{-1}, which was 14% lower.

However, viability remained >99% for all experiments until day 4. Glucose consumption and product formation also showed a slight increase for the above-mentioned experiments. For example, glucose concentration in B-P03_d7 on day 2 was already 17% lower than the control, and the IgG titer was 17% higher. The specific consumption and production rates can be seen in Appendix A (Figure A3). Nevertheless, comparable peak VCDs were achieved, from 11.5×10^6 cells mL^{-1} (B-P03-d6_2) to 12.4×10^6 cells mL^{-1} (B-P03-d7_2). The final IgG titers also showed only minor differences. The lowest value was measured for B-P03-d3_2 (included in the control group) with 254 mg L^{-1}, and the highest value for B-P03-d6_1 with 284 mg L^{-1}. Although high viability can be assumed even with higher VCDs, since higher VCDs are not necessary for freezing a sufficiently large WCB, a cell density of about 100×10^6 cells mL^{-1} at the time of freezing was aimed for to establish the UHCD-WCB. Perfusion cultivation P04 was run for 6 d and afterward used to freeze the UHCD-WCB at 260×10^6 cells mL^{-1} (Section 2.9).

3.3. Evaluation of a Perfusion-Based UHCD-WCB

3.3.1. Batch Experiments in Shake Flasks

As described in Section 2.4, cryovials from the UHCD-WCB were also thawed and used to directly inoculate shake flasks. Here, a triplicate with three cryovials was performed. Batch experiments inoculated directly from cryovials frozen with low VCD (15×10^6 cells mL^{-1}) served as the positive control, which were also performed in triplicate. The growth and viability curves as well as the courses of glucose and IgG concentration of the experiments are shown in Figure 7. The shake flasks inoculated directly from the UHCD-WCB had a slightly prolonged lag phase. On day 1, the growth rate was 0.0266 ± 0.0022 h^{-1} and therefore 23% lower compared to 0.0344 ± 0.0028 h^{-1} in the positive control. A viability drop on day 1 was also observed, but the viability remained high at $97.8 \pm 1.0\%$ and increased again to >99% in all experiments during the cultivation. The maximum VCD achieved in the batch experiments inoculated with the UHCD-WCB was $13.0 \pm 0.3 \times 10^6$ cells mL^{-1}, which was slightly lower (-9%) than in the control group ($14.2 \pm 0.2 \times 10^6$ cells mL^{-1}). According to the shorter lag phase in the positive control, the death phase had already begun on day 5 ($55 \pm 23\%$ viability), while at this time the viability of the batches from the UHCD-WCB was still $96 \pm 1\%$. Although product formation was also delayed in accordance with the delayed growth, the final product titers achieved were comparable: 225 ± 3 mg L^{-1} in the positive control and 222 ± 0 mg L^{-1} for the cells from the UHCD-WCB. On days 1 and 2, in contrast to the experiments from Parts 1 and 2 of the study, the glucose concentration was lower than in the control in all three shake flasks, although the VCDs were lower. The specific glucose consumption rates were 53% higher than for the control group on day 1 and 34% higher on day 2. The course of the glucose consumption rate can be seen in Appendix A (Figure A4). To check the morphology of the cells, microscopic images of the cells were taken after thawing a cryovial with 15×10^6 cells mL^{-1}, a cryovial with 260×10^6 cells mL^{-1}, and after one week of standard inoculum production. No abnormalities in cell morphology were observed (Figure A5).

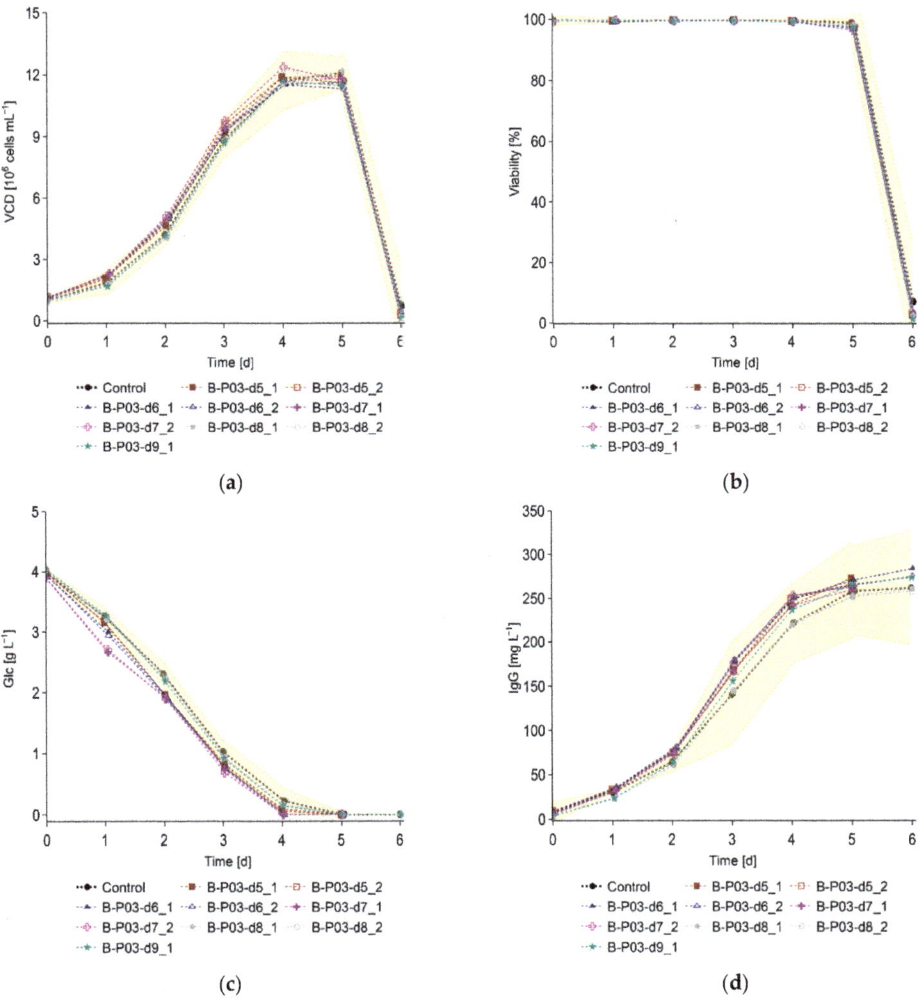

Figure 6. Course of (**a**) VCD, (**b**) viability, (**c**) glucose (Glc), and (**d**) IgG concentration during the batch experiments inoculated from perfusion cultivation (nomenclature: B = Batch, P03 = Source of inoculum, d5–d9 = day of cell harvest from P03). Control: batch experiments inoculated from P03 on days 3 and 4 ($n = 4$).

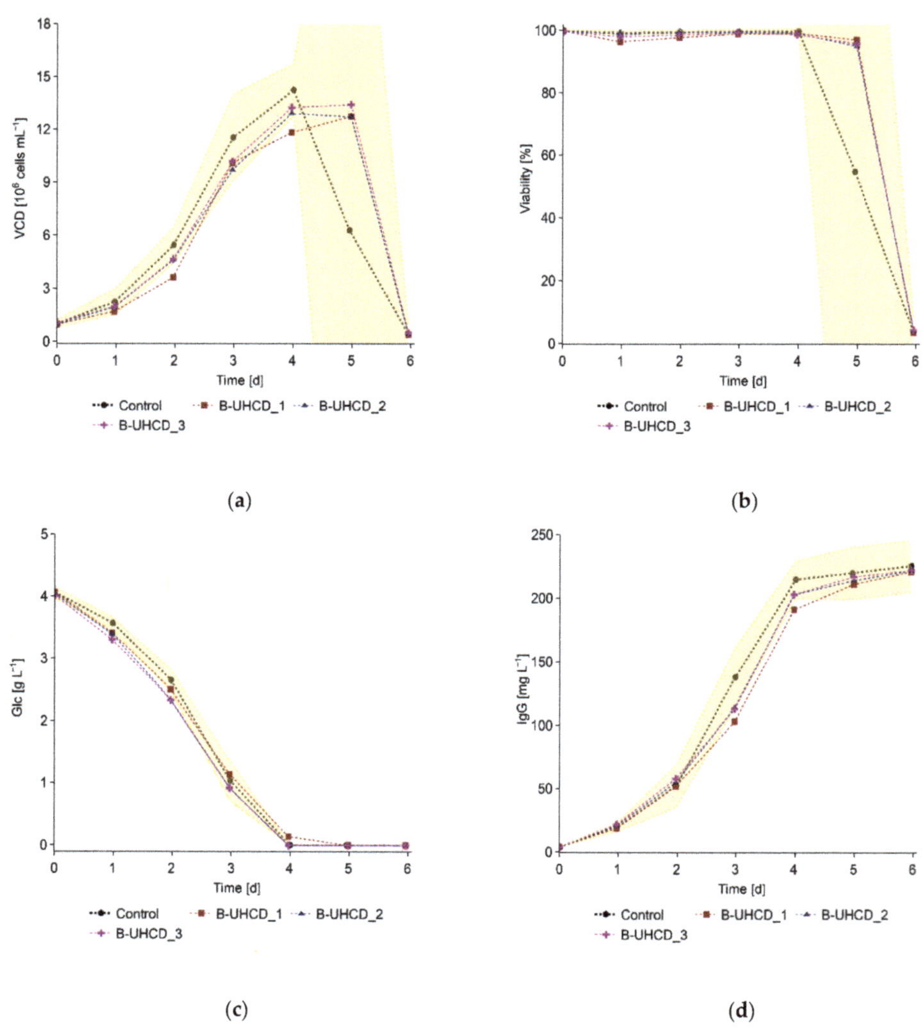

Figure 7. Course of (**a**) VCD, (**b**) viability (**c**) glucose (Glc), and (**d**) IgG concentration during the batch experiments inoculated directly from the UHCD-WCB (nomenclature: B = Batch, UHCD = inoculated from UHCD-WCB). Control: batch experiments inoculated directly from cryovials containing lower VCDs (15×10^6 cells mL^{-1}, $n = 3$).

3.3.2. Perfusion Experiment

After successfully confirming the suitability of the UHCD-WCB as the inoculum for batch experiments, a wave-mixed perfusion bioreactor with a 1 L working volume was inoculated with a starting VCD of 1.2×10^6 cells mL^{-1} from a 5 mL cryovial. The growth and viability curve as well as the course of the perfusion rate and CSPR are shown in Figure 8. As expected, the growth rate was only 0.0185 h^{-1} on the first day of cultivation due to the lag phase but subsequently increased to values >0.0300 h^{-1}. The viability drop, which was determined in the previous shake flask experiments, was only slight at 97.9% viability on day 2, and subsequently, the viability was >98% until the end of cultivation.

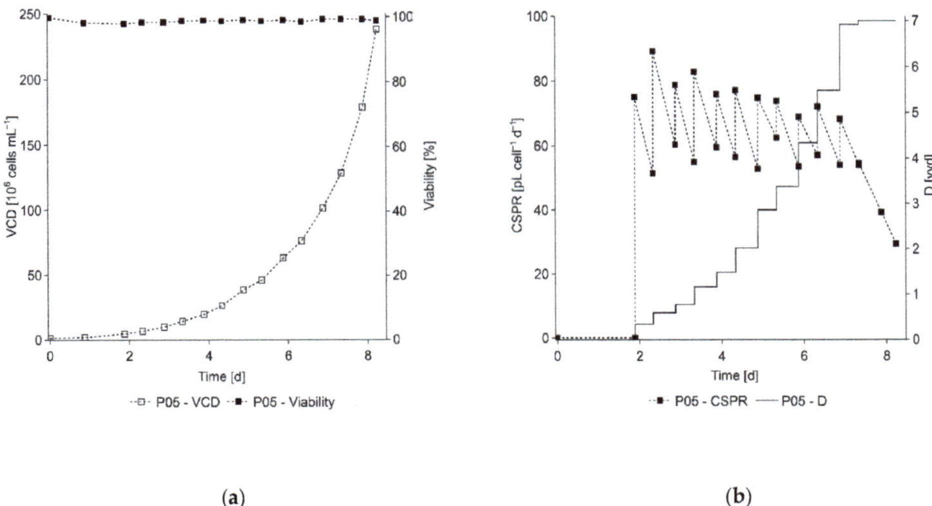

(**a**)   (**b**)

Figure 8. (**a**) Course of the VCD and viability of P05, and (**b**) course of D and the CSPR of P05, inoculated directly from the UHCD-WCB.

In contrast to P03, in which the perfusion rate was increased until the end of cultivation, a maximum perfusion rate of 7 vvd was set for this cultivation, in order to achieve higher final VCDs with the same amount of medium. As a result, the CSPR dropped to 29 pL cell^{-1} d^{-1} towards the end, but a maximum VCD of 238 × 10^6 cells mL^{-1} with a viability of 98.8% was achieved after 8.2 d. Despite the limitation of the perfusion rate on the last day of cultivation, the cells continued to grow exponentially until the end of cultivation; over the whole period, the growth rate was 0.0265 ± 0.0065 h^{-1}, comparable to 0.0259 ± 0.0096 h^{-1} in perfusion cultivation P03 (Section 3.2).

4. Discussion

The intensification of production processes with regard to the cryopreserved cell banks has so far been limited to high volumes through the use of cryobags and maximum freezing VCDs of up to 150 × 10^6 cells mL^{-1} [10,13–19]. Regarding N−1 perfusion processes, the production of cells in the perfusion mode has already been achieved up to >200 × 10^6 cells mL^{-1} [11,12], but the cells have so far only been used with cell densities of up to 100 × 10^6 cells mL^{-1} for inoculation in subsequent experiments [7–10]. In this work, both approaches, the intensification of the seed train by freezing UHCD and the use of perfusion processes for cell production, were pursued. In the first stage of the study, the use of direct inoculation from cryovials was investigated. If cryovials with 15 × 10^6 cells mL^{-1} and 10% DMSO are used for inoculation with a VCD of 1 × 10^6 cells mL^{-1}, it must be noted that almost 0.7% DMSO remain in the medium. Direct inoculation from cryovials with 15 × 10^6 cells mL^{-1} was found to have a lag phase compared to a seven-day inoculum production, but, nevertheless, comparable maximum VCDs and product titers were found. Kleman et al. described that up to 1% DMSO did not affect the growth of a HEK cell line, but greater than 0.3% did for a CHO-S cell line [25]. In this work, no growth inhibition was observed for the ExpiCHO-S cell line used. At higher freezing VCDs, the consideration of the remaining DMSO is negligible due to the large dilution effect. When using cryovials containing up to 250 × 10^6 cells mL^{-1} for direct inoculation, higher freezing VCDs resulted in an increased lag phase. However, this growth shift by a few hours had no influence on the maximum VCDs and IgG titers. The viability remained >99% until the beginning of the death phase. In previous publications, a viability drop often was reported in addition to a lag phase in the first days after inoculation [14,16,17].

A robust perfusion process was established, and it was shown that VCDs >200 $\times$ 10^6 cells mL^{-1}, as well as very high perfusion rates (9.4 vvd), are realizable with the wave-mixed perfusion bags with an internal filter. When using cells from a perfusion process for subsequent batch experiments, it was found that growth and product formation in the batch experiments were fastest between VCDs of 36 $\times$ 10^6 cells mL^{-1} (d6 of the perfusion process) and 102 $\times$ 10^6 cells mL^{-1} (d8) in the perfusion process, with a peak on day 7 (62 $\times$ 10^6 cells mL^{-1}). At lower as well as higher VCDs, the growth rates were slightly lower; nevertheless, the batch experiment inoculated at the highest VCD of 170 $\times$ 10^6 cells mL^{-1} also achieved a comparable IgG titer and peak VCD.

Another perfusion cultivation was performed, harvested at 90 $\times$ 10^6 cells mL^{-1} and frozen as UHCD-WCB with 260 $\times$ 10^6 cells mL^{-1}. Concentration of the cell density and media exchange with fresh medium containing DMSO was carried out in two centrifugation steps, as the viscosity also increases with increasing freezing cell density, making handling and centrifugation more difficult. Due to the pre-cooling of the bag, the transfer of the cell suspension to the centrifugation tubes, and the two-step medium exchange, the freezing process from a wave-mixed bag with 1 L working volume was more time-consuming than freezing from shake flasks, but Heidemann et al. described a time window of up to 2.5 h for the preparation of cryovials [26]. For the process described in this study, cells were transferred to the -80 °C freezer within 30 min of being removed from the controlled conditions of the wave-mixed bioreactor, compared to 15 min for freezing of cells grown in shake flasks.

The established UHCD-WCB was used for direct inoculation of batch experiments as well as a wave-mixed perfusion bioreactor. A low viability drop after inoculation was observed, as well as a 9% lower peak VCD in the batch experiments compared to the control group, with a comparable IgG titer. The perfusion process was comparable to previous experiments. Cells grew exponentially after a short lag phase on the first day to a maximum VCD of 238 $\times$ 10^6 cells mL^{-1} and a viability of 98.8% reached after 8.2 d.

This study demonstrates that high freezing cell densities of up to 260 $\times$ 10^6 cells mL^{-1} with cells taken from perfusion processes are suitable for establishing a UHCD-WCB with high viability and that only short lag phases occur after thawing. The UHCD-WCB is intended to be used to inoculate either the final inoculum production step (step N$-$1) as perfusion or even the production bioreactor (step N) directly from the cryovial. The successful perfusion cultivation with inoculation from a UHCD cryovial lays the foundation for further experiments. Direct inoculation of a 1 L bioreactor bag as N$-$1 perfusion would produce a sufficient number of cells within one week to inoculate a high-seed fed-batch at a 50 L scale with a starting VCD of 5 $\times$ 10^6 cells mL^{-1}. This would shorten inoculum production by more than 35% compared to the standard process (7 d shake flasks + 4 d wave-mixed bag, low-seed fed-batch process). In addition, the high-seed approach can shorten the production process for the ExpiCHO-S cell line and achieve more than 25% higher IgG titers within the same time (research article in preparation). Since a cryovial with 4.5 mL cell suspension is sufficient to inoculate the largest available wave-mixed perfusion bioreactor with an internal filter (max. working volume 25 L, available from Cytiva and Sartorius) at a minimum working volume of 5 L with about 0.25 $\times$ 10^6 cells mL^{-1}, even one-step inoculum production for bioreactors at production scale can be achieved. Here, a cubic meter scale bioreactor could be inoculated from the wave-mixed perfusion bag.

During the experiments of this study, the growth behavior, viability, and production performance of the cells were investigated as quality parameters of the cell suspensions. In further experiments for the development of IgG production processes based on the established UHCD-WCB, additional investigations, for example, of the IgG quality, can be carried out.

Author Contributions: Conceptualization, J.M., V.O., D.E. and R.E.; methodology, J.M. and V.O.; validation, J.M. and V.O.; formal analysis, J.M. and V.O.; investigation, J.M. and V.O.; resources, D.E. and R.E.; data curation, J.M.; writing—original draft preparation, J.M. and V.O.; writing—review and editing, R.E. and D.E.; visualization, J.M. and V.O.; supervision, D.E. and R.E.; project administration, D.E. and R.E. All authors have read and agreed to the published version of the manuscript.

Funding: This research received no external funding.

Data Availability Statement: The data presented in this study are available on request from the corresponding author.

Acknowledgments: We would like to thank Deniz Türkcan and Noémi Weiss for the support during the experimental part.

Conflicts of Interest: The authors declare no conflict of interest.

Appendix A

The glucose and IgG concentration courses of the batch experiments are shown and described in the results section. The influences of the different experimental approaches on the course of the glucose consumption rate and IgG production rate can be seen in the following Figures A1–A4.

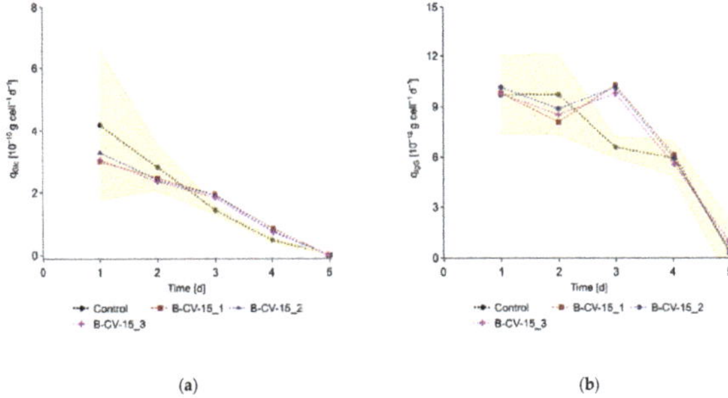

(a)

(b)

Figure A1. Course of (**a**) glucose consumption rate, and (**b**) IgG production rate during the batch experiments inoculated directly from cryovials (nomenclature: B = Batch, CV = inoculated out of cryovial, 15 = VCD in cryovial ($\times 10^6$ cells mL^{-1})). Control: batch experiments with standard inoculum production ($n = 3$).

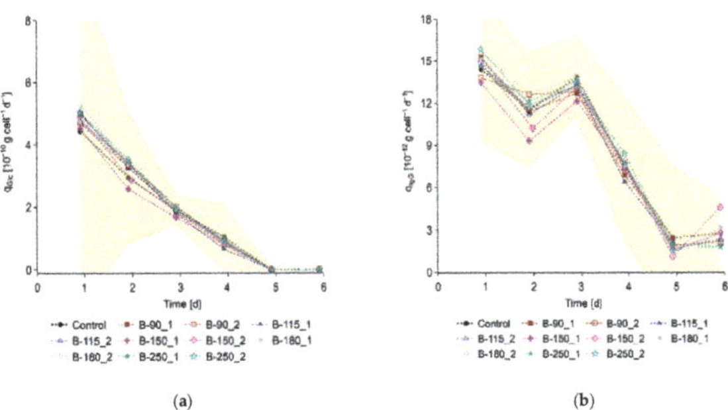

(a)

(b)

Figure A2. Course of (**a**) glucose consumption rate, and (**b**) IgG production rate during the batch experiments inoculated directly from cryovials containing high VCDs from 90 to 250 $\times$ 10^6 cells mL^{-1} (nomenclature: B = Batch, 90/115/150/180/250 = VCD in cryovial ($\times 10^6$ cells mL^{-1})). Control: batch experiments inoculated directly from cryovials containing lower VCDs (15–40 $\times$ 10^6 cells mL^{-1}, $n = 3$).

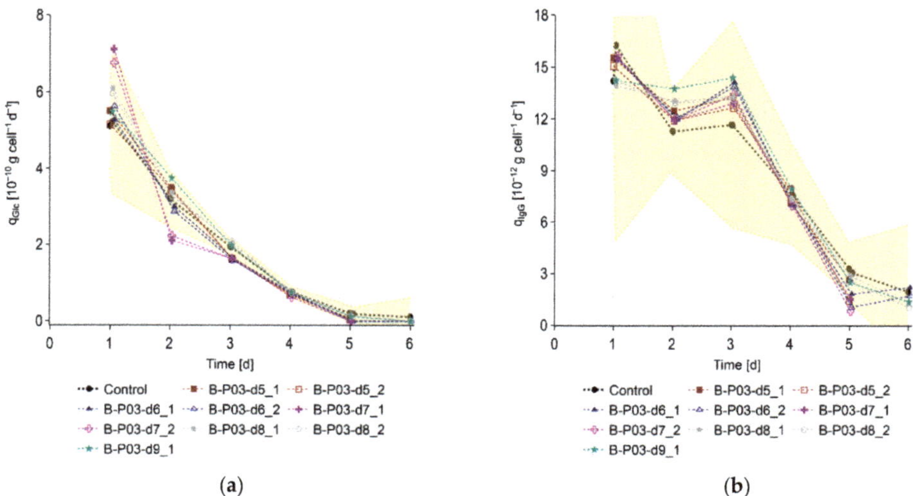

(a)   (b)

Figure A3. Course of (**a**) glucose consumption rate, and (**b**) IgG production rate during the batch experiments inoculated from perfusion cultivation (nomenclature: B = Batch, P03 = Source of inoculum, d5–d9 = day of cell harvest from P03). Control: batch experiments inoculated from P03 on days 3 and 4 ($n = 4$).

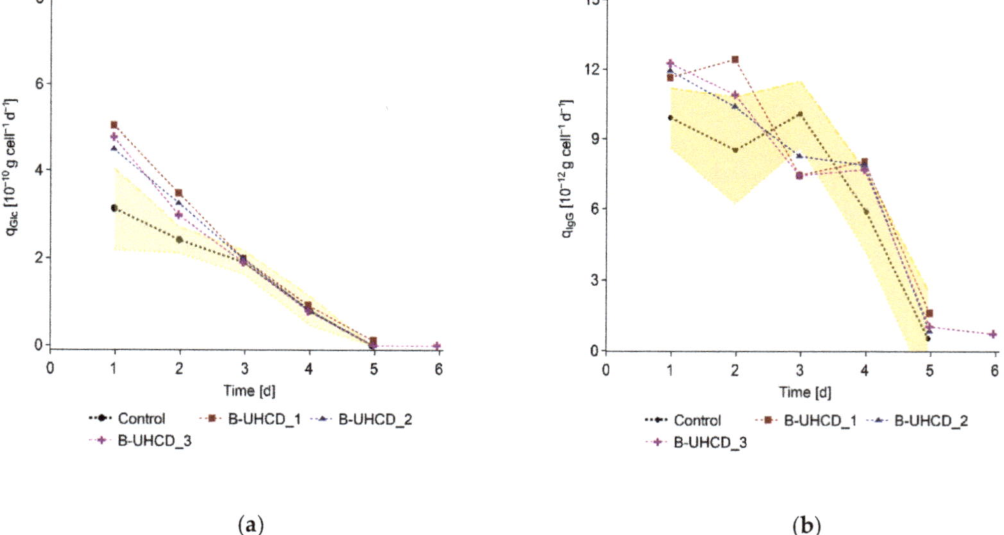

(a)   (b)

Figure A4. Course of (**a**) glucose consumption rate, and (**b**) IgG production rate during the batch experiments inoculated directly from the UHCD-WCB (nomenclature: B = Batch, UHCD = inoculated from UHCD-WCB). Control: batch experiments inoculated directly from cryovials containing lower VCDs (15×10^6 cells mL^{-1}, $n = 3$).

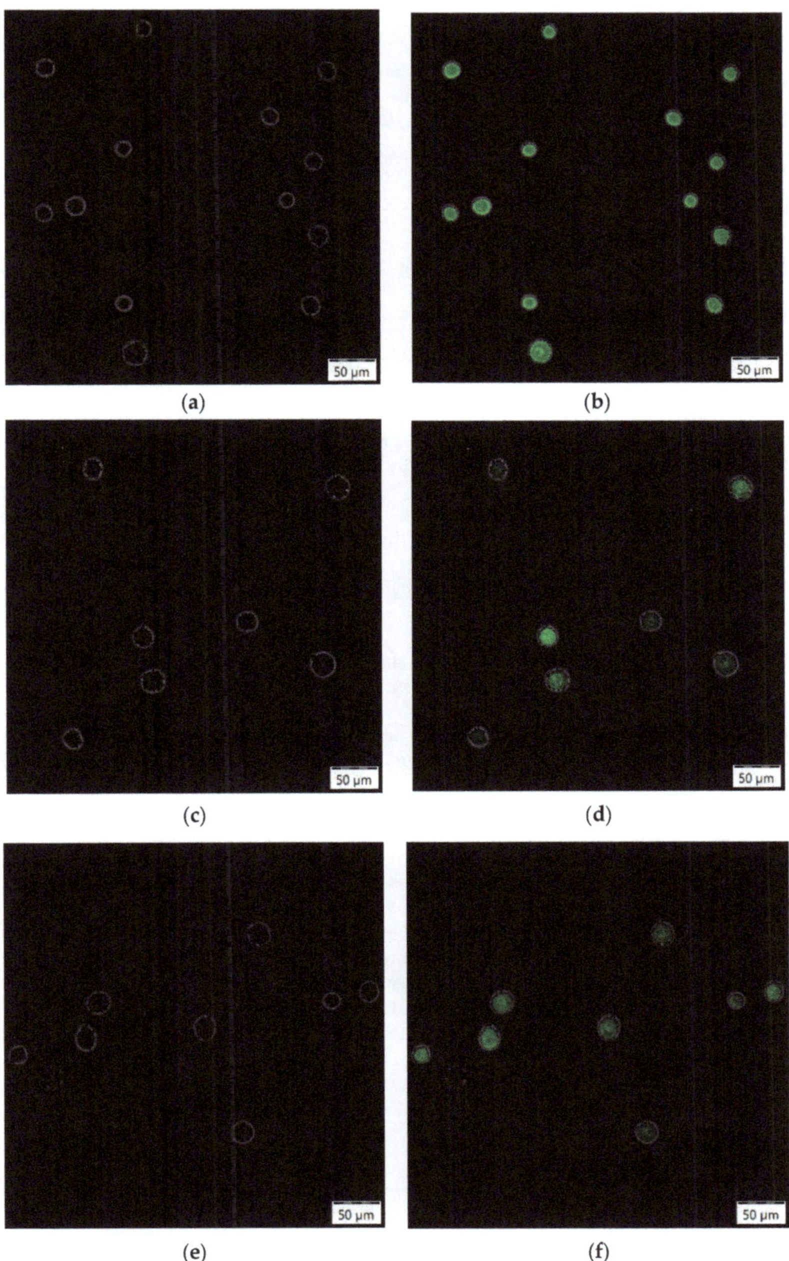

Figure A5. Microscopic images (200×) of (**a**,**b**) cells after one-week standard inoculum production, (**c**,**d**) cells from a cryovial containing 15×10^6 cells mL^{-1}, 1 d after thawing, and (**e**,**f**) cells from a cryovial containing 260×10^6 cells mL^{-1}, 1 d after thawing. (**a**,**c**,**e**) phase contrast images, (**b**,**d**,**f**) phase contrast and fluorescence images (fluorescein diacetate staining for cells with intact cell membrane).

References

1. Evaluate Pharma(R) WORLD PREVIEW 2021 Outlook to 2026. 2021, pp. 1–28. Available online: https://info.evaluate.com/rs/607-YGS-364/images/WorldPreviewReport_Final_2021.pdf (accessed on 4 December 2021).
2. Walsh, G. Biopharmaceutical Benchmarks 2018. *Nat. Biotechnol.* **2018**, *36*, 1136–1145. [CrossRef] [PubMed]
3. Kunert, R.; Reinhart, D. Advances in Recombinant Antibody Manufacturing. *Appl. Microbiol. Biotechnol.* **2016**, *100*, 3451–3461. [CrossRef] [PubMed]
4. Huang, Y.-M.; Hu, W.; Rustandi, E.; Chang, K.; Yusuf-Makagiansar, H.; Ryll, T. Maximizing Productivity of CHO Cell-Based Fed-Batch Culture Using Chemically Defined Media Conditions and Typical Manufacturing Equipment. *Biotechnol. Prog.* **2010**, *26*, 1400–1410. [CrossRef] [PubMed]
5. Kelley, B.; Kiss, R.; Laird, M. A Different Perspective: How Much Innovation Is Really Needed for Monoclonal Antibody Production Using Mammalian Cell Technology? In *New Bioprocessing Strategies: Development and Manufacturing of Recombinant Antibodies and Proteins*; Kiss, B., Gottschalk, U., Pohlscheidt, M., Eds.; Springer: Cham, Switzerland, 2018; pp. 443–462.
6. Kshirsagar, R.; Ryll, T. Innovation in Cell Banking, Expansion, and Production Culture. *Adv. Biochem. Eng. Biotechnol.* **2018**, *165*, 51–74. [CrossRef]
7. Schulze, M.; Lemke, J.; Pollard, D.; Wijffels, R.H.; Matuszczyk, J.; Martens, D.E. Automation of High CHO Cell Density Seed Intensification via Online Control of the Cell Specific Perfusion Rate and Its Impact on the N-Stage Inoculum Quality. *J. Biotechnol.* **2021**, *335*, 65–75. [CrossRef]
8. Stepper, L.; Filser, F.A.; Fischer, S.; Schaub, J.; Gorr, I.; Voges, R. Pre-Stage Perfusion and Ultra-High Seeding Cell Density in CHO Fed-Batch Culture: A Case Study for Process Intensification Guided by Systems Biotechnology. *Bioprocess Biosyst. Eng.* **2020**, *43*, 1431–1443. [CrossRef]
9. Xu, J.; Rehmann, M.S.; Xu, M.; Zheng, S.; Hill, C.; He, Q.; Borys, M.C.; Li, Z.J. Development of an Intensified Fed-Batch Production Platform with Doubled Titers Using N-1 Perfusion Seed for Cell Culture Manufacturing. *Bioresour. Bioprocess.* **2020**, *7*, 17. [CrossRef]
10. Wright, B.; Bruninghaus, M.; Vrabel, M.; Walther, J.; Shah, N.; Bae, S.-A.; Johnson, T.; Yin, J.; Zhou, W.; Konstantinov, K. A Novel Seed-Train Process Using High-Density Cell Banking, a Disposable Bioreactor, and Perfusion Technologies. *Bioprocess Int.* **2015**, *13*, 16–25.
11. Clincke, M.F.; Mölleryd, C.; Zhang, Y.; Lindskog, E.; Walsh, K.; Chotteau, V. Very High Density of CHO Cells in Perfusion by ATF or TFF in WAVE BioreactorTM: Part I: Effect of the Cell Density on the Process. *Biotechnol. Prog.* **2013**, *29*, 754–767. [CrossRef] [PubMed]
12. Müller, J.; Teale, M.; Steiner, S.; Junne, S.; Neubauer, P.; Eibl, D.; Eibl, R. Intensified and Continuous MAb Production with Single-Use Systems. In *Cell Culture Engineering and Technology*; Pörtner, R., Ed.; Springer: Cham, Switzerland, 2021; pp. 401–429.
13. Ninomiya, N.; Shirahata, S.; Murakami, H.; Sugahara, T. Large-Scale, High-density Freezing of Hybridomas and Its Application to High-density Culture. *Biotechnol. Bioeng.* **1991**, *38*, 1110–1113. [CrossRef] [PubMed]
14. Heidemann, R.; Mered, M.; Wang, D.Q.; Gardner, B.; Zhang, C.; Michaels, J.; Henzler, H.J.; Abbas, N.; Konstantinov, K. A New Seed-Train Expansion Method for Recombinant Mammalian Cell Lines. *Cytotechnology* **2002**, *38*, 99–108. [CrossRef] [PubMed]
15. Tao, Y.; Shih, J.; Sinacore, M.; Ryll, T.; Yusuf-Makagiansar, H. Development and Implementation of a Perfusion-Based High Cell Density Cell Banking Process. *Biotechnol. Prog.* **2011**, *27*, 824–829. [CrossRef]
16. Clincke, M.-F.; Mölleryd, C.; Samani, P.K.; Lindskog, E.; Fäldt, E.; Walsh, K.; Chotteau, V. Very High Density of Chinese Hamster Ovary Cells in Perfusion by Alternating Tangential Flow or Tangential Flow Filtration in WAVE BioreactorTM—Part II: Applications for Antibody Production and Cryopreservation. *Biotechnol. Prog.* **2013**, *29*, 768–777. [CrossRef] [PubMed]
17. Seth, G.; Hamilton, R.W.; Stapp, T.R.; Zheng, L.; Meier, A.; Petty, K.; Leung, S.; Chary, S. Development of a New Bioprocess Scheme Using Frozen Seed Train Intermediates to Initiate CHO Cell Culture Manufacturing Campaigns. *Biotechnol. Bioeng.* **2013**, *110*, 1376–1385. [CrossRef]
18. Bausch, M.; Brandl, M.; Horry, H. Seed Train Intensification Using High Cell Density Cryopreservation and Specially-Designed Expansion Medium. White Paper. 2020. Available online: https://www.sigmaaldrich.com/deepweb/assets/sigmaaldrich/product/documents/304/890/seed-train-intensification-using-hcdc-and-specially-designed-expansion-medium-wp5751en-mk.pdf (accessed on 29 July 2021).
19. Cytiva One-Step Seed Culture Expansion from One Vial of High-Density Cell Bank to 2000 L Production Bioreactor. 2016. Available online: https://cdn.cytivalifesciences.com/api/public/content/digi-17729-original (accessed on 10 November 2020).
20. Riesen, N.; Eibl, R. Single-Use Bag Systems for Storage, Transportation, Freezing, and Thawing. In *Single-Use Technology in Biopharmaceutical Manufacture*; Eibl, R., Eibl, D., Eds.; John Wiley & Sons, Inc.: Hoboken, NJ, USA, 2011; pp. 13–20. ISBN 9780470433515.
21. Bögli, N.C.; Ries, C.; Adams, T.; Greller, G.; Eibl, D.; Eibl, R. Large-Scale, Insect-Cell–Based Vaccine Development. *Bioprocess Int.* **2012**, *10*, 2–7.
22. Eibl, R.; Steiger, N.; Wellnitz, S.; Vicente, T.; John, C.; Eibl, D. Fast Single-Use VLP Vaccine Productions Based on Insect Cells and the Baculovirus Expression Vector System: Influenza as Case Study. *Adv. Biochem. Eng. Biotechnol.* **2014**, *138*, 99–125. [CrossRef]
23. Howe, W.G. Two-Sided Tolerance Limits for Normal Populations—Some Improvements. *J. Am. Stat. Assoc.* **1969**, *64*, 610–620. [CrossRef]

24. Guenther, W.C. *Sampling Inspection in Statistical Quality Control. Griffin's Statistical Monographs & Courses*; No. 37; C. Griffin: London, UK, 1977; ISBN 9780852642375.
25. Kleman, M.I.; Oellers, K.; Lullau, E. Optimal Conditions for Freezing CHO-S and HEK293-EBNA Cell Lines: Influence of Me$_2$SO, Freeze Density, and PEI-Mediated Transfection on Revitalization and Growth of Cells, and Expression of Recombinant Protein. *Biotechnol. Bioeng.* **2008**, *100*, 911–922. [CrossRef]
26. Heidemann, R.; Lünse, S.; Tran, D.; Zhang, C. Characterization of Cell-Banking Parameters for the Cryopreservation of Mammalian Cell Lines in 100-ML Cryobags. *Biotechnol. Prcg.* **2010**, *26*, 1154–1163. [CrossRef] [PubMed]

Article

Proof-of-Concept of Continuous Transfection for Adeno-Associated Virus Production in Microcarrier-Based Culture

Brian Ladd [1,2], Kevin Bowes [3], Mats Lundgren [4], Torbjörn Gräslund [2,5] and Veronique Chotteau [1,2,*]

[1] Department of Industrial Biotechnology, School of Engineering Sciences in Chemistry, Biotechnology and Health, Royal Institute of Technology (KTH), 11428 Stockholm, Sweden; ladd@kth.se
[2] AdBIOPRO, VINNOVA Competence Centre for Advanced Bioproduction by Continuous Processing, Royal Institute of Technology (KTH), 11428 Stockholm, Sweden; torbjorn@kth.se
[3] Cobra Biologics, Charles Rivers, Keele ST5 5SP, UK; kevin.bowes@crl.com
[4] Cytiva, 75323 Uppsala, Sweden; mats.lundgren@phase2phase.com
[5] Department of Protein Science, School of Engineering Sciences in Chemistry, Biotechnology and Health, Royal Institute of Technology (KTH), 11428 Stockholm, Sweden
*   Correspondence: veronique.chotteau@biotech.kth.se

Citation: Ladd, B.; Bowes, K.; Lundgren, M.; Gräslund, T.; Chotteau, V. Proof-of-Concept of Continuous Transfection for Adeno-Associated Virus Production in Microcarrier-Based Culture. *Processes* **2022**, *10*, 515. https://doi.org/10.3390/pr10030515

Academic Editor: Ralf Pörtner

Received: 14 February 2022
Accepted: 2 March 2022
Published: 4 March 2022

Publisher's Note: MDPI stays neutral with regard to jurisdictional claims in published maps and institutional affiliations.

Abstract: Adeno-associated virus vectors (AAV) are reported to have a great potential for gene therapy, however, a major bottleneck for this kind of therapy is the limitation of production capacity. Higher specific AAV vector yield is often reported for adherent cell systems compared to cells in suspension, and a microcarrier-based culture is well established for the culture of anchored cells on a larger scale. The purpose of the present study was to explore how microcarrier cultures could provide a solution for the production of AAV vectors based on the triple plasmid transfection of HEK293T cells in a stirred tank bioreactor. In the present study, cells were grown and expanded in suspension, offering the ease of this type of operation, and were then anchored on microcarriers in order to proceed with transfection of the plasmids for transient AAV vector production. This process was developed in view of a bioreactor application in a 200 mL stirred-tank vessel where shear stress aspects were studied. Furthermore, amenability to a continuous process was studied. The present investigation provided a proof-of-concept of a continuous process based on microcarriers in a stirred-tank bioreactor.

Keywords: Adeno-associated virus; transfection; PEI; continuous; gene therapy; microcarriers; bioreactor; transient expression

1. Introduction

Gene therapy has the potential to be one of the next great revolutions in medicine. It allows for not only treatment but also potential cures for many debilitating diseases. Through the choice of vector, different tissues can be targeted making the treatment highly specific with few off-target effects. Interest in gene therapy has intensified with more than 1680 new drug trials being conducted from 2004 to 2017 [1].

One of the most promising vectors for the targeted delivery of therapeutic genes is the Adeno-associated virus (AAV) due to its highly specific targeting and lack of an immune response [2]. Most of the current AAV based therapies, are for rare diseases with low patient populations or have targets that require low doses. The current roadblock for the application of AAV based therapies for more prevalent diseases, or ones that require a high dose, is the limited manufacturing capacity [3]. An efficient scalable manufacturing platform for AAV vectors would allow for larger trials which would accelerate drug development and provide treatment options for diseases that affect larger populations.

Production methods of AAV based on transient transfection require plasmids encoding the viral proteins and DNA, which are required for vector assembly. Typically, the two open reading frames encoding the Rep proteins and the capsid proteins are placed on the same plasmid (pRepCap) and the adenoviral genes necessary for AAV replication is placed on a second plasmid (pHelper); the gene of interest (GOI), flanked by the inverted terminal repeat sequences, is placed on a third (pGOI). However, there are designs that can reduce the number of plasmids [1–4]. The producer cells are grown either adherently or in suspension. Adherent cells are often able to provide a higher cell-specific production of AAV [2]. The popularity of the transient transfection method is due to the versatile production and rapid process development that it facilitates. The plasmids are transfected at the time of production and the specific serotype and GOI can be readily changed with only minor alterations to the process. This versatility allows for a quick transition between the production of different serotypes and GOIs.

The transfection agent that is most widely used is the cationic polymer, polyethylenimine (PEI). The positively charged PEI facilitates transfection by forming a complex with the negatively charged plasmid DNA; this complex is called the polyplex. When added to the cells, the polyplex is taken up and, through endosomal escape due to the proton sponge effect of the PEI, the PEI-DNA complex is released within the cytosol where it can be transported into the nucleus. It has been shown that the size and properties of these complexes have a large influence on transfection efficiency [5–7], with smaller complexes being more readily taken up by the cell but having a lower chance of reaching into the nucleus and larger particles being taken up more slowly but are more effective in delivering their DNA payload.

Depending on the specific cell line and transfection conditions, a larger or smaller polyplex will lead to a more effective transfection. Therefore, it is necessary to reoptimize the transient transfection conditions when the mode of production changes or if a new cell line is used. In addition to the size of the polyplex, shear forces acting on the cells play an important role in the effectiveness of the transfection [8,9].

It has been shown that shear forces influence the uptake of nanoparticles including PEI-DNA complexes [8–10]. During transfection, the integrity of the cell membrane is compromised, increasing their susceptibility to damage by shear, meaning that conditions that previously did not affect cells could become damaging after transfection [9].

The cell-specific yields and vector quality are up to 15-fold higher in adherent systems compared to single-cell suspension systems [11]. While it has been shown that AAV vectors can be produced by cells in suspension with volumetric yields similar to processes with adherent cells, this is however, only achieved after a long and labor-intensive optimization [12,13]. Two-dimensional production systems, such as roller bottles or T-flasks, provide high specific productivity of AAV but suffer from scalability, making large scale production not economical [3]. The solution is to use a three-dimensional system that maintains the high cell-specific productivity of adherent cells but allows the process to scale with the volume rather than the surface area. If an adherent system could have the same cell density as a suspension platform, around 2×10^6 cells mL^{-1}, the volumetric yields would be significantly higher. Potentially, a microcarrier-based transient transfection process would combine the scalability of suspension cultures with the high specific productivity of adherent production, which could result in an extremely efficient process [14].

Microcarriers have been used since the 1960s for the cultivation of anchorage-dependent cells in a suspension system. Cultivations using microcarriers have been successfully used at scales in excess of 2000 L for the production of vaccines [15]. Microcarriers thus provide a possible solution to scale up the bottleneck of anchorage-dependent AAV production.

While being able to provide a close to suspension-like scalability, microcarriers have their own set of limitations. Among them is the increased sensitivity of the cells to shear due to the increase in effective size of the cell to the hydrodynamic environment [16]. Vortices, also called eddies, are formed in a turbulent fluid with the size of these eddies being proportional to the power transferred to the fluid. The smallest of these eddies and power input to the fluid are related by the Kolmogorov eddy length equation (1). When the eddy size decreases in a turbulent flow, the size between a particle and an eddy becomes comparable and the shear acting on the surface of the particle increases. If the Kolmogorov eddy length is of comparable size to the microcarrier, high shear is experienced on the surface, damaging the attached cells [16].

The generation of a large volume culture with microcarriers is an issue for manufacturing. To enable an increase in the number of adherent cells, cells anchored on microcarriers need to be detached and reinoculated to a larger amount of microcarriers in a larger volume. This detaching and reattaching procedure presents a loss in efficiency compared to suspension cell culture.

Depending on the cell line, the anchorage dependency of the cells can be modified based on the culture medium used. If the anchorage dependency would be only required at a certain stage in the culture, during transfection and AAV production, for example, it could be more efficient and economically attractive to keep the cells in suspension during the previous stages, i.e., cell expansion, and then shift the culture to anchorage dependency on microcarriers at the production stage. Such a system lends itself very well to the production of AAV at scale because it can make use of the increased product quality and yield of adherent cells while maintaining the efficiency and economics of suspension culture to build the cell mass.

In the present study, it was hypothesized that HEK293T cells could be grown in suspension, offering the easiness of this type of operation and that the cells would then be anchored on microcarriers in order to proceed with the triple plasmid transient transfection in adherent cells for AAV expression. This latter process to benefit from a higher specific AAV yield, is often reported higher for adherent cell systems than cells in suspension. Bearing in mind that this process was developed in view of scaled-up commercial application in a bioreactor, the culture on microcarriers was studied in terms of limitations brought by shear forces in stirred tank vessels and the feasibility of continuous transfection. Finally, the present investigation aimed at providing a proof-of-concept of a process based on microcarriers in view of a readily scalable solution for the production of AAV at a commercial scale.

2. Materials and Methods

2.1. Culture Medium for Cell Expansion and Passaging

HEK293T cells (Cobra Biologics, Charles Rivers, ST55SP Keele, UK) were cultured as single cells in suspension and routinely passaged twice a week in Hyclone CDM4HEK293 medium, cat. No. SH30858.02 (Cytiva, 75323, Uppsala, Sweden). This medium was selected among four culture media for its ability to support the growth of cells in suspension. Using this medium, the specific growth rate, calculated using nonlinear least squares fitting of the total cell density, was 0.80 day^{-1} against 0.61 to 0.70 day^{-1} for the other tested media.

2.2. Medium for Transfection

A selection of the medium used for the triple transfection (presented in Section 2.3) among five media, A, ... , E, was carried out in static tissue culture T-25 flasks with a total end volume of 7 mL and a total DNA mass of 80 μg. The relative AAV titer was measured via CHO cell transduction (presented in Section 2.7). Supplementary Figure S1 shows the average transduction results normalized to the maximal observed expression obtained in medium D. From the transduction, the relative GFP expression indicated that the HEK293T cells transfected in medium D generated a greater production of the active vector than in the other media. Based on this result, it was decided that medium D would be used for all the transfection experiments. Medium D is a proprietary serum-free medium.

2.3. Adeno-Associated Virus Triple Transfection

The three plasmids for AAV production; pHelper, pRepCap, and pGOI (Cobra Biologics, Charles Rivers, ST55SP Keele, UK), were at a DNA concentration of 1 mg mL^{-1} in TE buffer. The pGOI encoded green fluorescent protein (GFP) and the pRepCap were derived from AAV serotype 9. The transfection reagent was polyethyleneimine PEIpro (Polyplus), at a stock concentration of 1 mg mL^{-1}. The three plasmids, pHelper, pGOI encoding GFP, and pRepCap, were mixed at a volume ratio of 2:1:1 with an end DNA concentration of 80 μg mL^{-1} in the cell culture medium. The PEIPro was also diluted to a concentration of 80 μg mL^{-1} in the cell culture medium. The DNA mix was then added to the PEIpro at a volume ratio of 1:1 (DNA: PEIpro ratio 1:1), briefly vortexed, and then left to incubate for 5 min before transfection in HEK293T cells. The number of passages before transfection did not exceed 30 passages. Prior to transfection, a 70% medium exchange was performed by stopping the agitation and allowing the microcarriers to sediment; after which, the appropriate amount of supernatant was removed and fresh medium was added. In the spinner flask transfections, the volume was reduced by 50%, from 50 mL to 25 mL. The volume was increased back to 100% 2 h post-transfection.

2.4. Cell Count and Viability Measurements

For all the transfection experiments and cell passaging/back-up maintenance, the density and viability of HEK293T and CHO cells were measured by BioProfile FLEX Analyzer (Nova Biomedical, Waltham, MA, USA) which uses the trypan blue exclusion method. For all the transduction experiments, the cell density and viability were measured by Norma XS (Iprasense, Clapiers, France), which is based on holographic imaging. All cell counts were made from samples taken directly from the culture without dilution.

2.5. Flow Cytometry

A Gallios flow cytometer (Beckman Coulter, Brea, CA, USA) was used for the quantification of the fraction of cells expressing GFP. The 488 nm excitation laser was used along with a 525 nm detector, channel 1, and a 575 nm detector, channel 2. An initial forward and side scatter gate was used to isolate the cell population. The fluorescence channels 1 and 2 were used to gate the cells expressing GFP by the constant ratio observed between these channels. All gating was done in the software Kaluza (Beckman Coulter, Brea, CA, USA); the data were exported to MATLAB (MathWorks, Natick, MA, USA) for further analysis.

2.6. Bioreactor and Spinner Flask Cultures

A DASBox system (Eppendorf, Hamburg, Germany) was used for the bioreactor experiment with 150 mL working volume, under feedback controls of pH, dissolved oxygen concentration, impeller speed, and temperature. The bioreactor was equipped with two marine impellers with a diameter of 3 cm [17]. The control software regulated the dissolved oxygen by varying the flow rate and proportion of air or pure oxygen at atmospheric pressure into the head space of the vessel. The pH was controlled by varying the flow rate of CO_2 into the head space (no upregulation of the pH was required). The agitation rate was controlled via an onboard tachometer. In the perfusion operation, a sedimentation tube with a diameter of 1 cm was used to retain the microcarriers.

Cultures in spinner flasks were performed using Bellco spinner flasks (Bellco Glass, Vineland, New Jersey USA), magnetically agitated at speeds specified in the text, in an incubator (37 °C, 5% pCO$_2$). The spinner flasks were equipped with a dual blade impeller, diameter 5.1 cm, height 2.27 cm.

The microcarriers Cytodex 3 (Cat. No. 17-5487-01) and Cytodex 1 (Cat. No. 17-5488-01) (Cytiva, 75323, Uppsala, Sweden) were prepared according to the manufacturer instructions. The spinner flask cultivations used gamma sterilized Cytodex 3 and for the bioreactor cultivations. Cytodex 1 or 3 microcarriers were prepared and then autoclaved for 20 min at 121 °C. Cytodex 1 and 3 microcarriers have a dry mass specific surface area of 4400 cm^2 g^{-1} and 2700 cm^2 g^{-1}, respectively. A concentration of 3 g L^{-1} was targeted, corresponding to a volume specific surface area of 13.2 cm^2 mL^{-1} and 8.1 cm^2 mL^{-1} for Cytodex 1 and 3. A target inoculation density of 50 cells per microcarrier, or 7.9 $\times$ 10^4 cells cm^{-2} and 4.5 $\times$ 10^4 cells cm^{-2} for Cytodex 1 and 3, was used unless otherwise specified.

2.7. Transduction Assay

For the determination of the amount of AAV produced, CHO cells were used for transduction. In a Corning 24 well deep well plate or tissue culture treated Corning 96 well plate, CHO cells were seeded to a final density of 0.1×10^6 cells mL^{-1} in 1.5 mL or 150 µL of Ex-cell 302 medium from Sigma-Aldrich. 1 mL or 0.1 mL of AAV containing the sample was added to this culture and the plate was sealed with an adhesive filter and placed in a 37 °C, 5% pCO$_2$ incubator and shaken at 300 RPM with an orbital diameter of 2.5 cm or was statically incubated. After 48 h, the cells were analyzed by flow cytometry or by a fluorescence plate reader with an excitation wavelength of 500 nm, a dichroic filter of 520 nm, and an emission wavelength of 540 nm.

3. Results

3.1. Microcarrier Cultivation of HEK293T Cells

Medium D was selected among five media for its potential to support transfection and AAV production. This medium was also able to support the cell attachment and growth of microcarriers. A major benefit was that the medium used for growth on the microcarriers did not need to be changed to a different medium before transfection, which gave a great benefit to reduce the operations, a factor important for the scale-up of this process.

A preliminary study was dedicated to the characterization of the culture of the cells on microcarriers. Cells were inoculated in a Bellco 125 mL spinner flask with 3 mg mL^{-1} Cytodex 3 microcarriers, at a cell density of 0.64×10^6 cells mL^{-1} in a final culture volume of 50 mL at two different agitation rates. It was observed that a non-homogeneous distribution of cells on the microcarriers occurred for certain agitation rates. In Figure 1, the distribution of cells on the microcarriers is shown for two different agitation rates, the minimum speed allowing suspension, 50 RPM, and a higher rate, 70 RPM.

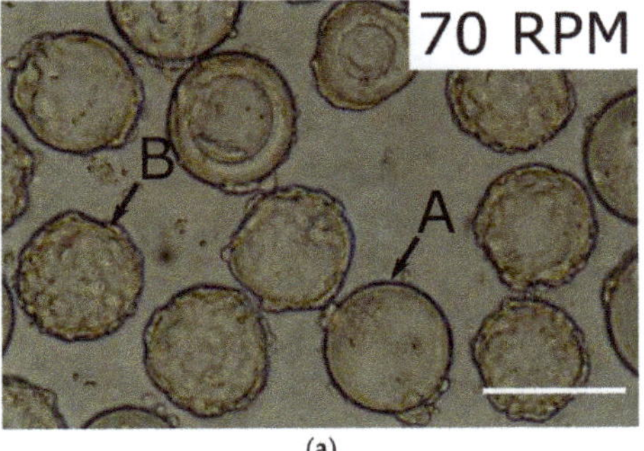

(a)

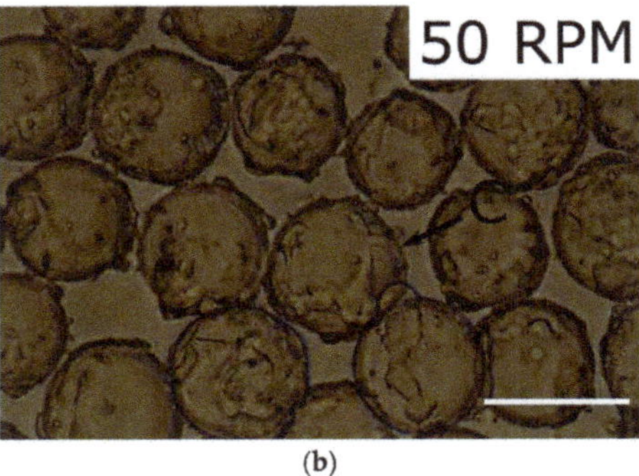

(b)

Figure 1. Pictures comparing the cell distribution on microcarriers of HEK293T cells inoculated at different agitation rates and a density of 0.64×10^6 cells mL^{-1}; the scale bar represents 200 µm. Arrows point at microcarrier bare from cells (A), fully confluent (B), and with homogenous cell distribution (C). (a) 70 RPM, (b) 50 RPM.

The cell distribution on the microcarriers was more homogenous when using 50 RPM agitation than 70 RPM. This was likely due to the fact that while a higher agitation rate led to greater mixing, it also brought more shear, which could detach cells from the microcarriers. Confluent microcarriers are better able to withstand this shear and are therefore preferentially sustaining cell growth compared to more sparsely populated carriers, accentuating the inhomogeneity. To minimize the effect of shear while maintaining the homogenization, the lowest agitation rate able to maintain adequate mixing should be used.

The agitation rate can greatly affect the attachment and growth of the cells on microcarriers [18]. The minimum rate able to homogenously suspend the microcarriers is known as the just suspended agitation rate or N_{js} [19]. The minimum agitation rate to maintain the microcarriers in suspension is affected by the reactor and impeller geometries as well as the medium properties and can be calculated with the Zwietering correlation [19].

$$N_{js} = S \times \nu^{0.1} \times d_p{}^{0.2} \times \left(\frac{g \times (\rho_s - \rho_l)}{\rho_l} \right)^{0.45} \times \frac{X^{0.13}}{D^{0.85}} \tag{1}$$

with N_{js}: Just suspended stirring speed (s^{-1}), S: Zwietering coefficient (–), ν: kinematic viscosity (m^2 s^{-1}), X: solid loading fraction (kg solid kg^{-1} fluid), d_p: microcarrier diameter (m), D: impeller diameter (m), g: gravitational constant (m s^{-2}), ρ_s: density of the microcarrier (kg m^{-3}), ρ_l: density of the fluid (kg m^{-3}).

In the present study, the minimum agitation rate for a 3 mg mL^{-3} suspension of Cytodex 3 microcarriers was measured in both a 125 mL Bellco spinner flask and a DASBox reactor with dual marine impellers. The just suspended stirring speed in the spinner flask was measured to be 50 RPM and the Zwietering coefficient was calculated to 4.8, a value in line with the literature [20]. In the DASBox, the just suspended stirring speed was measured to be 150 RPM generating a Zwietering coefficient of 9.2. This larger Zwietering coefficient indicated that the DASBox system was less efficient in suspending microcarriers than the spinner flask for the present configurations of these vessels.

3.2. Hydrodynamic Comparison between Spinner Flasks and the DASBox Bioreactor

To determine the maximum shear limit acceptable for the cells on microcarriers for the bioreactor system, the agitation was slowly increased after the cells had attached to the microcarriers. This was applied from 150 RPM to 400 RPM by steps of 50 RPM in duplicate. Figure 2 shows photographs taken from both reactors, where the rows correspond to the reactor number, and columns a and b show samples taken from the 250 RPM and 300 RPM conditions. The attached cells were sheared off the microcarriers when the agitation rate increased from 250 RPM, column a, to 300 RPM, column b. This is most clearly seen by observing aggregated microcarriers; at 250 RPM cells were seen on the perimeter of the aggregates but at 300 RPM, column b, these perimeter cells had been sheared off.

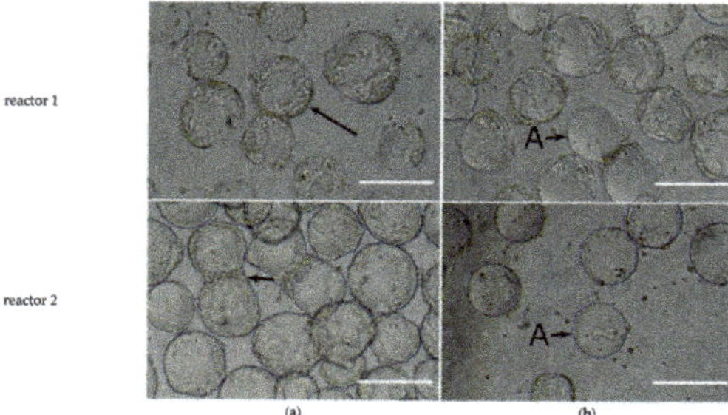

Figure 2. Determination of the maximum shear limit for HEK293T cells attached to Cytodex 3 microcarriers for the bioreactor system DASBox with dual marine impellers—experiment performed

in duplicate: reactor 1 (row 1) and reactor 2 (row 2); the scale bar represents 200 µm. The microcarriers were inoculated at 150 RPM; 24 h after inoculation, the impeller speed was slowly increased from 150 RPM to 400 RPM in steps of 50 RPM intervals. Images of the microcarriers are shown for ((a)—left column) 250 RPM where the cells were attached to the microcarriers, and ((b)—right column) 300 RPM where cell detachment from the microcarriers was observed. The unlabeled arrows represent cells well attached to microcarriers and the arrows labeled with A represent microcarriers where the cells have been sheared off by the hydrodynamic forces.

To make a detailed comparison between the spinner flasks and the DASBox; key hydrodynamic parameters were calculated for each vessel. The most important of these parameters is the Kolmogorov eddy length shown in Equation (2).

$$\sigma = \left(\varepsilon \times \left(\frac{\eta}{\rho} \right)^3 \right)^{\frac{1}{4}} \tag{2}$$

with σ: eddy size (m), ε: energy dissipation per mass (W kg^{-1}), η: dynamic viscosity of the fluid (Pa s), ρ: density of the fluid (kg m^{-3}).

Equation (2) requires the energy dissipation per mass, which can be calculated by Equation (3).

$$\varepsilon = \frac{Ne \times N^3 \times D^5}{V} \tag{3}$$

with Ne: dimensionless Newton number for the impeller (–), N: rotational frequency (s^{-1}), D: impeller diameter (m), V: reactor volume (m^3).

The Newton number, Ne, was calculated for the spinner flask at agitation rates 50 RPM and 70 RPM using the Nagata correlation [21], see Table 1. The Newton number for the DASBox bioreactor was assumed to be identical to the value for a marine impeller as given in [22], also listed in Table 1.

Table 1. Newton numbers for the Bellco spinner flask and the DASBox bioreactor.

	Spinner Flask 50 RPM	Spinner Flask 70 RPM	DASBox
Ne	0.52	0.46	0.36 [20]

Using Equations (2) and (3) and the values in Table 1, the Kolmogorov eddy lengths were calculated for the spinner flask and the DASBox bioreactor and graphically represented as shown in Figure 3. Note that the power input for the DASBox was doubled to account for the dual marine impellers.

Croughan et al. determined that the area of cell damage begins when the eddy length decreases below 2/3 of the microcarrier diameter and an eddy length below 100 µm leads to rapid cell death [16]. These regions are marked in orange and red, respectively, in Figure 3. The DasBox was inoculated using an agitation rate of 150 RPM, allowing for even coverage and providing a low shear environment for them to develop a more robust attachment. In contrast, inoculation at 70 RPM in the spinner flask, an agitation point located within the cell damage region, leads to an inhomogeneous distribution of cells because of the increased shear. While the eddy length for 250 RPM in the DASBox bioreactor lies within the damage zone, the cells were able to withstand the shear. However, these conditions would most likely lead to negative effects if maintained for prolonged periods of time. The 300 RPM condition in the DASBox, another agitation point within the cell death region, leads to rapid cell loss. This is not surprising because at this agitation rate the eddy length is below 100 µm which is associated with cell death on microcarriers [16].

These values provided the limiting operating configurations for the spinner flasks and the bioreactor systems. Namely, inoculation should be at an agitation rate below the cell damage zone, 70 RPM for the spinner flask, and 233 RPM for the DASBox. Increases in agitation up to 250 RPM in the DASBox can be tolerated, but prolonged exposure would most likely be detrimental. Finally, an agitation rate of 300 RPM and greater in the DASBox would shear cells off the microcarriers and result in cell death.

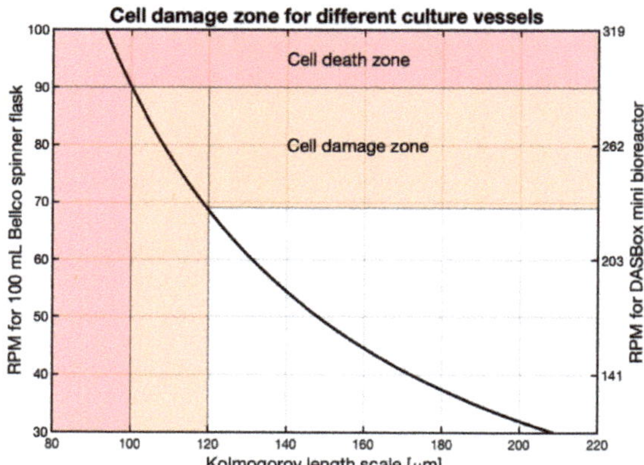

Figure 3. The Kolmogorov eddy length (x-axis) is represented as a function of the agitation speed in the Bellco spinner flask, left y-axis, and in the DASBox bioreactor, right y-axis, where the length scale associated with cell damage is marked in orange and the area associated with rapid cell death in red.

3.3. Triple Transfection of Cells Adherent on Microcarriers for AAV Production

3.3.1. Mass Transfer Considerations for Microcarrier Based Transfections

Mixing during transfection improves the mass transfer between the bulk culture medium and the cells in comparison with static transfection. Thanks to the agitation, the PEI-DNA polyplexes only need to diffuse through a small film layer surrounding a particle instead of the liquid height to reach the cell surface. This causes a local increase in the concentration of PEI and DNA on the cell surface and could be a possible explanation of the improved nanoparticle uptake by cells under shear [10]. This local concentration can be calculated using the results from the shear studies performed in Section 3.

The Sherwood number represents the ratio of the convective mass transfer to the diffusive mass transfer and can be used to compare the relative effects of the different transport phenomena. Using the correlation in Equation (4), the Sherwood number, Sh, can be calculated for the case of microcarriers [23].

$$Sh = 2 + 0.4\left(\varepsilon \times \frac{d_p^4}{v^3}\right)^{\frac{1}{4}} \times Sc^{\frac{1}{3}} \text{ where } Sc = \frac{v}{D_f} \tag{4}$$

ε: energy dissipation per mass (W kg^{-1}), v: kinematic viscosity of the fluid (m^2 s^{-1}), d_p: microcarrier diameter (m), D_f: diffusion coefficient of the PEI-DNA complex (m^2 s^{-1}), Sc: Schmidt number (–).

Using the Sherwood number from Equation (4), the normalized concentration of a chemical species on the surface of a microcarrier can be estimated for different cellular uptake rates [24] as follows;

$$X_N = \frac{X_b - X_c}{X_b} = \frac{\psi \times R_c \times d_p}{D_f \times X_b \times Sh} \tag{5}$$

X_N: normalized surface concentration (–), X_b: concentration in the bulk (mol L^{-1}), X_c: concentration on the surface (complexes m^{-3}), ψ: surface coverage of cells per unit area (cells m^{-2}), R_c: cell-specific reaction rate (complexes cell^{-1} s^{-1}).

The concentration of PEI-DNA complexes was estimated by assuming a polyplex density similar to that of a protein, as studied in [25], and a polyplex radius of 300 nm, which was the average radius for a 5 min incubation reported in [26]. Equation (5) shows the normalized polyplex concentration as a function of the cell-specific complex uptake rate for a spinner flask operating at 50 RPM. Here, the cell-specific uptake rate was an estimate based on observations in several reports that the PEI-based transfection is complete after a maximum of 4 h, corresponding to an uptake rate of 0.01 complexes cell^{-1} s^{-1} [26–29].

The normalized concentration, see Equation (4), indicates the percentage decrease from the bulk concentration, where a normalized concentration of 5% means that the surface concentration is 95% of the bulk (100% − 5% = 95%). The normalized concentration gives an indication of where the majority of the mass transfer resistance lies. This is best illustrated by looking at the extreme case where the normalized concentration is either 1 or 0. When the normalized concentration is 1, this means that the surface concentration is 0, which can only be the case when the uptake rate of the cell greatly exceeds the transfer rate, i.e., the cell instantly takes up any polyplex that reaches the cell surface. Conversely, when the normalized concentration is 0, the concentration at the cell surface is equal to the bulk concentration, i.e., the transfer to the cell surface greatly exceeds the cells' ability to take up the polyplex Figure 4 shows that at the relevant uptake rates of between 0.01 and 0.05 complexes cell^{-1} s^{-1}, the surface concentration (100%—normalized concentration) of the PEI-DNA complexes is between 75% to 95% of the bulk concentration. This means that the transport rate to the surface exceeds the uptake rate.

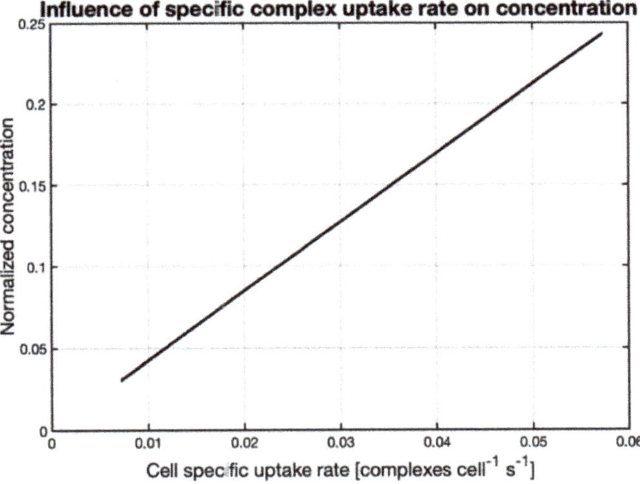

Figure 4. Theoretical normalized concentration of PEI-DNA complexes, calculated by Equation (5), as a function of the cell-specific polyplex uptake rate for a polyplex radius of 300 nm in a spinner flask stirred at 50 RPM. The cell-specific uptake rate was estimated based on PEI transfection completion obtained after a maximum of 4 h, corresponding to an uptake rate of 0.01 complexes cell^{-1} s^{-1}. The PEI-DNA complexes concentration was estimated by assuming a polyplex density similar to that of a protein, and a polyplex radius of 300 nm, which was the average radius for a 5 min incubation.

The internalization mechanism mediated by heparin sulfate proteoglycans (HSPG) has been shown to affect the transfection efficiency of CHO cells, with lower numbers of HSPG binding sites resulting in lower transfection efficiency [30]. In addition, Mozley et al. showed that the rate of polyplex internalization was linked with the regeneration rate of HSPGs [30]. Therefore, if the rate of polyplex delivery to the cell exceeds the rate of this uptake mechanism, the polyplexes could either be internalized through another mechanism or be altered by the conditions in the culture; both will affect the transfection. It is then important to consider not only the overall amount of DNA that is added but also the concentration of DNA at transfection.

3.3.2. Transfection in Spinner Flasks

HEK293T cells immobilized on Cytodex 3 microcarriers were transfected at two different DNA concentrations. Both DNA concentrations were chosen based on the preliminary transfection experiments in a static T-flask. The first condition was a final DNA concentration of 5.72 µg mL^{-1} and the second concentration was taken identical to the static experiment, 11.44 µg mL^{-1}. The cell density at the time of transfection was 1.4×10^6 cells mL^{-1} two hours post-transfection (2 hpt) the volume was increased from 25 mL to 50 mL, diluting the cell density down to 0.7×10^6 cells mL^{-1}. Both vessels were sampled daily, and the supernatant was used to transduce CHO cells to evaluate the production of biologically active AAV vectors. Figure 5a shows the fraction of transduced CHO cells as a function of the hours post-transfection (hpt).

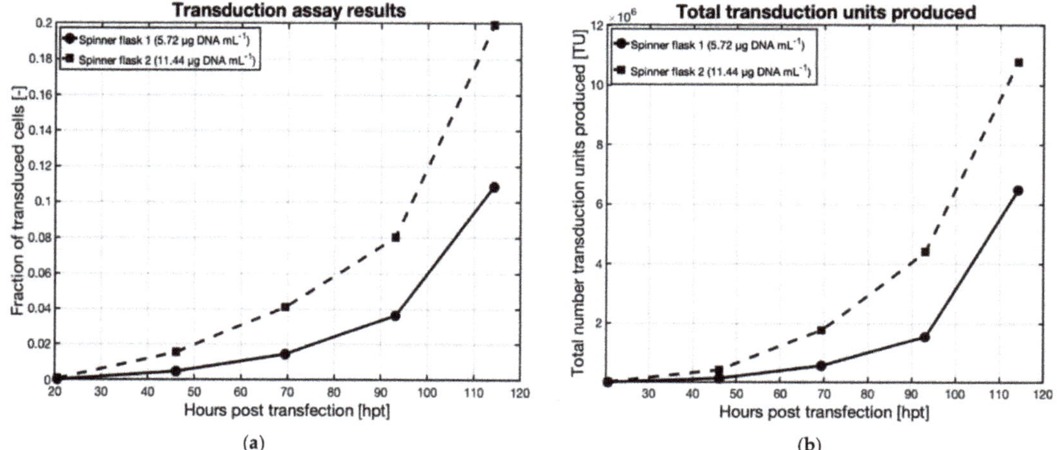

Figure 5. Effect of DNA concentration on the transfection of HEK293T cells immobilized on Cytodex 3 microcarriers in 125 mL Bellco spinner flasks, measured by transduction assays of CHO cells using supernatant samples harvested at various times from the spinner flasks with a cell density at transfection of 1.4×10^6 cells mL^{-1}; spinner flask 1 and 2 had final DNA concentrations of 5.72 µg DNA mL^{-1} and 11.44 µg DNA mL^{-1}, respectively. (**a**) Fraction of transduced CHO cells from supernatant samples at selected points post-transfection and (**b**) Total number transduction units produced in both spinner flasks, calculated from Equation (6).

It can be seen in Figure 5a that twice as many CHO cells were transduced when using the supernatant from the highest DNA concentration, indicating that doubling the DNA concentration generated production of twice the amount of active AAV vectors. This suggests that at a concentration of 11.44 µg DNA mL^{-1} the polyplex uptake mechanisms were not saturated. As a matter of fact, in case the uptake mechanisms had been saturated, the increase in the transduced fraction would have been smaller than the increase in DNA. It also indicates that the other mechanisms for AAV production were not saturated for this same reason. The latter condition, providing a superior outcome, was selected for further studies.

To quantify the number of active virus particles obtained at 114 h post-transfection the supernatants from both spinner flasks 1 and 2 were used to perform a Tissue Culture Infectious Dose 50 assay, TCID$_{50}$ assay, in which the supernatant samples were serially diluted and used to transduce CHO cells. Figure 6a,b represent, for transfected DNA concentrations of 5.72 µg mL^{-1} and 11.44 µg mL^{-1}, the number of transduced CHO cells as a function of supernatant volume used in the serial dilution. When volumes up to 0.25 mL of supernatant i.e., high dilutions, were used in the transduction assay, the amount of transduced cells increased linearly with the supernatant volume. In Figure 6a,b, lines fitted for this range of volume supernatant were drawn. The slopes of the values in the linear range of Figure 6a,b represent the change in the number of transduced CHO cells given the corresponding increase in the volume of supernatant added, in other words, it is a direct quantification of the concentration of active AAV in the supernatant. These slopes can be expressed in transduction units per milliliter [TU mL^{-1}], where one transduction unit is able to infect and cause one CHO cell to become GFP positive, and are 1.358×10^5 [TU mL^{-1}] and 2.143×10^5 [TU mL^{-1}] for spinner flasks 1 and 2 respectively.

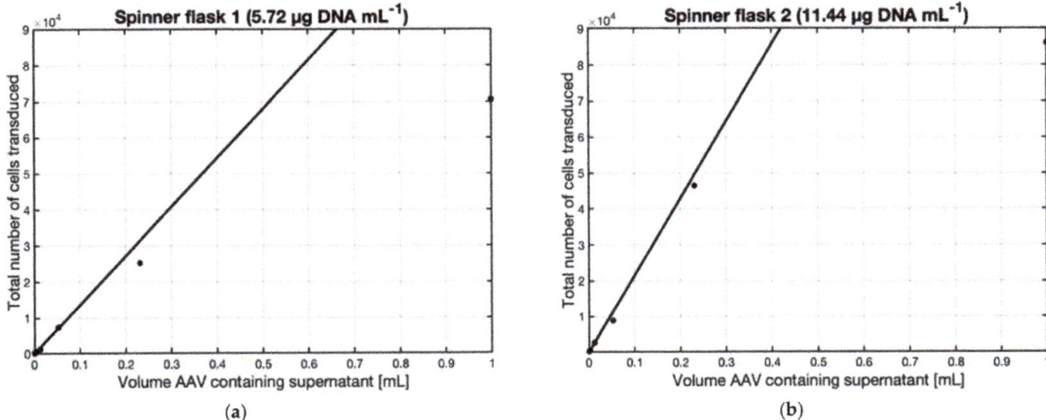

Figure 6. TCID$_{50}$ assay of supernatant samples harvested 114 h post-transfection from HEK293T cells immobilized on Cytodex 3 microcarriers transfected in 125 mL Bellco spinner flasks at a final concentration of (**a**) 5.72 µg DNA mL^{-1} or (**b**) 11.44 µg DNA mL^{-1} and a cell density of 1.4×10^6 cells mL^{-1}, corresponding to spinner flasks 1 or 2 of Figure 5a; supernatant samples were serially diluted and used to transduce CHO cells. The number of transduced cells increased with the supernatant volume until the number of cells used for the transduction was reached, leading to asymptotic behavior.

As can be seen in Figure 6a,b, the total amount of transduction units, measured by the TCID$_{50}$ assay, increased with the total number of transduced cells. This increase was linear for the lower values but approached an asymptote when the total number of transduced cells (y-axis) approaches the number of cells used for the assay. This behavior is expected as it would be impossible to transduce more than the number of cells in the assay. In Figure 7, the titers calculated in the TCID$_{50}$ assay were used to plot the total amount of transduced CHO cells vs. the total number of transduction units given in the transduction assay for both DNA conditions.

The data in Figure 7 fits a model with the function

$$y = \frac{b \times x}{(a - x)} \tag{6}$$

where x is the total number of transduced cells and y is the total number of transduction units with the parameters $a = 128{,}900$ and $b = 107{,}100$ determined by linear regression.

This relationship can be used to evaluate the number of transduction units from a CHO transduction assay culture, providing the titer of active AAV's. Figure 5b shows the total amount produced for both DNA conditions. The trend is very similar to the fraction of transduced cells, which is expected given that the same assay protocol was used for both assays.

This equation is specific for the transduction protocol used in this work and will most likely need to be recalibrated if parameters such as the medium containing the AAV, the cell number for the transduction, or the cell type are changed. It remains, however, extremely useful to monitor the titer of AAV during process changes.

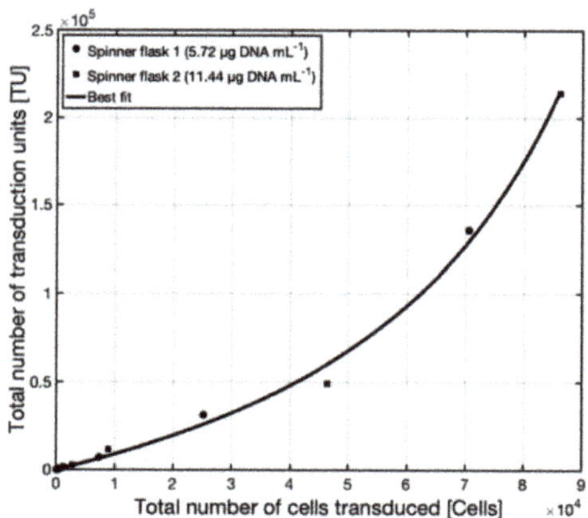

Figure 7. The total number of transduction units calculated from the titer determined by the $TCID_{50}$ assay, represented as a function of the total number of transduced CHO cells; approximation by model $y = b\,x\,(a - x)^{-1}$, coefficients $a = 128{,}900$ and $b = 107{,}100$ determined by linear regression.

3.3.3. Transfection in Bioreactor

The successful transfection of cells on microcarriers in spinner flasks showed that microcarrier cultures can be an effective tool to scale up AAV vector production for adherent cells. However, it was observed that shear could play a critical role in this process and therefore had to be taken into account in the up-scaling. To evaluate the importance of this factor on AAV production, a set-up mimicking a large-scale reactor was used. In the present study, a 150 mL DASbox bioreactor was operated with a perfusion system based on a sedimentation tube from which supernatant was pumped to the harvest tank while the microcarriers were retained in the bioreactor.

To limit the damage caused by shear, a rate of 175 RPM was used to closely mimic the conditions in the spinner flask at 50 RPM. While the DASBox operated slightly above the N_{js}, according to Figure 3, the Kolmogorov eddy length for the DASBox at 175 RPM was well below the cell damage zone and was similar to the shear environment in the spinner flask at 50 RPM. In these conditions, the Sherwood numbers (Equation (4)), or the ratio of convective mass transfer to diffusive mass transfer, were 35.71 for the DASBox (150 mL) and 33.00 for the spinner flask (50 mL). These values are similar, indicating that the mass transfer of the polyplexes to the surface of the cells was similar in these systems. In addition to the Sherwood numbers, the other variables of Equation (5) used to determine the normalized concentration gradient of the PEI-DNA complex were also similar in both systems. It was thus expected that the normalized surface concentrations of polyplex were similar between the spinner flask and DASBox.

Cells at a density of 1.5×10^6 cells mL^{-1} were transfected in the DASbox bioreactor with a final DNA concentration of 11.44 µg mL^{-1}. Before transfection, the agitation was stopped, allowing the microcarriers to sediment; after which, 113 mL of supernatant was removed and 43 mL of polyplex mixture was added under agitation. Two hours post-transfection, 75 mL of fresh medium were added to restore the volume in the reactor to 150 mL. After transfection, samples were taken daily and the supernatant was used to evaluate the AAV production by transducing CHO cells, see Figure 8.

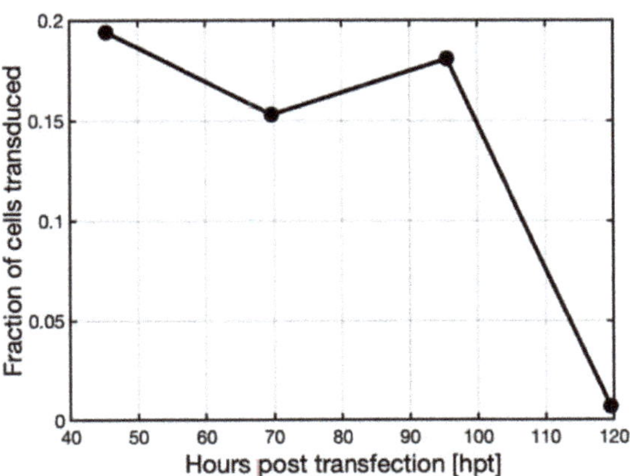

Figure 8. Transduction assay results of CHO cells using supernatant samples harvested at various time points from HEK293T cells immobilized on Cytodex 3 microcarriers after transfection at a final DNA concentration of 11.44 µg DNA mL^{-1} in a 150 mL DASBox bioreactor operated in perfusion with a cell density at the time of transfection of 1.5×10^6 cells mL^{-1}; results indicated a stable AAV titer until day 4 followed by a decrease due to an increase of the perfusion rate.

It was observed that about 15% to 20% of the cells produced GFP when transduced with supernatant taken at days 2, 3, or 4 post-transfection, after which, the titer decreased steeply. The decrease in AAV titer on day 5 of the experiment was due to the perfusion rate being increased in the experiment and thus diluting the AAV.

Using the correlation of Equation (6) between the total number of CHO cells transduced and the total amount of transduction units, it is possible to calculate the accumulated volumetric productivity of the reactor, shown in Figure 9.

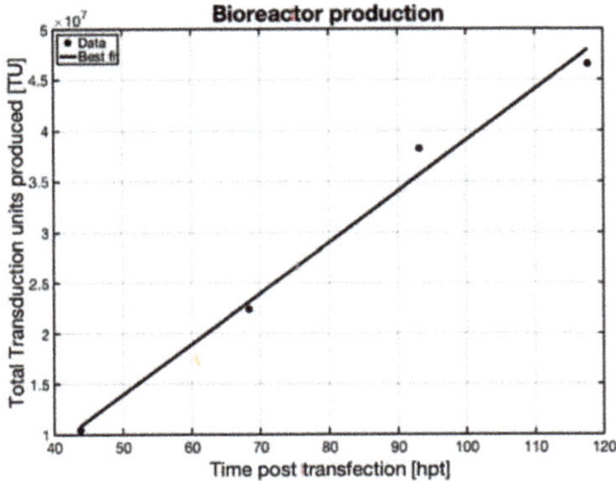

Figure 9. Total accumulated transduction units produced in the DASBox bioreactor operated in perfusion with a working volume of 150 mL and transfected with 11.44 µg DNA mL^{-1}, indicating that AAV was produced until the last day, day 5.

The total accumulated AAV production in the reactor was continuously increasing until day 5, after which the culture was terminated. The spinner flasks started at a low level of expression that increased exponentially over time, Figure 5b; whereas in the DASBox, the production increased linearly. The different production behaviors could be due to the differing environments in the DASBox compared to the spinner flasks, however, the productivity of the bioreactor was consistent with the spinner flasks. The total number of transduction units produced in the bioreactor was four times larger than the amount produced in the spinner flask at the same DNA load, and used four times the volume of media, resulting in equal volumetric productivity.

3.4. Proof of Concept Continuous Transfection

On a large scale, the transient transfection process based on PEI requires transferring large volumes of mixed PEI and DNA which cannot be achieved in the same time frames that small- or micro-scale operations allow. The PEI and DNA mixture must incubate for a fixed amount of time to achieve the desired polyplex size. The time required to transfer large volumes of liquid will affect the delivered polyplex size; as such, this critical parameter will not be the same between the start and end of transfection in a batch reactor. Additionally, the use of a larger tank to prepare the polyplexes will take longer to homogenously mix and a larger distribution of polyplex sizes could occur because of the rapid condensation reaction. To counteract this, a continuous mode was adopted in the present work for the transfection, where a defined incubation time independent of the volume of the PEI and DNA was set.

A plug flow reactor is suited for a continuous transfection due to its narrow residence time distribution. To explore this concept, a continuous mixture of PEI and DNA followed by continuous transfection is proposed as follows. At the inlet of the reactor, the PEI and DNA are mixed, if the proper flow parameters are met, this can be achieved with a static mixer. This mixture then flows through the reactor while the PEI:DNA complexes form. At the outlet, the complexes exit after having incubated for a given time. This leads to a fixed complex formation time which is equal to the residence time of the reactor and is independent of the volume.

Furthermore, the continuous system needs to take into account the fact that the cells are adhering to microcarriers which requires two systems.

1. One system that can continually generate microcarriers with cells with a specific cell to bead ratio.
2. Another system to continuously transfect the cells on the inoculated microcarriers with the PEI:DNA complex.

System (1) was achieved with a combination of a static mixer and an intermediate stirred tank. Figure 10 shows a block flow diagram of this process.

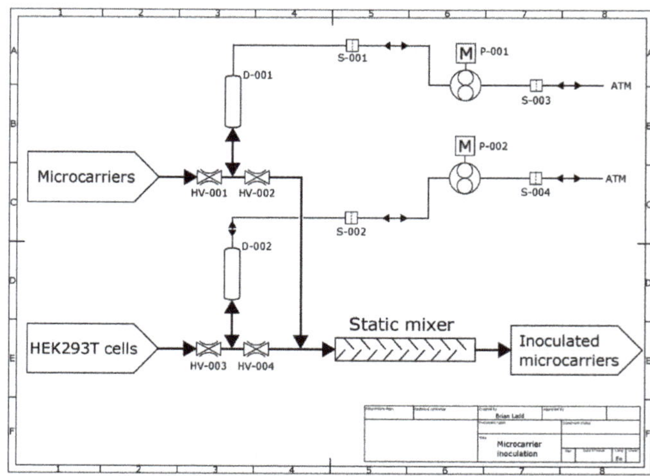

Figure 10. Process flow diagram of the continuous microcarrier inoculation system; D-001 and D-002 are 10 mL vessels; HV-001 to HV-004 are pinch valves; P-001 and P-002 are air pumps; S-001 to S-004 are 0.2 μm sterile air filters; ATM denotes atmosphere; DASBox bioreactor vessels are connected to labels "Microcarriers", "HEK293T cells" and "Inoculated microcarriers".

The microcarriers and cell suspension enter the diagram at the corresponding labels and were drawn by suction into vessels D-001 and D-002, respectively, and then the pinch valves HV-001 and HV-003 were closed. After the vessels D-001 and D-002 were filled, pinch valves HV-002 and HV-004 were opened and air pumped by pumps P-001 and P-002 was used to push the fluids through the static mixer into the intermediate reactor, denoted by the label "Inoculated microcarriers". The vessels D-001 and D-002 could hold up to 10 mL of liquid each and dispense it at a constant rate.

The static mixer shown in Figure 10 was built in-house with nine inserts made from curved stainless-steel inserts placed in a 5 mm ID silicon tube, similar in design to a Kenics type static mixer.

The static mixer was connected to three DASBox reactors, one contained sterile hydrated Cytodex 1 microcarriers at a concentration of 10 mg mL^{-1} in medium (labeled "Microcarriers" in Figure 10), the second reactor had a suspension culture of HEK293T cells at 8×10^6 cells mL^{-1} (labeled "HEK293T cells"), and the third reactor was the intermediate reactor connected to the outflow of the static mixer (labeled "Inoculated microcarriers"). A cell to microcarrier ratio of 100 was chosen which resulted in a microcarrier to cell volumetric flow ratio of 1:1. In two cycles, a total of 20 mL of both cells and microcarriers, for a total volume of 40 mL, was mixed at a total flow rate of 10 mL mL^{-1}. This flow rate was chosen to match the maximum power input for the DASBox using a marine impeller at 250 RPM, the highest agitation rate the cells were able to withstand on microcarriers. While this power input exceeded the limit for inoculation, the cells only experienced this stress for a brief amount of time. The maximum power input per unit mass for the DASBox was calculated using Equation (7) and the power input from a pipe flow was calculated using Equation (8), derived from the Darcy-Weisbach equation.

$$\varepsilon_{max,impeller} = \frac{20 \times Ne \times \rho \times N^3 \times D^5}{V} \tag{7}$$

$$\varepsilon_{tube\ flow} = \frac{512 \times \dot{V}^2 \times \eta}{\pi^2 \times d^6 \times \rho} \tag{8}$$

With $\varepsilon_{max,impeller}$: Maximum energy dissipation per mass for an impeller (W kg^{-1}), $\varepsilon_{tube\ flow}$: energy dissipation per mass for pipe flow (W kg^{-1}), Ne: Newton number, 0.36 for a marine impeller (–), N: rotational frequency (s^{-1}), D: impeller diameter (m), V: Fluid volume (m^3), $\dot{V}$: fluid flow rate (m^3 s^{-1}), η: dynamic viscosity of the fluid (Pa s^{-1}), d: diameter of the tube (m), and ρ: Density of the fluid (kg m^{-3}).

Figure 11 shows a sample from the outlet of the static mixer directly after the first cycle of 10 mL microcarriers and cells.

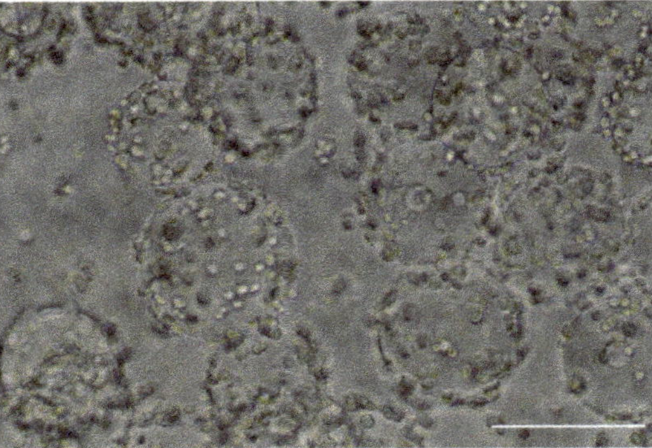

Figure 11. Image of the cell-loaded microcarriers taken at the exit of the static mixer of the continuous microcarrier inoculation process as depicted in Figure 10; the scale bar represents 200 μm.

It can be seen that the cells have attached well and demonstrate that a continuous stream of inoculated microcarriers can be obtained with a highly uniform distribution of cells on microcarriers.

At the small scales used here, a fully continuous transfection would require extremely low flow rates, on the order of μL min^{-1}, and would thus require a specialty microfluidic static mixer and very small diameter tubing. To alleviate this, a semi-continuous process can be implemented; Figure 12 shows a block flow diagram of the semi-continuous process that was developed.

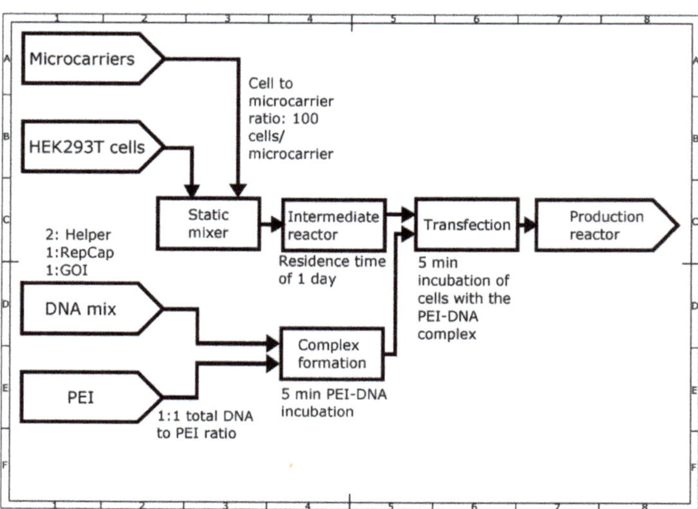

Figure 12. Block flow diagram of a semi-continuous transfection process. Microcarriers and HEK293T cells enter the process at the corresponding labels, where they are mixed in the static mixer; after which, the cells attach to the microcarriers. The inoculated microcarriers flow into the intermediate reactor which has a residence time of one day. The DNA mix and PEI enter the process at the corresponding labels where they are mixed at the given ratio in the complex formation step. The HEK293T cells attached to the microcarriers are then transfected with the formed complex, after which they are pumped into the production reactor.

In this process, the transfection reagents were mixed in a 100 mL Duran flask and allowed to incubate for 5 min; then, 50 mL of inoculated microcarriers were added and allowed to incubate for a further 5 min under gentle mixing. After this incubation, the transfected microcarriers were pumped into the production reactor using positive pressure from an air pump. The volume in the production reactor was kept below 200 mL. Prior to the addition of the transfected microcarriers, 50 mL was harvested from the reactor to maintain the volume. This process was repeated seven times over 20 days.

The long running time of this experiment would generate a large number of samples; to cope with this increased analytical demand, an adaptation was done to the transduction protocol presented above. The previous transduction protocol used suspension CHO cells and analyzed the GFP expression through flow cytometry; in the modified version, the same CHO cell type was used, but in a statically incubated 96 well plate measuring GFP expression through a plate reader. This modified protocol benefits over the previous transduction assay in two aspects; (i) the 96 well plate format is more conducive to the use of conventional multichannel pipettes, increasing the throughput; (ii) using a plate reader over a flow cytometer saves on sample preparation and running time, reducing the time needed to analyze many samples from hours to minutes. For these reasons, most of the samples of the semi-continuous transfection experiment were analyzed with the modified transduction protocol, however, at selected time points the transduction protocol used in previous experiments was performed to harmonize the results.

The AAV production in the supernatant as measured by the previously used transduction protocol and the modified protocol can be seen in Figure 13a,b, respectively.

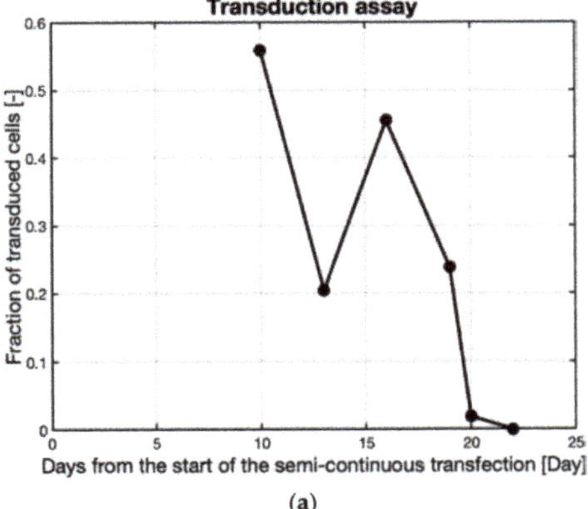

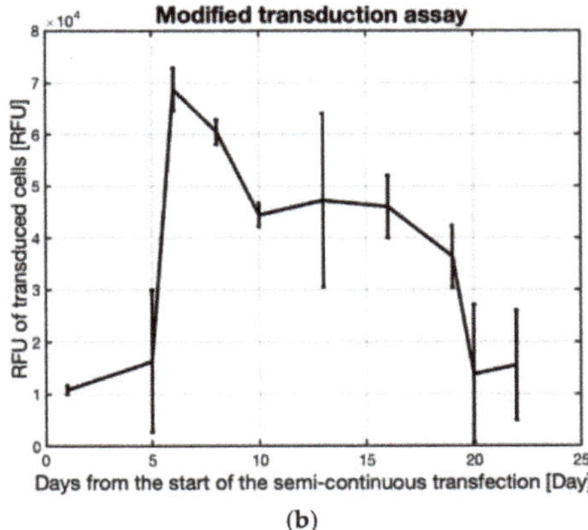

Figure 13. (a) Transduction assay results of CHO cells using supernatant samples harvested at various times during the continuous transfection, Figure 12, of HEK293T cells immobilized on Cytodex 1 microcarriers. (b) Modified transduction assay results of CHO cells using supernatant samples harvested at various times during the continuous transfection, Figure 12, of HEK293T cells immobilized on Cytodex 1 microcarriers.

The results of both transduction assays in Figure 13 show that the production of the active virus was maintained in the reactor for at least 16 days. When compared to the previous bioreactor transfection, the fraction of transduced cells is higher, indicating a higher titer. A drop in production was seen in the modified transduction assay after day 6 followed by a further decrease after day 16. This latter decrease could be due to a reduction of the transfection efficiency as the same mixing flask was used during the whole process, possibly subjecting the transfections to an accumulation of interfering components as some medium remained in the transfection flask after transferring the cells to the production reactor, influencing the composition of the complexation medium and leading to a less efficient transfection. The earlier decrease in production could be due to medium limitations in the production reactor. Fresh medium was added to the reactor after each transfection, but it is possible that in the altered state of virus production the 50 mL or 0.25 reactor volumes medium exchange was not sufficient. The presence of virus production until day 16 would suggest the latter decrease in AAV production was due to a change in transfection efficiency rather than medium limitation because medium limitation should have caused a decrease in production at an earlier time. This experiment shows that continuous transfection can be used to extend the production of AAV over a longer time than a traditional batch process.

4. Discussion

Gene therapy is set to revolutionize medicine. At the forefront of viral vectors, AAV is a leading candidate. The current roadblock for AAV is the production capacity needed for more extensive clinical trials and, later, supplying the commercial demand. A flexible production process is needed that can produce AAV vectors at different scales, depending on the demand. An ideal process would need to be easily scalable from vector screening all the way to large scale production. Adherent transfections, in addition to being a widely used AAV production strategy at the discovery phase, provide cell-specific titers up to 15 times that of suspension cells [11,31]. Current methods based on 2D cultures suffer from the scalability needed to take an AAV therapy from discovery to commercial approval. Microcarriers provide the solution to this problem. A process using microcarriers can be readily implemented to existing transfection protocols without modification and thus does not need extensive optimization [14]. The work presented here shows that a 2D transfection process using T-flasks can be successfully scaled up to a 200 mL bioreactor with an effective surface area of over 1600 cm^2. A proof-of-concept continuous transfection is shown to be possible which will be able to increase the production capacity of a microcarrier-based system without increasing the footprint of the reactor.

The present system is a proof-of-concept showing that it is possible to continuously transfect cells adhering to microcarriers, themselves generated by continuous microcarrier inoculation. A fully automated transfection process could reduce the variability seen with manual transfections. Through a greater process control, the concentration of DNA, PEI, and perhaps even the size of the complex, could be dynamically changed to yield the most efficient transfection. A continuous transfection system could also be used to better understand the effect that certain medium components have on the production of AAV, e.g., by comparing the production rates in steady states of two different media compositions, allowing a direct comparison between both media. On the contrary, in batch mode, the dynamic nature can convolute the effects of certain components. In addition, and probably the most important issue for scale-up, at large-scale, a truly homogenous batch transfection is impossible since the homogenization cannot be instantaneous. The only way to generate a narrow distribution of PEI—DNA mixing time and therefore polyplex size, is to use a continuous transfection.

Looking at a scaled-up process of a 2000 L reactor scale performed in batch mode, the volume of PEI-DNA complex liquid needed to be transferred using the present transfection protocol is at least 500 L. To transfer the entire polyplex volume under a minute requires a pump flow of 30 m^3 h^{-1} from a mixing tank of at least 500 L placed in close proximity to the culture bioreactor. Importantly, mixing in a 500 L size tank is potentially not evenly distributed, with areas of low mixing, e.g., near the impeller shaft, bioreactor wall, and bottom and liquid surface. DNA and PEI are large molecules, with a very low diffusion coefficient, which implies that mixing would mainly happen in the intense mixing zone, near the impeller. In batch mixing and operation, this localized mixing adds a great deal of heterogeneity into the amount of time the polyplexes have to react with each other and new PEI or DNA molecules, thus leading to a wide distribution in polyplex size.

As demonstrated here, the process can instead be carried out in continuous mode. When the involved liquid volume flow rates are larger than μL min^{-1}, the process can be transitioned to be fully continuous which provides easier operation and most likely more controlled transfection conditions for the cells. At a large scale, a homogenous single batch transfection leads to very large uncertainty, and the only way to have a narrow distribution of PEI:DNA complexation time is to use a continuous transfection. Furthermore, the continuous stream of inoculated microcarriers obtained here can also be scaled-up regardless of the amount needed, achieving a highly uniform distribution of cells on microcarriers.

Supplementary Materials: The following supporting information can be downloaded at: https://www.mdpi.com/article/10.3390/pr10030515/s1.

Author Contributions: Conceptualization, B.L., V.C. and T.G.; methodology, B.L.; formal analysis, B.L.; investigation, B.L.; resources, V.C., K.B. and M.L.; data curation, B.L.; writing—original draft preparation, B.L.; writing—review and editing, B.L., V.C. and T.G.; visualization, B.L.; supervision, V.C., K.B. and T.G.; project administration, V.C., K.B. and T.G.; funding acquisition, V.C. All authors have read and agreed to the published version of the manuscript.

Funding: This research was supported by the Competence Centre for Advanced BioProduction by Continuous Processing, AdBIOPRO, funded by Sweden's Innovation Agency VINNOVA (diaries nr. 2016-05181), by the *Marie Skłodowska-Curie European Industrial Doctorate programme EID, STACCATO* *grant No 813453*, Cobra Biologics, Charles Rivers, for the cell line and the plasmid, and Cytiva for the culture medium.

Institutional Review Board Statement: Not applicable.

Informed Consent Statement: Not applicable.

Data Availability Statement: Not applicable.

Conflicts of Interest: The authors declare no conflict of interest.

References

1. Ginn, S.L.; Amaya, A.K.; Alexander, I.E.; Edelstein, M.; Abedi, M.R. Gene therapy clinical trials worldwide to 2017: An update. *J. Gene Med.* **2018**, *20*, e3015. [CrossRef]
2. Naso, M.F.; Tomkowicz, B.; Perry, W.L., 3rd; Strohl, W.R. Adeno-Associated Virus (AAV) as a Vector for Gene Therapy. *BioDrugs* **2017**, *31*, 317–334. [CrossRef] [PubMed]
3. Clément, N. *Large-Scale Clinical Manufacturing of AAV Vectors for Systemic Muscle Gene Therapy*; Springer: Cham, Switzerland, 2019. [CrossRef]
4. Robert, M.A. Manufacturing of recombinant adeno—associated viruses using mammalian expression platforms. *Biotechnol. J.* **2017**, *12*, 1600193. [CrossRef] [PubMed]
5. Huang, X.; Hartley, A.V.; Yin, Y.; Herskowitz, J.H.; Lah, J.J.; Ressler, K.J. AAV2 production with optimized N/P ratio and PEI-mediated transfection results in low toxicity and high titer for in vitro and in vivo applications. *J. Virol. Methods* **2013**, *193*, 270–277. [CrossRef] [PubMed]
6. Ulasov, A.V.; Khramtsov, Y.V.; Trusov, G.A.; Rosenkranz, A.A.; Sverdlov, E.D.; Sobolev, A.S. Properties of PEI-based polyplex nanoparticles that correlate with their transfection efficacy. *Mol. Ther.* **2011**, *19*, 103–112. [CrossRef] [PubMed]
7. Sang, Y.; Xie, K.; Mu, Y.; Lei, Y.; Zhang, B.; Xiong, S.; Chen, Y.; Qi, N. Salt ions and related parameters affect PEI–DNA particle size and transfection efficiency in Chinese hamster ovary cells. *Cytotechnology* **2013**, *67*, 67–74. [CrossRef]
8. Chen, P.; Demirji, J.; Ivleva, V.B.; Horwitz, J.; Schwartz, R.; Arnold, F. The transient expression of CHIKV VLP in large stirred tank bioreactors. *Cytotechnology* **2019**, *71*, 1079–1093. [CrossRef]
9. Rawat, J.; Gadgil, M. Shear stress increases cytotoxicity and reduces transfection efficiency of liposomal gene delivery to CHO-S cells. *Cytotechnology* **2016**, *68*, 2529–2538. [CrossRef]
10. Shurbaji, S.; Anlar, G.G.; Hussein, E.A.; Elzatahry, A.; Yalcin, H.C. Effect of Flow-Induced Shear Stress in Nanomaterial Uptake by Cells: Focus on Targeted Anti-Cancer Therapy. *Cancers* **2020**, *12*, 1916. [CrossRef]
11. Emmerling, V.V.; Pegel, A.; Milian, E.G.; Venereo-Sanchez, A.; Kunz, M.; Wegele, J.; Kamen, A.A.; Kochanek, S.; Hoerer, M. Rational plasmid design and bioprocess optimization to enhance recombinant adeno-associated virus (AAV) productivity in mammalian cells. *Biotechnol. J.* **2016**, *11*, 290–297. [CrossRef]
12. Grieger, J.C.; Soltys, S.M.; Samulski, R.J. Production of Recombinant Adeno-associated Virus Vectors Using Suspension HEK293 Cells and Continuous Harvest of Vector From the Culture Media for GMP FIX and FLT1 Clinical Vector. *Mol. Ther.* **2016**, *24*, 287–297. [CrossRef] [PubMed]
13. Zhao, H.; Lee, K.J.; Daris, M.; Lin, Y.; Wolfe, T.; Sheng, J.; Plewa, C.; Wang, S.; Meisen, W.H. Creation of a High-Yield AAV Vector Production Platform in Suspension Cells Using a Design-of-Experiment Approach. *Mol. Ther. Methods Clin. Dev.* **2020**, *18*, 312–320. [CrossRef] [PubMed]
14. Fliedl, L.; Kaisermayer, C. Transient gene expression in HEK293 and vero cells immobilised on microcarriers. *J. Biotechnol.* **2011**, *153*, 15–21. [CrossRef] [PubMed]
15. GE. *Microcarrier Cell Culture Principles and Methods*; Pensoft: Sofia, Bulgaria, 2008; Volume 18-1140-62 AC.
16. Croughan, M.S.; Hamel, J.F.; Wang, D.I.C. Hydrodynamic effects on animal cells grown in microcarrier cultures. *Biotechnol. Bioeng.* **2006**, *95*, 295–305. [CrossRef]
17. Schwarz, H.; Zhang, Y.; Zhan, C.; Malm, M.; Field, R.; Turner, R.; Sellick, C.; Varley, P.; Rockberg, J.; Chotteau, V. Small-scale bioreactor supports high density HEK293 cell perfusion culture for the production of recombinant Erythropoietin. *J. Biotechnol.* **2020**, *309*, 44–52. [CrossRef]
18. Nienow, A.W. Reactor engineering in large scale animal cell culture. *Cytotechnology* **2006**, *50*, 9–33. [CrossRef]
19. Zwietering, T.N. Suspending of solid particles in liquid by agitators. *Chem. Eng. Sci.* **1958**, *8*, 244–253. [CrossRef]
20. Ibrahim, S.; Nienow, A.W. Suspension of Microcarriers for Cell Culture with Axial Flow Impellers. *Chem. Eng. Res. Des.* **2004**, *82*, 1082–1088. [CrossRef]
21. Xie, M.-H.; Zhou, G.-Z.; Xia, J.-Y.; Zou, C.; Yu, P.-Q.; Zhang, S.-L. Comparison of Power Number for Paddle-Type Impellers by Three Methods. *J. Chem. Eng. Jpn.* **2011**, *44*, 840–844. [CrossRef]
22. Hall, S. 16-Blending and Agitation. In *Branan's Rules of Thumb for Chemical Engineers*, 5th ed.; Hall, S., Ed.; Butterworth-Heinemann: Oxford, UK, 2012; pp. 257–279. [CrossRef]
23. Sano, Y.; Yamaguchi, N.; Adachi, T. Mass Transfer Coefficients for Suspended Particles in Agitated Vessels and Bubble Columns. *J. Chem. Eng. Jpn.* **1974**, *7*, 255–261. [CrossRef]

24. Croughan, M.S. *Hydrodynamic Effects on Animal Cells in Microcarrier Bioreactors*; Massachusetts Institute of Technology: Cambridge, MA, USA, 1983.
25. Erickson, H.P. Size and shape of protein molecules at the nanometer level determined by sedimentation, gel filtration, and electron microscopy. *Biol. Proced. Online* **2009**, *11*, 32–51. [CrossRef] [PubMed]
26. Han, X.; Fang, Q.; Yao, F.; Wang, X.; Wang, J.; Yang, S.; Shen, B.Q. The heterogeneous nature of polyethylenimine-DNA complex formation affects transient gene expression. *Cytotechnology* **2009**, *60*, 63. [CrossRef] [PubMed]
27. Bertschinger, M.; Schertenleib, A.; Cevey, J.; Hacker, D.L.; Wurm, F.M. The kinetics of polyethylenimine-mediated transfection in suspension cultures of Chinese hamster ovary cells. *Mol. Biotechnol.* **2008**, *40*, 136–143. [CrossRef] [PubMed]
28. Blackstock, D.J.; Goh, A.; Shetty, S.; Fabozzi, G.; Yang, R.; Ivleva, V.B.; Schwartz, R.; Horwitz, J. Comprehensive Flow Cytometry Analysis of PEI-Based Transfections for Virus-Like Particle Production. *Research* **2020**, *2020*, 1387402. [CrossRef]
29. Godbey, W.T.; Wu, K.K.; Mikos, A.G. Tracking the intracellular path of poly(ethylenimine)/DNA complexes for gene delivery. *Proc. Natl. Acad. Sci. USA* **1999**, *96*, 5177–5181. [CrossRef]
30. Mozley, O.L.; Thompson, B.C.; Fernandez-Martell, A.; James, D.C. A mechanistic dissection of polyethylenimine mediated transfection of CHO cells: To enhance the efficiency of recombinant DNA utilization. *Biotechnol. Prog.* **2014**, *30*, 1161–1170. [CrossRef]
31. Blessing, D.; Vachey, G.; Pythoud, C.; Rey, M.; Padrun, V.; Wurm, F.M.; Schneider, B.L.; Deglon, N. Scalable Production of AAV Vectors in Orbitally Shaken HEK293 Cells. *Mol. Ther. Methods Clin. Dev.* **2019**, *13*, 14–26. [CrossRef]

Article

Large-Scale Production of Size-Adjusted β-Cell Spheroids in a Fully Controlled Stirred-Tank Reactor

Florian Petry and Denise Salzig *

Institute of Bioprocess Engineering and Pharmaceutical Technology, University of Applied Sciences Mittelhessen, Wiesenstrasse 14, 35390 Giessen, Germany; florian.petry@lse.thm.de
* Correspondence: denise.salzig@lse.thm.de; Tel.: +49-641-309-2630; Fax: +49-641-309-2553

Abstract: For β-cell replacement therapies, one challenge is the manufacturing of enough β-cells (Edmonton protocol for islet transplantation requires $0.5–1 \times 10^6$ islet equivalents). To maintain their functionality, β-cells should be manufactured as 3D constructs, known as spheroids. In this study, we investigated whether β-cell spheroid manufacturing can be addressed by a stirred-tank bioreactor (STR) process. STRs are fully controlled bioreactor systems, which allow the establishment of robust, larger-scale manufacturing processes. Using the INS-1 β-cell line as a model for process development, we investigated the dynamic agglomeration of β-cells to determine minimal seeding densities, spheroid strength, and the influence of turbulent shear stress. We established a correlation to exploit shear forces within the turbulent flow regime, in order to generate spheroids of a defined size, and to predict the spheroid size in an STR by using the determined spheroid strength. Finally, we transferred the dynamic agglomeration process from shaking flasks to a fully controlled and monitored STR, and tested the influence of three different stirrer types on spheroid formation. We achieved the shear stress-guided production of up to $22 \times 10^6 \pm 2 \times 10^6$ viable and functional β-cell spheroids per liter of culture medium, which is sufficient for β-cell therapy applications.

Keywords: spheroid strength; β-cells; diabetes; shear stress-guided production; hydrodynamic stress

Citation: Petry, F.; Salzig, D. Large-Scale Production of Size-Adjusted β-Cell Spheroids in a Fully Controlled Stirred-Tank Reactor. *Processes* **2022**, *10*, 861. https://doi.org/10.3390/pr10050861

Academic Editor: Alok Kumar Patel

Received: 21 March 2022
Accepted: 24 April 2022
Published: 27 April 2022

Publisher's Note: MDPI stays neutral with regard to jurisdictional claims in published maps and institutional affiliations.

1. Introduction

Diabetes involves the selective autoimmune destruction or dysfunction of insulin-producing β-cells, located within the islets of Langerhans in the pancreas. The large and growing number of patients living with diabetes [1] has generated interest in the promise of β-cell therapy to restore lost β-cell mass [2]. However, β-cells have unique characteristics and must be transplanted as spheroids to be able to exert their full biological activity in the recipient. The β-cell spheroids are living drugs, and proper manufacturing is an important step to bring these therapeutics into clinics. It is difficult to manufacture sufficient β-cell numbers. The total estimated number of β-cells in the human pancreas is $\sim 10^9$ [3], which is the benchmark for the manufacturing process. Islets of Langerhans from deceased donors show a high functionality, but there is a shortage of donor material, the adult β-cells lose their functionality over time, and they cannot be expanded in vitro [4,5]. Induced pluripotent stem cells (iPSCs) can help to address the material shortage, but iPSCs lack functionality compared with native islets and require a complex and laborious differentiation protocol [4–6]. Various mouse and rat models have been used for diabetes research, which led to the development of rodent β-cell lines such as RIN (rat), INS-1 (rat) and MIN6 (mouse) [4]. However, the development of human β-cell lines has been hampered by their insufficient functionality and xenotropic viral contamination [5,7–9]. Even so, β-cell lines can be expanded in vitro over multiple passages without decreasing in functionality, therefore, β-cell lines serve as good models for diabetes research, and the development of manufacturing protocols for cell therapy [10,11]. The 3D cultivation of β-cell lines as spheroids or agglomerates (these terms are used synonymously) can enhance glucose-dependent insulin secretion, and β-cells cultured as re-aggregated 3D structures (pseudo-islets) show a greater viability, proliferation and functionality than individual cells [12,13]. Such agglomerates facilitate the reconstitution of the native pancreatic microenvironment, including cell–cell interactions, the extracellular matrix (ECM), cell coupling and tight junctions, cell polarization, and changes in gene expression triggered by bioactive molecules and forces acting on the cytoskeleton [14–16].

A manufacturing process for cell therapy must produce uniform spheroids in sufficient numbers. As mentioned above, the benchmark for β-cell therapy is to manufacture $\sim 10^9$ cells [3] or $1–3.2 \times 10^6$ spheroids, following the cell and islet amount within the human pancreas [15,17]. Although the mean islet size is 140 μm [3], a major portion of islets ranges between 20 and 50 μm (37%) [18], or 10 and 50 μm (60%) [19]. Even though the contribution of the smaller islets to the total islet mass is small, they might be crucial for the outcome of the transplants, as smaller islets are more robust against hypoxic conditions [20]. Ultimately, this results in an increased functionality, as shown for islets < 100 μm in comparison with islets > 250 μm [21–23], therefore, we aim to manufacture spheroids between 20 and 100 μm. Small-scale systems, such as microtiter plates, can generate enough spheroids for research purposes, but they are labor-intensive and vulnerable to contamination, they lack process control, and they are difficult to scale up to produce sufficient cell quantities for clinical applications. Some groups address the scale-up issue by using spinner flasks [24–27] or rotating wall vessel reactors (RWVRs) [22,28], which enable a limited degree of process control and increase the manufacturing scale up to 500 mL. However, these systems achieve improper mixing and have limited possibilities to adjust the spheroid size. Our literature search did not reveal any reports regarding the large-scale (1 L or more) or stirred-tank bioreactor (STR) related expansion of β-cell spheroids, but STR results were found for other cell types (mainly iPSCs). Table 1 summarizes the critical aspects of the dynamic production of spheroids with different cell types in relation to our presented theoretical background (see Supplementary). STRs are fully controlled bioreactor systems, which allow the establishment of robust, larger-scale manufacturing processes. STRs also facilitate the online monitoring and control of process parameters such as O_2, pH, temperature, and biomass (the latter by dielectric spectroscopy). STR processes can also be automated and are scalable up to 6000 L (single-use and multi-use systems). Particularly, the scale-up of β-cell manufacturing is a crucial step. When transferring bioprocesses to larger scales, mass transport problems often occur and become process-limiting, which could not be observed or were negligible on a small scale. To avoid process limitations, a correct scale-up strategy is needed, for which the similarity theory provides good service. This theory uses dimensionless numbers (such as the stirrer power number, N_P, or the Reynolds number, Re) for the description of a physical–technical behavior of the manufacturing system (e.g., the STR), which is to be maintained constant during the transition to another scale, so that the physical similarity remains [29]. In the context of β-cell spheroid manufacturing, we see STRs as a potent choice of scalable bioreactor systems, as STRs provide geometric similarity (d_s/D_T, H_L/D_T, d_S/d_H,) at different scales, and the development of a turbulent flow, which enables a constant N_P and the admissible application of Kolmogorov's theory of isotropic turbulence (see theoretical background or [30]).

Although we consider the STR as a proper system for β-cell spheroid manufacturing, we must be aware of the fact that the formation of cell agglomerates differs from the formation in static culture systems. Agglomeration in static cultures involves the self-assembly of cells in hanging drops or parabolic wells with cell-repellent surfaces [31]. In contrast, in an STR, the cell agglomeration requires collisions between cells to establish cell–cell connections, facilitated by protein-mediated adhesion forces on the cell surface. Cell agglomeration progresses, until the hydrodynamic forces become too high to support further cell attachments to the spheroid. A steady state is reached when the spheroid strength (the total of all cell adhesion forces within the spheroid) equals the hydrodynamic forces in the culture medium.

Table 1. Overview of dynamic aggregate/spheroid processes with different cell types, applications, bioreactor types, and the respective scalability in accordance with the similarity theory. We evaluated the production scale, the stirrer type (or similar), and consequently determined the flow range based on own calculations regarding the Reynolds number, Re. Additionally, we listed the seeding density, and the size range of the aggregates and yield per batch. Finally, we compared the described agglomeration techniques, depending on the kind of regulating force, the fluid dynamic, and the ability to adjust the aggregate size. We defined the production of spheroids as controlled when the cells aggregate under isotropic conditions, i.e., at Re > 10,000. This leads to a small size distribution of the aggregates and prevents further agglomeration of the spheroids themselves. Moreover, the applied hydrodynamic forces can restrict the size of the formed spheroids. At ratios of $d_{Sph}/\lambda < 3$, the size restriction is facilitated by surface erosion (for further information, see Supplementary or [30]).

Cell Type	Cell Name	Application	Bioreactor Type; Scalability	Production Scale [mL]	Stirrer Type	Reynolds Number Re [−] and Flow Range **	Seeding Density [Cells mL^{-1}]	Size Range [µm]	Yield	Agglomeration Technique	Reference
Islets/ β-cell lines	INS-1	Bioprocess model for diabetes therapy	STR (Infors HT); scalable	1000	30-SPB; 45-SPB; Rushton	Re = 11,000–18,000; turbulent	5×10^5	40–50	22×10^3 spheroids mL^{-1}; 1.1×10^3 IEQs mL^{-1}; 22×10^6 spheroids batch^{-1}; 1.1×10^6 IEQs batch^{-1};	Controlled: isotropic conditions; surface erosion ($d_{Sph}/\lambda < 3$); narrow size distribution; size restricted	This study
	MIN6	Research studies and clinical application	Spinner flask (ProCulture, Corning); semi-scalable	Not specified	Straight blade paddle impeller *	Re = 2300 (d = 40 mm, 60 rpm); non-turbulent	0.2–0.4×10^5	100–400	$\sim 0.7 \times 10^6$ cells mL^{-1}	Uncontrolled: non-isotropic; broad size distribution; no size restriction	[26]
	Primary neonatal porcine pancreatic islet cells	Scalable process to expand pancreatic endocrine tissue for cell therapy	Spinner flask (Corning); semi-scalable	100	Magnetic stir bars	-	1.3×10^5	Not specified	1×10^6 cells mL^{-1}; 1×10^8 cells batch^{-1}	Uncontrolled: non-isotropic; broad size distribution; no size restriction	[25]
ESC	CyT49	Diabetes therapy	6-well plates; not scalable	5.5	No stirrer	Non-turbulent	10×10^5	100–200	~ 1000 aggregates mL^{-1}; ~ 5500 aggregates batch^{-1}	Uncontrolled: non-isotropic; broad size distribution; no size restriction	[32]
	Royan H5 and H6	Bioprocess development for production of aggregates	Spinner flask (Cellspin; Integra Biosciences); semi-scalable	100	Magnetic pendulums *	-	2–10×10^5	140–200	2×10^6 cells mL^{-1}; 2×10^8 cells batch^{-1}	Uncontrolled: non-isotropic; broad size distribution; no size restriction; addition of shear protectant	[24]
iPSC	hCBiPSC2	Development of suspension culture for iPSCs	STR (cellferm®, DASGIP AG); scalable	100	45° 0°, 45° 60°, 60° 60° impeller	Re ≈ 2500 (d = 40 mm, 60 rpm); Non-turbulent	4–5×10^5	50–150	2×10^6 cells mL^{-1}; 2×10^8 cells batch^{-1}	Uncontrolled: non-isotropic; broad size distribution; no size restriction	[33]
	hHSC_1285i_iPS2	Development of high-density bioprocessing	STR (DASbox, Eppendorf AG); scalable	150–500	Eight-blade impeller (60° pitched)	Re = 1200–2600 (d = 30 mm); non-turbulent	5×10^5	50–400	2.9×10^6 cells mL^{-1}; 4.6–14.5×10^8 cells batch^{-1}	Uncontrolled: non-isotropic; broad size distribution; no size restriction	[34]
	hiPSC1 and hiPSC4	Bioprocess development for production of iPSCs aggregates	Spinner flask (Cellspin; Integra Biosciences); semi-scalable	100	Magnetic pendulums *	-	2–10×10^5	140–200	1.8×10^6 cells mL^{-1}; 1.8×10^8 cells batch^{-1}	Uncontrolled: non-isotropic; broad size distribution; no size restriction; addition of shear protectant	[24]
Tumor cells	MCF7, BT474, HCC1954, HCC1806, A549, H460, H157, H1650 and HT29	3D cancer models	Baffled spinner flasks (Corning® Life Sciences); semi-scalable	125–500	Straight blade paddle impeller	Re = 2300–3800 (d = 40 mm, 60–100 rpm); non-turbulent	2×10^5	81–298 (depending on cell type)	1000–1500 spheroids mL^{-1}; 0.16–0.63×10^6 spheroids batch^{-1}	Uncontrolled: non-isotropic; broad size distribution; no size restriction	[27]

* We assumed the stirrer type and size, based on the product information of the bioreactor system. ** The flow range is based on own calculations regarding the Reynolds number, Re (Re < 10,000 corresponds to non-turbulent and Re > 10,000 to turbulent).

We aim to develop a large-scale STR-based production process for viable and functional β-cell spheroids, allowing the production of large numbers of spheroids within a defined size range of 20 to 100 μm. The term "large-scale" has no precise definition, but we aimed to scale up our process to a working volume of 1 L, which is 100-times higher than the volume of RWVRs, and repeatedly higher than comparable processes with iPSCs (Table 1). We used the INS-1 as our model β-cell line to develop this process. We used shaking flasks as a screening platform to characterize the behavior of INS-1 β-cells in a turbulent flow regime, and to determine the minimal seeding density and spheroid strength. We then used the spheroid strength (equaling the sum of the bonding forces between adjacent cells mediated by membrane proteins [30]) to predict the spheroid size in the STR. Furthermore, we investigated the influence of the stirrer design by testing three stirrer types with different pumping directions (axial, axial/radial and radial), stirrer-swept volumes, and maximum energy dissipation to mean energy dissipation ratios. Finally, we scaled up from the shaking flasks to the STR by keeping the energy dissipation constant.

2. Materials and Methods

2.1. Cells and Culture Medium

We used the rodent β-cell line INS-1 (kindly provided by Sebastian Hauke from European Molecular Biology Laboratory, Heidelberg, Germany), which originated from a Simian virus-induced rat insulinoma. Pre-cultures, prepared in RPMI 1640 medium (Biochrom, Berlin, Germany) and supplemented with 10% (v/v) fetal bovine serum (FBS) and 0.05 M 2-mercaptoethanol (Carl Roth, Karlsruhe, Germany), were incubated at 37 °C in a 5% CO_2 atmosphere. The cells were seeded with 5×10^4 cells cm^{-2} into 25–175 cm^2 T-flasks (Sarstedt, Nuembrecht, Germany) and cultured to a confluence of 80–90%. Before passaging, the cells were observed by microscopy to ensure the absence of morphological defects and contamination. The cells were washed once with 0.3 mL cm^{-2} phosphate-buffered saline (PBS, Biochrom) before detachment with 0.012 mL cm^{-2} trypsin (Biochrom) for 5–7 min at 37 °C. After detachment, premature agglomeration was avoided by omitting centrifugation. Although the INS-1 cell line was robust, we did not exceed a passage number of 35.

2.2. Static Spheroid Formation

Static 3D cultures as spheroids were generated in 96-well plates (U-bottom) with a cell-repellent surface (Greiner, Kremsmünster, Austria). The INS-1 were seeded at 10^3 cells per well, where the cells were forced into agglomeration by the parabolic shape. The working volume was 200 μL, and a 50% medium exchange was performed every other day. Daily imaging of the spheroids was used to determine the diameter and circularity, while staining with calcein/ethidium was used to determine the viability. Each experiment was performed in 12-fold biological replicates.

2.3. Shaking Flask Cultivation

Preliminary experiments were carried out in shaking flasks at 37 °C in a 5% CO_2 atmosphere. We used 100-mL shaking flasks with a working volume of 20 mL. The shaking flasks had an inner diameter of 0.064 m, and four baffles. For all experiments, we used the same Celltron shaking plate (Infors HT, Bottmingen, Switzerland) with an eccentricity of 2.5 cm and varying rotational frequencies. The cells were inoculated with a seeding density of 5×10^5 cells mL^{-1}. Daily samples were stained with calcein/ethidium to test viability, and an image-based analysis of the particle distribution was carried out as described below.

2.4. Production of Spheroids in a STR

For the large-scale production of β-cell spheroids, we used the Labfors 5 Bioreactor system (Infors HT). The bioreactor tank had an inner diameter of 0.115 m and a dished bottom. The working volume V_L was 1 L, resulting in a ratio of liquid height to tank diameter of $H_L/D_T = 1$. We used three different stirrer types: a 30° three-segment pitched-blade stirrer (30°-3-SPB), a 45° three-segment pitched-blade stirrer (45°-3-SPB), and a Rushton turbine (Table S1). To achieve a constant mean energy dissipation of $\bar{\varepsilon} = 35$ kW kg^{-1} in each STR run, the stirrer speed for the 30°-3-SPB was 183 rpm, for the 45°-3-SPB it was 141 rpm, and for the Rushton turbine it was 162 rpm. The STR was equipped with process analytical technology, including a temperature probe, a pH probe, a dissolved oxygen (DO) probe, and a dielectric spectroscopy probe (all from Hamilton Germany). During cultivation, the temperature was maintained at 37 °C, the pH was regulated by gassing with CO_2 or the addition of 1 M NaOH, and the DO concentration was kept above 40% by discontinuous submersed gassing (sparger). A DO of 100% represents oxygen-saturated culture medium (without cells) at 37 °C, the corresponding stirrer speed, and 100 mL min^{-1} continuous submersed gassing with air. Every 12 min, the permittivity (correlating with viable biomass) was measured inline at 1000 kHz, combined with a frequency scan of 300–10,000 kHz. The four installed probes, the shaft guide, the sparger, and the pipes for sampling and harvest were used as baffles, and this was sufficient to achieve complete baffling [35]. After the sterilization of the bioreactor, pre-cultured cells in multiple T-175 flasks were harvested, using trypsin. The cells were in the exponential growth phase and not overgrown. As above, we avoided centrifugation to prevent a premature agglomeration. The seeding density in the STR was 5×10^5 cells mL^{-1}. The cells were stained with calcein/ethidium to test their viability, and an image-based analysis of the particle distribution was carried out as described below.

2.5. Calcein/Ethidium Staining to Analyze Particle Count, Size, Circularity, and Areas of Green and Red Particles

We used the LIVE/DEAD Viability/Cytotoxicity Kit (Invitrogen (Waltham, MA, USA), Thermo Fisher Scientific (Waltham, MA, USA)) containing calcein AM and ethidium homodimer 1. Calcein AM is a true live stain that detects intracellular esterase activity, and ethidium homodimer 1 intercalates into DNA to detect dead cells. The cells were stained without washing steps to prevent cell/spheroid loss. After sampling, five technical replicate (100-µL) samples were transferred to flat-bottom 96-well plates. The staining solution was added directly to the sample (final concentration = 2 µM calcein/ethidium) and incubated for 30 min at 37 °C. No washing steps were applied to prevent cell/spheroid loss. The samples were analyzed using a Cytation3 (BioTek Instruments, Winooski, VT, USA). The fluorescent cells and spheroids in entire wells were captured by acquiring multiple images at 10× magnification. The particle count, size, circularity, and areas of green and red particles were determined, using Gen5 v2.07.17 to assess cell viability.

2.6. Particle Size Distribution

Image analysis was restricted to the size range 0–300 µm. To distinguish between single cells and spheroids, we set the threshold to 20 µm. All counts <20 µm were defined as single cells, whereas all counts >20 µm were evaluated as spheroids. The size of the particle distribution was expressed using the Sauter diameter d_{32} as shown in Equation (1):

$$d_{32} = 6 \cdot \frac{\sum V_{Sph}}{\sum S_{Sph}} \tag{1}$$

where S_{Sph} represents the surface of the spheroids. We found that the Sauter diameter d_{32} described the size distribution of the spheroids very well compared with the mean diameter, median and modus.

We compared the progression of the single cell and spheroid counts of the culture period in addition to the spheroid distribution (20–300 µm) by constructing box plots, showing the median, mean, minimum (1%) and maximum values (99%). The span width of the spheroid distribution within the margin of 1 to 99% was used to compare the width of the particle distributions from different cultures.

2.7. Assessment of Viability

The viability of the whole sample was determined from the respective green A_{green} and red A_{red} areas. Hereby, we discriminated between the areas according to our determined size ranges to determine the viability of the single cells (areas of the size range: 0–20 µm), the spheroids (areas of the size range: 20–300 µm) or the total viability (areas of the size range: 0–300 µm) using the following Equation (2):

$$Via = \frac{A_{green}}{A_{green} + A_{red}} * 100 \tag{2}$$

2.8. Determination of Growth Rate and Expansion Factor

The size of the viable spheroids was used to determine the volume growth rate μ_{Vol} of the spheroids. We assumed that the spherical volume of the spheroids increases due to the exponential cell growth within the spheroids, and that the cells are constant in volume. This leads to the correlation shown in Equation (3):

$$\mu_{Vol} = \frac{ln\left(V_{Sph,2}\right) - ln\left(V_{Sph,1}\right)}{t_2 - t_1} \tag{3}$$

where V_{Sph} corresponds to the spheroid volume at time point t.

The resulting volume doubling time $t_{D,Vol}$ can be calculated as shown in Equation (4):

$$t_{D,Vol} = \frac{ln(2)}{\mu_{Vol}} \tag{4}$$

The volume expansion factor V_{Ex} can be expressed as shown in Equation (5):

$$V_{Ex} = \frac{V_{Sph,2}}{V_{Sph,1}} = \frac{d^3_{Sph,2}}{d^3_{Sph,1}} \tag{5}$$

2.9. Assessment of β-Cell Functionality

The glucose-dependent insulin secretion of the β-cell spheroids was measured by exposing the cells to varying glucose concentrations, and measuring the secreted insulin. After sampling, three technical replicates of 500 µL were washed twice with PBS to remove the culture medium and possible insulin residues present in FBS. The cells were then incubated in 500 µL medium lacking FBS, but containing 1.1 mM glucose for 40 min at 37 °C. The supernatant was removed, centrifuged, and stored at −20 °C for analysis (sample for basal insulin secretion). We then added 500 µL of medium lacking FBS, but containing 16.7 mM glucose for 20 min at 37 °C. The supernatant was removed, centrifuged, and stored at −20 °C for analysis (sample for acute insulin secretion). The insulin in the supernatants was analyzed, using a rat insulin ELISA kit (DRG Instruments, Marburg, Germany). The samples were measured in duplicate and evaluated within the ELISA working range, using a four-parameter logistic curve.

2.10. Statistical Analysis

If not stated otherwise, all experiments were performed as three independent runs, and presented as mean value ± standard deviations (STDV). For statistical analysis, the following were applied: (1) two groups: Student's t-test, (2) for more than two groups: a one-way ANOVA followed by a post-hoc analysis using Bonferroni correction. Intervals of significance were indicated as follows: $* p < 0.05$, $** p < 0.01$, and $*** p < 0.001$.

3. Results and Discussion

3.1. Differences in Static and Dynamic Formation of INS-1 Spheroids

To gain an initial insight into the differences of the static and the dynamic formation of spheroids, we used cell-repellent 96-well plates as a static system, and baffled shaking flasks as a simple dynamic system, compatible with established models of power consumption [36,37]. Under static conditions, we determined INS-1 agglomerates after 24 h, having 181 ± 3 µm in diameter with a circularity of 0.92 ± 0.05. Under dynamic conditions, we determined the agglomeration of INS-1 cells within 24 h, producing spheroids with a Sauter diameter $d_{32} = 94 \pm 12$ µm and a circularity of 0.65 ± 0.02 (Figure 1). Dynamic and static spheroid formation occurred at similar rates, but the circularity of the static spheroids was significantly (*** $p < 0.001$) higher. The volume growth rate μ_{Vol} of 0.14 ± 0.04 d^{-1} and $t_{D,Vol} = 5.0 \pm 1.4$ d for dynamic spheroid formation was significantly (** $p < 0.01$) lower than the corresponding values for static cultures ($\mu_{Vol} = 0.327 \pm 0.013$ d^{-1}, $t_{D,Vol} = 2.12 \pm 0.08$ d). The dynamic growth of spheroids is likely to be restricted by surface erosion, as growing cells in the outer layer are sheared off and/or collide with/adhere to suspended single cells or smaller agglomerates. Spheroids in the size range of 59–269 µm were reported to form after 7 d in shaking cultures of the β-cell line RIN-5F, while maintaining β-cell functionality [38]. The method was not characterized in detail. Based on the information provided, we assume that the cells were not limited by hydrodynamic forces, thus allowing a continuous spheroid growth. Our static spheroids developed a necrotic core and dropped to ~50% viability after 24 h due to mass transfer limitations, whereas the viability of spheroids formed in a dynamic culture was always close to 100% (except some single cells within the spheroids) regardless of their size (up to 200 µm). Similarly, agglomerates of the rodent β-cell line MIN6 were produced after 2 d in static culture dishes and spinner flasks. Although the spheroid concentration remained constant while they grew from 100 to 400 µm, a dense necrotic core formed in the static spheroids (viability ~65%), whereas those in the spinner flasks maintained ~85% viability [26]. The spheroids in the dynamic culture were also characterized by a lower lactate dehydrogenase and caspase activity, and lower detected levels of fragmented DNA, using the TUNEL assay [26].

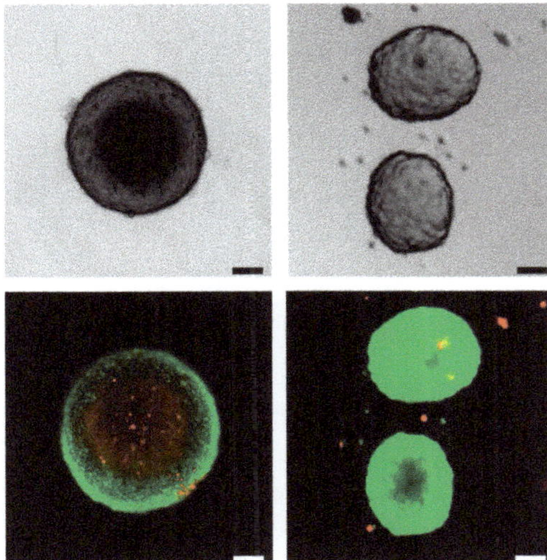

Figure 1. Comparison of spheroid structures. Left panels show INS-1 spheroids produced in static 96-well plates with cell-repellent surfaces. These spheroids developed a necrotic core, seen as a denser area in the bright field image (**upper**), and the corresponding red area by live/dead staining (**lower**). Right panels show highly viable spheroids produced in dynamic shaking flask cultures. Scale bar = 100 µm.

3.2. Influence of Seeding Density on INS-1 Spheroid Formation in Shaking Flasks

Next, we investigated relevant process parameters such as seeding density, turbulence working range, and spheroid strength in shaking flasks. We tested seeding densities from 1×10^4 to 1×10^6 cells mL^{-1}. At values between 1×10^5 and 6×10^5 cells mL^{-1}, we observed a plateau of a spheroid size with a narrow size distribution ($d_{32} = 63 \pm 5$ μm) and a rapid spheroid formation (within 24 h) and, therefore, defined 1×10^5 to 6×10^5 cells mL^{-1} as our working range. At values below 1×10^5 cells mL^{-1}, agglomeration was inhibited ($d_{32} = 29 \pm 7$ μm, ** $p < 0.01$), whereas values above 6×10^5 cells mL^{-1} generated larger spheroids ($d_{32} = 82 \pm 3$ μm, ** $p < 0.01$) due to further visible agglomeration of the spheroids themselves. The seeding density is not only important for a dynamic spheroid formation, but also affects the functionality of the β-cells, which is dependent on the cell number in each spheroid. For example, when spinner flasks were used to form spheroids from primary neonatal porcine pancreatic islet cells at different seeding densities, low values (6.3×10^3 and 5×10^4 cells mL^{-1}) reduced the number of aggregates, whereas the highest value (1.3×10^5 cells mL^{-1}) increased the number of islet-like aggregates containing insulin-positive cells [25]. The positive effect reflected the increase in cell–cell interactions at higher seeding densities. Our chosen working range also matches the values reported for other cell types that grow as spheroids (Table 1), including iPSCs [39,40]. We therefore carried out all subsequent experiments with a seeding density of 5×10^5 cells mL^{-1}.

3.3. Influence of Power Input on INS-1 Spheroid Formation in Shaking Flasks

The development of a turbulent flow regime is necessary for the scale-up of the INS-1 spheroid formation process in a STR. Based on earlier models [36,37], we calculated a Reynolds number of Re = 10^4 (fully developed turbulence) in the shaking flasks at a frequency of 100 rpm. We investigated the spheroid formation at shaking frequencies in the range between 90–130 rpm, and correlated the spheroid size to the power input. We observed a linear decrease in the spheroid size between 100–120 rpm (corresponding to 35–60 W m^{-3}), combined with a narrow span width of the spheroid distribution of 85 ± 28 μm, where we anticipated the turbulent flow regime. Spheroids produced with an increased power input (and thereby increasing hydrodynamic forces) showed the anticipated and significant (** $p < 0.01$) decrease in size from $d_{32} = 86 \pm 6$ μm at 100 rpm to $d_{32} = 44 \pm 5$ μm at 120 rpm (Figure 2). An increasing stirrer speed also reduced the size of hPSCs spheroids [34] and, similar to our results, the viability of the spheroids remained high, although increasing shear forces led to a surface erosion and the loss of cells from the spheroid surface. Therefore, we determined an increasing amount of dead single cells with an increasing power input (35–60 W m^{-3}). In relation to 100 rpm, spheroids produced under non-turbulent conditions (90 rpm, 28 W m^{-3}) were twice as large on average ($d_{32} = 168 \pm 4$ μm, ** $p < 0.01$) and were distributed over a broader size range (span width: 232 ± 24 μm), whereas no spheroids were formed at 130 rpm (=74 W m^{-3}) and the size distribution represented the profile of a single-cell suspension ($d_{32} = 10 \pm 3$ μm, ** $p < 0.01$). Moreover, the efficiency of spheroid formation, represented as the number of formed spheroids in relation to the total cell number (Figure 2), fell to low values of 6% and 5% at 90 and 130 rpm, respectively. Within the turbulent range (100–120 rpm), the spheroid formation efficiency increased to 27% (110 rpm). These data agreed with the spheroid formation theory discussed in the supplementary, which we adapted from particle systems such as clay, latex, and glass [41–46]. Based on the ratio d_{Sph}/λ, we concluded that the INS-1 spheroid formation process reached a steady-state diameter that reflected the shear forces of the corresponding eddies acting on the spheroid surface. This assumption is supported by the decreasing spheroid size as the power input increases. The spheroid size is, therefore, limited by the Kolmogorov eddy size, which can also be described as the equilibrium between kinetic energy in the culture medium and the bonding energy of the spheroids.

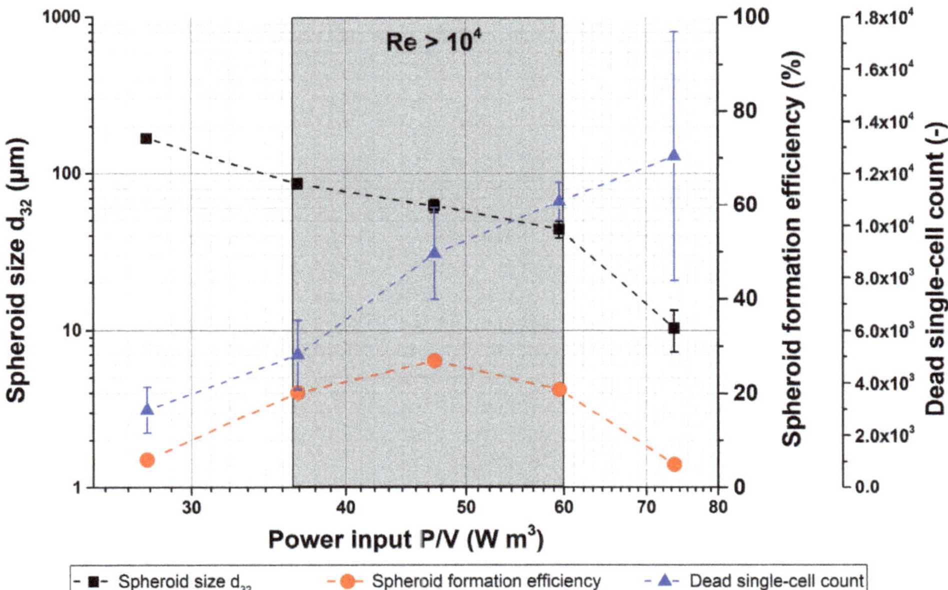

Figure 2. Spheroid size (d_{32}, black squares) from dynamic cultivation in shaking flasks ($n = 3$, error bars represent STDV). Spheroids produced with a power input lower than the gray highlighted working range (35–60 W m^{-3}; 100–120 rpm) were twice as large, whereas increasing shear forces reduced the steady-state size until agglomeration was prevented at a P/V of $\approx$75 W m^{-3} (130 rpm). An increase in power input displaced cells from the spheroid surface, leading to greater numbers of suspended dead singe-cells (blue triangles) in the supernatant. The spheroid formation efficiency (the number of formed spheroids in relation to the total cell number; red dots) supported the turbulent range from 100 to 120 rpm with an increased efficiency above 20%.

3.4. Determination of INS-1 Spheroid Strength in Shaking Flasks

The spheroid strength is an essential parameter in our process, so we estimated the value in shaking flask experiments. The spheroid strength is influenced by the strength and number of bonds (F_{ad}) between cells, the size and shape of the cells, and the compaction of the spheroid. The particle analysis and wastewater treatment literature provide multiple approaches to determine spheroid strength [41,44,47,48], which we adapted for the INS-1 spheroids. These methods are based on the relationship between the power input/energy dissipation and the corresponding floc/spheroid diameter, thus distinguishing between agglomerate splitting by tensile forces, and surface erosion due to shear forces. However, we are aware that the agglomeration of non-biological particle systems is enabled by polymers or electrostatic interactions, whereas the agglomeration of cells is also based on the interaction between surface proteins (such as integrins and cadherins) on adjacent cells [49]. This is a highly variable precondition for different cell types and even for the same cell type, because the expression of surface proteins on cells can change during growth, senescence, and differentiation, and can be influenced by cultivation and harvesting methods.

Because there is no standardized procedure to evaluate spheroid strength, we produced spheroids within the turbulent working range under increasing power input, and correlated the steady-state spheroid size (after 24 h) with energy dissipation (Figure 2). We applied the following method [41,50] to describe the steady-state diameter d using Equation (6):

$$d = C_{Agg} \cdot G^{-\gamma} \tag{6}$$

where C_{Agg} is the strength of the agglomerate (here, spheroid strength), γ is the stable agglomerate size exponent, and the shear rate G is defined using Equation (7):

$$G = \sqrt{\frac{\epsilon}{\vartheta_L}} \tag{7}$$

We calculated for the exponent γ a value of 2.8 ($R^2 = 0.99$) over the linear region of the equation. The exponent γ provides insight into the agglomeration behavior of the INS-1 spheroids: γ values of ~0.5 indicate fragmentation by tensile forces, whereas $\gamma > 2$ suggests surface erosion [50]. This fits to our data shown above that the increasing amount of dead single cells with increasing power input (35–60 W m^{-3}) is the result of surface erosion and the loss of cells from the spheroid surface.

We finally calculated a strength of 325 ± 5 N m^2 for the INS-1 spheroids. Our spheroid strength was similar in magnitude to monolayer cells interacting with planar surfaces. Here, the cells also attached to the surface via surface proteins such as integrins, and the sum of F_{ad} corresponded to the attachment strength. A single fibroblast displaying ~2×10^5 integrins on its surface, required a force of 400 N m^2 to detach the cell from the surface [51]. Notably, the spheroid strength we calculated was only valid for the first 24 h, and probably increased during cultivation due to the buildup of ECM components, and the rearrangement of cells led to spheroid compaction. The spheroid strength we determined was also only valid for the INS-1 under our specific experimental set-up, as the spheroid strength is highly dependent on the cell type and culture conditions. For example, we observed completely different spheroid strengths when comparing β-cells cultivated in a serum-free and in a serum-supplemented medium.

3.5. INS-1 Spheroid Formation in a Stirred-Tank Bioreactor Using Different Stirrer Types

Our investigations in shaking flasks served as a basis for the 1 L scale production of INS-1 spheroids in a fully controlled and monitored STR. We therefore chose the power input/mean energy dissipation from the shaking flask experiments (35 W m^{-3}), where we achieved a good INS-1 spheroid formation and kept that value constant in the STR process. We then investigated the influence of three different stirrer types. Each stirrer has a specific power number N_P that reflects the momentum of resistance, and, therefore, the reinforcement of the power input into the culture medium. Although higher N_P values are often associated with greater particle stress, several investigations have shown that the same power input for an axial flow stirrer with a lower N_P can cause more particle disintegration [52–55]. This may reflect the so-called energy dissipation circulation function (EDCF), which describes the particle stress in relation to the particle residence time and frequency within the stirrer-swept volume V_S, where the greatest particle stress occurs. Logically, axial pumping stirrers with low N_P values must increase the stirrer frequency to achieve the same power input, and this increases particle stress. Furthermore, the trailing vortex behind the stirrer blade is the region with ε_{max}, and is thus responsible for the destructive effects [54]. To investigate these effects, we selected three different stirrer types, varying in N_P and V_S, based on axial, axial/radial and radial pumping orders (Table S1).

We ran each stirrer type with the same power input/mean energy dissipation. After 24 h, we recorded values of $d_{32} = 94 \pm 12$ µm for the shaking flask (reference) cultures, $d_{32} = 51.5 \pm 1.2$ µm for the 30°-3-SPB stirrer, $d_{32} = 40 \pm 3$ µm for the 45°-3-SPB stirrer, and $d_{32} = 50.6 \pm 1.6$ µm for the Rushton turbine (Table 2). The spheroid size was significantly (*** $p < 0.001$) decreased in the STR, while the 30°-3-SPB and Rushton turbine produced a similar spheroid size. Spheroids generated with the 45°-3-SPB were significantly (* $p < 0.05$) smaller. Although shaking flasks achieve a homogenous energy dissipation, whereas energy dissipation in the STR is heterogeneous due to the power input of the centrally-installed stirrer, the higher value for d_{32} and the larger span width of the size distribution (Table 2) may reflect the manual manufacturing process and variations in shaking flask geometry. Although we aimed to transfer the energy dissipation from the shaking flasks to the STR to reproduce the results of the shaking flask experiments, we anticipated differences in the steady-state size of the spheroids after 24 h. It was challenging to measure the power input in baffled shaking flasks due to variations in shape, size, and number of baffles, hence the corresponding models may overestimate the energy dissipation. The overestimation of hydrodynamic forces was supported by the significantly (*** $p < 0.001$) larger d_{32} and a lower and decreasing single-cell count, given there is less stress on the spheroid surface. Furthermore, there were substantial differences in power input between the shaking flasks and STR. The power input in shaking flasks was reinforced by friction between the culture medium and the vessel wall, and the redirection of the tangential flow into radial flow by the baffles, whereas the power input in STRs was generated by the stirrer. Even so, we found that shaking flasks were suitable for preliminary experiments to guide the manufacture of spheroids, if turbulent conditions were used to generate homogenous stress.

Table 2. Overview of relevant stirrer properties such as the power number N_P, energy dissipation circulation function (EDCF), stirrer tip speed, and the ratio of maximum energy dissipation to the mean $\varepsilon_{max}/\bar{\varepsilon}$ (calculated by [56]). Further, we summarized important particle parameters such as the Sauter diameter d_{32}, the spheroid formation efficiency, and the span width of the spheroid distribution from 1 to 99% after 24 h.

Stirrer Type	Np [−]	d_{32} [μm]	Spheroid Formation Efficiency [%]	Span Width of Distribution [μm]	EDCF [kW m^{-3} s^{-1}]	Stirrer Tip Speed [m s^{-1}]	$\varepsilon_{max}/\bar{\varepsilon}$ [−]
Shaking flask	-	94 ± 12	7.7 ± 1.4	134 ± 11	-	-	-
30°-3-SPB	1.1	51.5 ± 1.2	60 ± 3	66 ± 7	0.9	0.62	5
45°-3-SPB	2.4	40 ± 3	8.7 ± 1.4	51 ± 6	0.6	0.49	6
Rushton	4.0	50.6 ± 1.6	33 ± 2	61 ± 3	1.6	0.46	13

The three stirrer types are compared in Figure 3, which shows the counts for single cells (0–20 μm) and spheroids, in addition to the spheroid size distribution as box plots. The desirable outcome would be low single-cell counts and a high spheroid count, while maintaining a narrow spheroid size distribution. The 30°-3-SPB stirrer achieved a continuous low single-cell count (maximum 60,000 cells mL^{-1}), indicating that most of the cells formed spheroids, which is also reflected by the high spheroid formation efficiency of $60 \pm 3\%$ after 1 d (Table 2). The increasing d_{32}, while maintaining a narrow span width of the distribution of 66 ± 7 μm on day 1 to 100 ± 6 μm on day 4, confirmed efficient spheroid growth with low surface erosion, and therefore, few single cells in suspension. In contrast, the 45°-3-SPB stirrer was associated with a high single-cell count (up to 150,000 cells mL^{-1}) throughout the cultivation period, while maintaining a similar span width from 51 ± 5 μm on day 1 to 79 ± 21 μm on day 4. The low spheroid formation efficiency of ~10% and the high single-cell count indicated destructive hydrodynamic forces. The Rushton turbine performance was midway between the other stirrers. The highest d_{32} was achieved after 24 h, and the hydrodynamic forces prevented further spheroid growth and even reduced the spheroid size while increasing the single-cell count. The span width of the spheroid distribution ranged from 61 ± 2 μm on day 1 to 88 ± 3 μm on day 4. The spheroid formation efficiency was ~35%. Although the increased single-cell count suggested a negative effect, it was a consequence of the surface erosion, and this confirmed our adapted model showing that spheroid formation and size are restricted by surface erosion, as reported for other particles [41]. Although we observed increasing single-cell counts for all three stirrers over time, spheroid growth was not limited in the STR with the 30°-3-SPB stirrer. The overall low single-cell count and high agglomeration efficiency of 60% indicate that the stress caused by hydrodynamic forces was acceptable with this device, in contrast to the size-limiting surface erosion resulting in higher single-cell counts with the 45°-3-SPB stirrer and Rushton turbine.

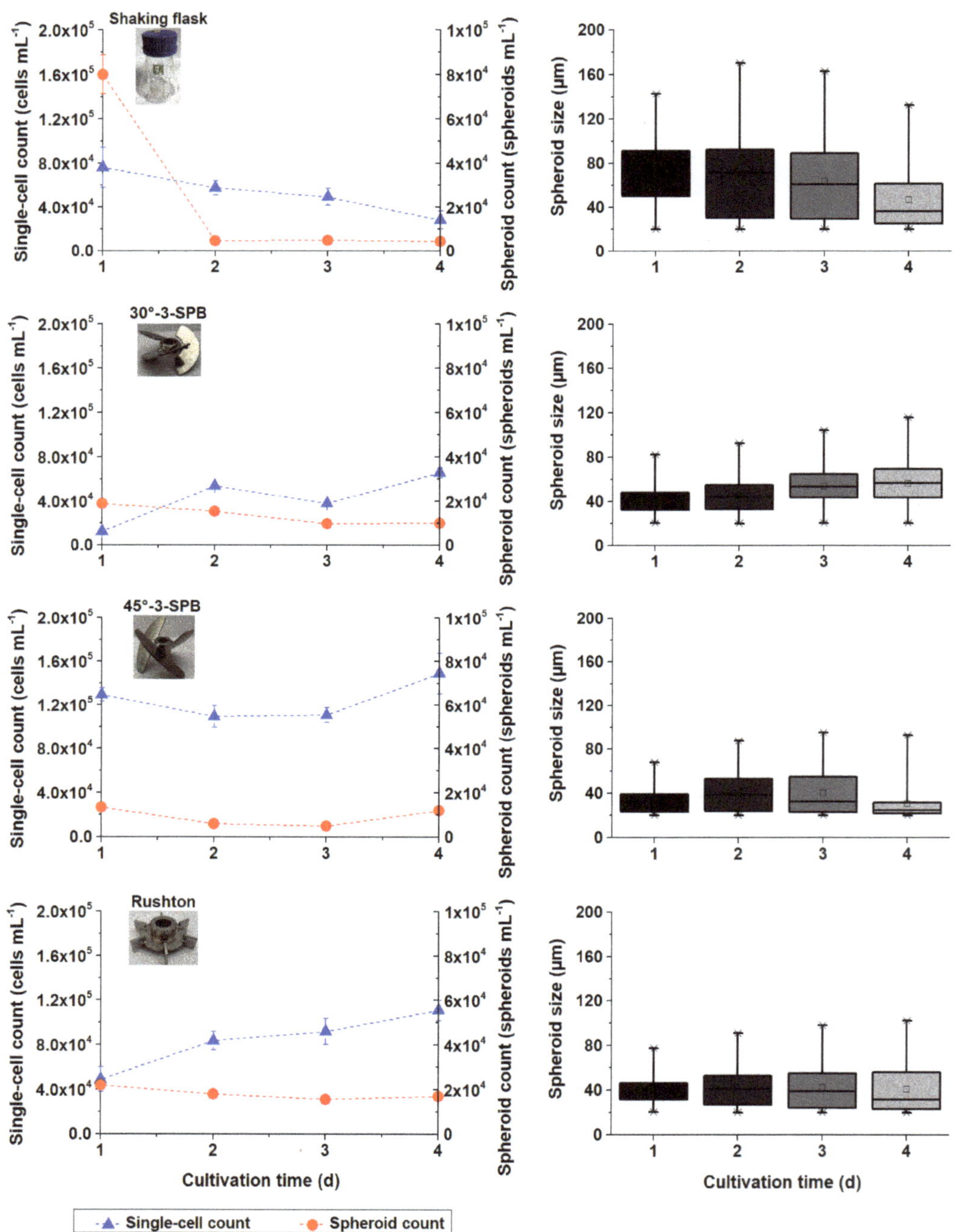

Figure 3. Particle counts for the shaking flask cultures (*n* = 3) and the three different stirrer types (*n* = 2) during cultivation for 4 d. Left column: single-cell count (blue triangles, size range: 0–20 μm, error bars = STDV) and spheroid count (red circles, size range: 20–300 μm, error bars = STDV). Right column: box plots of the spheroid size distribution showing the median (straight line), mean (square), minimum value 1% (due to our threshold at 20 μm), and maximum value 99%, each as cross.

Given the power numbers N_P of the three stirrers, our results support the statement that low N_P values (30°-3-SPB) cause less particle stress and encourage spheroid growth, whereas high N_P values (Rushton) produce smaller spheroids. This was also confirmed by the higher EDCF of the Rushton turbine (1.6 kW m^{-3} s^{-1}), compared with the lower value for the 30°-3-SPB stirrer (0.9 kW m^{-3} s^{-1}). The 45°-3-SPB stirrer had the lowest EDCF (0.6 kW m^{-3} s^{-1}), reflecting the 33% increase in vs. and, thus, the lowest stirrer frequency (Table S1). Although the stirrer tip speed is often used to describe the hydrodynamic forces during cell cultivation, in our case the stirrer tip speed provided an insufficient correlation. Whereas the Rushton turbine tip speed was 0.46 m s^{-1}, that of the 30°-3-SPB stirrer was 26% higher (0.62 m s^{-1}) and should have generated higher shear forces, which disagrees with our observations (Table 2). The 30°-3-SPB stirrer showed a slightly lower $\varepsilon_{max}/\bar{\varepsilon}$ ratio of 5 (assuming a more homogenous energy dissipation in the STR) than the 45°-3-SPB stirrer with a value of 6 (Table 2), which may help to explain the latter's poor performance, although this was nevertheless unexpected. An evaluation of multiple axial flowing stirrers, used to suspend microcarriers for the expansion of anchorage-dependent cells under low shear-stress conditions, resulted in the selection of an elephant ear impeller with a similar design to our 45°-3-SPB device [57]. The large vs. was found to facilitate the establishment of a microcarrier suspension while maintaining low shear rates and collisions between microcarriers (expressed as the turbulent collision severity) compared with axial stirrers with lower vs. and smaller blades to reinforce the power input (such as pitched-blade turbines). The reduced collision rate of this stirrer type could explain the poor performance of the 45°-3-SPB. Further, the angle of the stirrer blades, which affects the suspension efficiency, may also have contributed to our observed results. The 45°-3-SPB stirrer uses different mechanisms to suspend large and fine particles [58] and this could affect the agglomeration process, hence the high single-cell count associated with this stirrer from the start of the process.

3.6. Correlation between Predicted and Measured INS-1 Spheroid Size in STR

To correlate our experimental data with our theoretical background of the spheroid formation, we calculated the spheroid size in the STR after 24 h using our determined spheroid strength C_{Agg} (from the shaking flasks experiments) in Equation (8) [43]:

$$d = C_{Agg}^{l} \cdot d_{cell}^{m} \cdot \varepsilon^{-n} \tag{8}$$

where d_{cell} is the single-cell diameter, ε is the mean energy dissipation, and the exponents are $l = 0.5$, $m = 0$ (in the case of surface erosion) and $n = 0.25$.

Figure 4 shows the calculated spheroid size for each stirrer type in relation to the power input. For the calculations, we assumed that the spheroid strength is constant during the initial 24-h agglomeration phase. Our calculation of the spheroid size in the STR under different power inputs resulted in a good fit for the 45°-3-SPB stirrer, but for the 30°-3-SPB stirrer and Rushton turbine our calculation was off by ~10 μm. This deviation may reflect the different stirrer designs, because the 30°-3-SPB and the Rushton devices showed a similar behavior in terms of particle distribution (reduced single-cell count and higher spheroid formation efficiency) in contrast to the 45°-3-SPB stirrer. Although macroscale eddies are directly affected by the stirrer design and setup (e.g., d_S/D_T and baffling), eddies in the dissipation range are dependent on the fluid viscosity. However, the design and setup of the stirrer influence the magnitude of the energy level of λ, thus the sensitivity of our model is dependent on the precise characterization and perhaps adaption of the power consumption of the STR/stirrer combination. The same applies for the determination of C_{Agg}. Furthermore, the initial agglomeration process may differ for each stirrer type, and the 45°-3-SPB device could promote the formation of smaller, denser spheroids and, consequently, a different C_{Agg}. Our model, therefore, requires further empirical verification and assessment for robustness. We must still verify that spheroids are really affected by surface erosion (and not by tensile forces) and that similar conditions are found in the STR, to allow precise measurement of the effect.

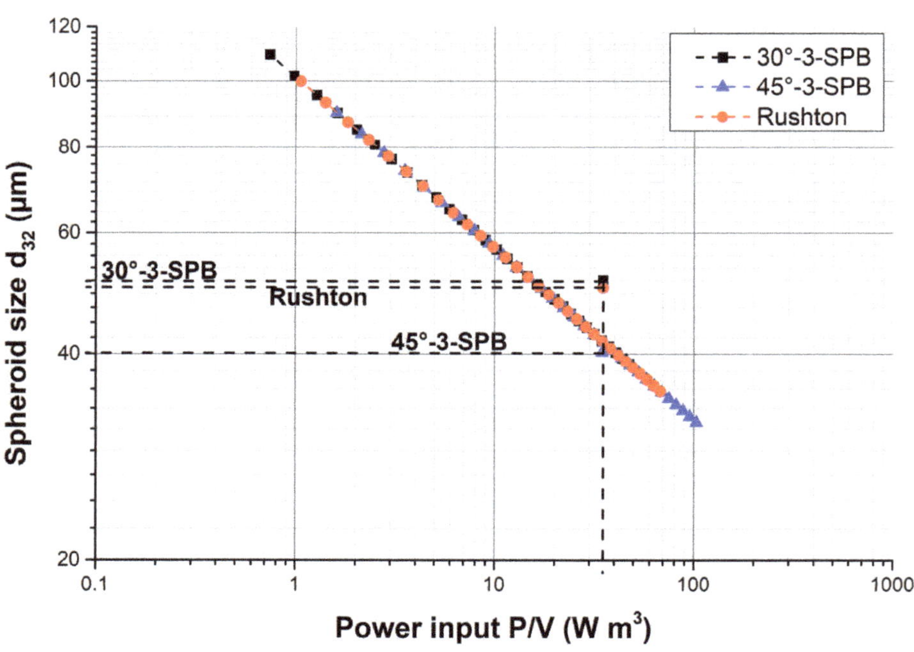

Figure 4. The relationship between spheroid size and power input. We determined the spheroid strength (C_{Agg}) in shaking flasks to predict the spheroid size in the STR at different power inputs (P/V) after 24 h. The relationship for each stirrer type shows the anticipated characteristic of a declining spheroid size with increasing P/V. With a constant P/V of 35 W m^{-3} (vertical dashed line), the measured spheroid size (horizontal dashed lines) for the 45°-3-SPB stirrer (blue triangle) fit the predicted spheroid size, whereas the 30°-3-SPB stirrer (black square) and the Rushton turbine (red circle) were offset by ~10 µm from the predicted value.

3.7. Growth and Viability of the INS-1 Spheroids Produced in the STR

The three different stirrer types resulted in the production of INS-1 spheroids, but the key objective was to produce viable and functional β-cell spheroids. We therefore evaluated the influence of each stirrer type on spheroid growth and viability for a further insight into INS-1 behavior under dynamic conditions. As described for the shaking flasks, we observed volume expansion either due to the attachment of single cells to the spheroids or cell growth within them. We calculated similar volume-based growth rates μ_{Vol} and volume doubling times $t_{D,Vol}$ for each stirrer type: 30°-3-SPB μ_{Vol} = 0.38 ± 0.018 d^{-1} ($t_{D,Vol}$ = 1.81 ± 0.09 d); 45°-3-SPB μ_{Vol} = 0.4 ± 0.2 d^{-1} ($t_{D,Vol}$ = 1.9 ± 1.0 d); and Rushton μ_{Vol} = 0.27 ± 0.072 d^{-1} ($t_{D,Vol}$ = 2.7 ± 0.7 d) (Table 3). The volume expansion of the spheroids depends on the culture system and the connected hydrodynamic conditions, as [59] could show that the fold expansion of hPSCs spheroids was highest in spinner flasks, compared with well plates and a ring-shaped culture vessel. Interestingly, the low fold expansion of the ring-shaped culture vessel was associated with a higher cell injury, which is attributed to increased shear forces. We hypothesize that this spheroid growth mainly reflected internal cell growth or an increase in spheroid strength due to compaction and stronger adhesion, thus, the extended buildup of an ECM. If we consider cell division within the spheroids, there should be no increase in size when one cell divides into two daughter cells, but if both daughter cells increase their volume or/and become surrounded by an extended ECM, the spheroid should expand. Other studies conclude an increase in spheroid size by cell divisions within the spheroids [26,34]. If the cells in the spheroid are protected by the ECM, the shear forces acting on the surface would need to be stronger than the force required to dislodge single cells that have adhered to the surface. Therefore, the viability and the count of single cells in the supernatant gives additional information about the performance of each stirrer type (Figure 5). The viability of the spheroids remained close to 100%, whereas we determined an accumulation of dead single cells over the cultivation period. Here, using the 30°-3-SPB stirrer, the continuous low single-cell count during the culture period was also connected to a high total viability (equaling single-cell viability and spheroid viability combined) of 90%. In contrast, high single-cell counts produced by the 45°-3-SPB were connected to low total viabilities down to 58%. In coherence with increasing single-cell counts, the Rushton turbine started with a high total viability of 91%, but revealed a reduction in total viability to 62% after day 2. In agreement with most of the dynamic cultivations, the increased mass transfer prevented the development of a necrotic spheroid core and resulted in an overall high viability of cells within the spheroids [26,33,34,59,60].

Table 3. Overview of relevant biological properties of the spheroids produced with each stirrer type, such as the volume-based growth rate μ_{Vol}, and the corresponding minimal time for doubling the volume $t_{D,Vol}$, the acute insulin secretion, and the insulin stimulation index SI (except for the 45°-3-SPB, due to poor performance). Further, we added the yield of spheroids $Y_{Spheroids}$, IEQs Y_{IEQs}, and cells Y_{cells} for each 1-L process.

Stirrer Type	μ_{Vol} [d^{-1}]	$t_{D,Vol}$ [d]	Acute Insulin Secretion [µg h^{-1} L^{-1}]	SI [−]	$Y_{Spheroids}$ [Spheroids L^{-1}]	Y_{IEQs} [IEQs L^{-1}]	Y_{cells} [Cells L^{-1}]
30°-3-SPB	0.38 ± 0.018	1.81 ± 0.09	32 ± 3	3.2 ± 1.2	$18 \times 10^6 \pm 2 \times 10^6$	$1.4 \times 10^6 \pm 0.6 \times 10^6$	$2 \times 10^8 \pm 0.1 \times 10^8$
45°-3-SPB	0.4 ± 0.2	1.9 ± 1.0	-	-	$14 \times 10^6 \pm 4 \times 10^6$	$0.3 \times 10^6 \pm 0.2 \times 10^6$	$0.8 \times 10^8 \pm 0.4 \times 10^8$
Rushton	0.27 ± 0.07	2.7 ± 0.7	17 ± 4	2.7 ± 1.0	$22 \times 10^6 \pm 0.6 \times 10^6$	$1.1 \times 10^6 \pm 0.4 \times 10^6$	$2.2 \times 10^8 \pm 0.2 \times 10^8$

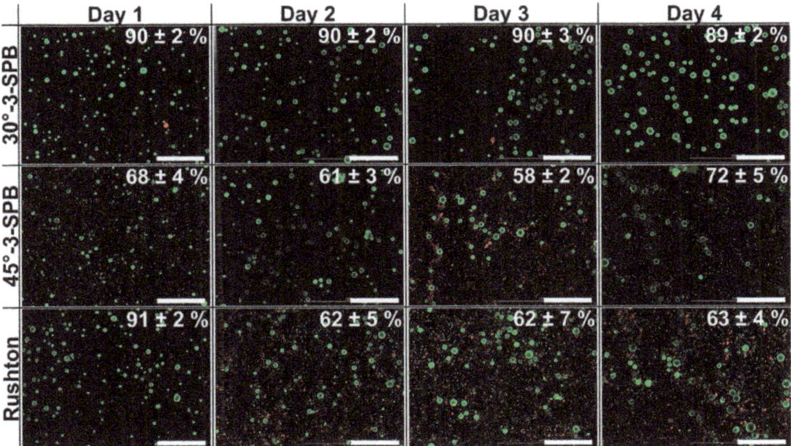

Figure 5. Analysis of the total viability of the STR process with each stirrer type over the culture period of 4 d (*n* = 2). Live/dead staining revealed high viabilities for the spheroids, whereas increasing amounts of dead single cells decreased the total viability, shown as mean ± STDV. The scale bar represents 500 µm.

We used the STR and stirrer-associated forces to regulate the size of the INS-1 spheroids, whereas others perform the contrary by using a rotational microgravity cell culture system to minimize any mechanical forces, thereby prolonging the shelf-life of primary β-cells [28]. As we conducted our proof-of concept with a β-cell line, we must consider that islet-derived β-cells are more sensitive to environmental changes and the process conditions. Islets cultivated with simulated microgravity maintained their structural integrity, were more potent, and functional for a longer time, whereas islets under static conditions developed necrotic cores, lost their exocrine mantle, and started to disintegrate. This reflected the dependence of β-cells on a sufficient diffusive supply of nutrients to maintain their complex microenvironment and functionality. The total viability of the STR process could be maintained (30°-3-SPB) or decreased (45°-3-SPB and Rushton turbine) over time due to surface erosion, but effective mixing in the STR increased the mass transfer, maintaining the viability of the spheroids close to 100%, despite the major impact of hydrodynamic forces (Figure 5).

3.8. Online Monitoring of Spheroid Growth and Destruction

The fact that with increasing power input, the spheroids were attacked by surface erosion, was supported by our inline dielectric spectroscopy data (Figure 6). The dielectric spectroscopy probe only detects the viable biomass (dead cells cannot be polarized) and does not distinguish between single cells and cells incorporated into spheroids, which was why we also measured spheroid size offline. Figure 6 shows the biomass signal relative to the offline data. We cultured the INS-1 cells with a low volumetric power input of 5.5 W m^{-3} (75 rpm) for 4 d, and then increased the P/V stepwise to up to 104 W m^{-3} (200 rpm) to induce spheroid destruction. Until day 4, we observed the exponential growth of INS-1 cells, which was demonstrated by the rapidly increasing spheroid size (up to d_{32} = 182 ± 9 μm) and the increasing biomass signal. More importantly, increasing the stirrer speed to 200 rpm reduced the spheroid size and inline biomass signal, and increased the single-cell counts, while maintaining high spheroid viability. This indicated that shear stress, acting on the spheroid surface due to surface erosion ($d_{Sph}/\lambda < 3$), and causing the disruption of ECM-embedded cells, leads to permanent cell damage, as evidenced by the declining biomass signal. The dielectric spectroscopy was also implemented for the expansion and hepatic differentiation of hiPSC spheroids, and showed a good correlation between offline data and online biomass [61].

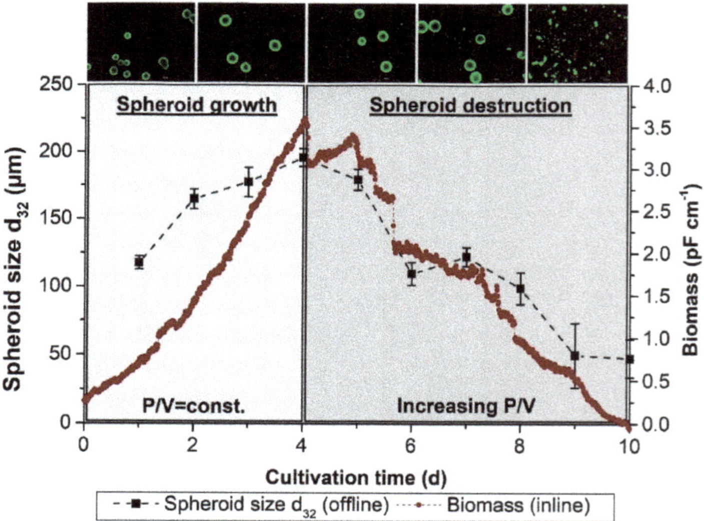

Figure 6. Representative spheroid production in an STR with the 45°-3-SPB stirrer. Dielectric spectroscopy showed an increase in biomass (red line) during the first 4 d of cultivation with a constant power input per volume (P/V). As dielectric spectroscopy only measures the volume of viable cells and does not distinguish between suspended individual cells and cells within spheroids, we also plotted the spheroid growth measured offline (black squares and dashed line; mean of 5 technical replicates, error bars = STDV). Upper lane: live/dead images of progressing spheroid growth and destruction. Increasing the P/V led to the disruption of the spheroids, resulting in single cells and smaller agglomerates, which were permanently affected by the hydrodynamic forces, clearly indicated by the decreasing biomass signal and the offline measurement.

3.9. Glucose-Stimulated Insulin Secretion

For the direct analysis of functionality, we tested the glucose-stimulated insulin secretion of the INS-1 spheroids. The responsiveness of β-cells to varying glucose levels provides more insight into the health of the cells, than the testing of the basal insulin secretion only. Although we observed wide deviations within the samples, reflecting the manual sample preparation method and variation among the spheroids in each sample, the INS-1 spheroids from all STR runs achieved a higher insulin secretion during the acute phase compared with basal secretion (Figure 7). These results supported the particle analysis and viability data. The 30°-3-SPB stirrer and the Rushton turbine promoted an efficient insulin secretion: 17 ± 4 µg h^{-1} L^{-1} with a stimulation index (SI) of 2.7 ± 1.0, and 32 ± 3 µg h^{-1} L^{-1} with a SI of 3.2 ± 1.2, respectively. The 45°-3-SPB stirrer promoted low levels of insulin secretion: 0.8 ± 0.3 µg h^{-1} L^{-1} (SI = 13 ± 7, solely for completeness, not significant). The poor performance of the 45°-3-SPB stirrer aligned with the accumulation of dead single cells (Figure 3), providing evidence of excessive stress. For the insulin secretion rate per spheroid, the 30°-3-SPB stirrer and the Rushton turbine achieved similar rates of 12 ± 6 and 12 ± 2 pg h^{-1} spheroid^{-1}, respectively, compared with the low value of 0.4 ± 0.5 pg h^{-1} spheroid^{-1} for the 45°-3-SPB device. We used an empirically determined conversion factor to approximate the number of cells contained in a spheroid and, thus, compensated for differences in spheroid size for each stirrer type, as previously recommended (Table 3) [21,62]. The Rushton turbine achieved the highest level of insulin secretion per 10^6 cells (97 ± 46 ng (h 10^6 cells)$^{-1}$), followed by the 30°-3-SPB stirrer (69 ± 43 ng (h 10^6 cells)$^{-1}$), and finally the 45°-3-SPB stirrer (2 ± 2 ng (h 10^6 cells)$^{-1}$). The insulin profiles of the dynamic cultured INS-1 as spheroids using the 30°-3-SPB and the Rushton stirrer were similar to the insulin secretion of the INS-1 as a monolayer. INS-1, challenged as a monolayer, secreted 47 (basal) and 95 (acute) ng (h 10^6 cells)$^{-1}$, which corresponded to SI = 2.1. In contrast, static produced spheroids (in cell-repellent 96-well plates) secreted 0.5 ng (h 10^6 cells)$^{-1}$ within the acute phase and thereby secreted much lower insulin amounts.

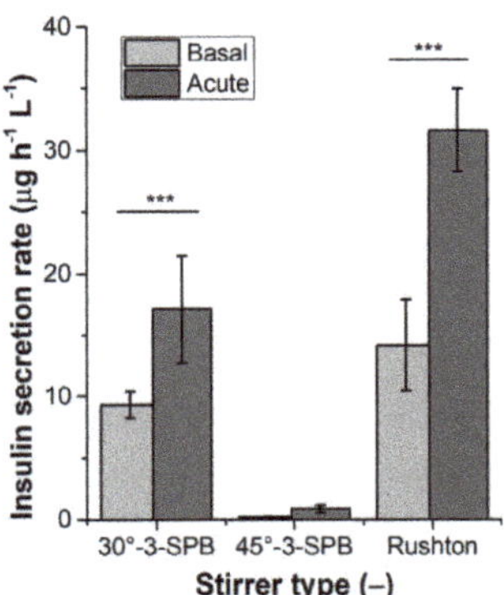
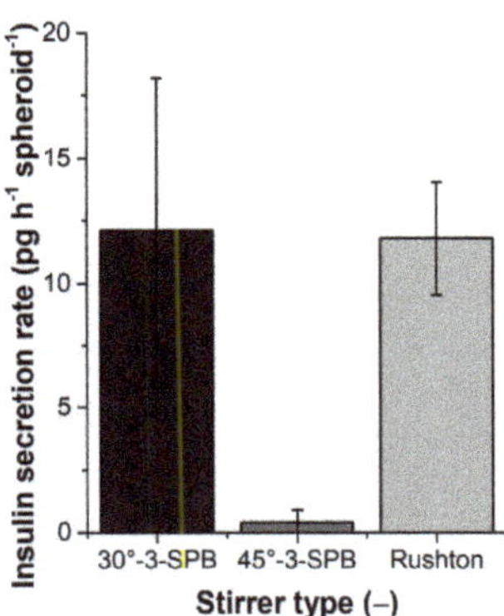
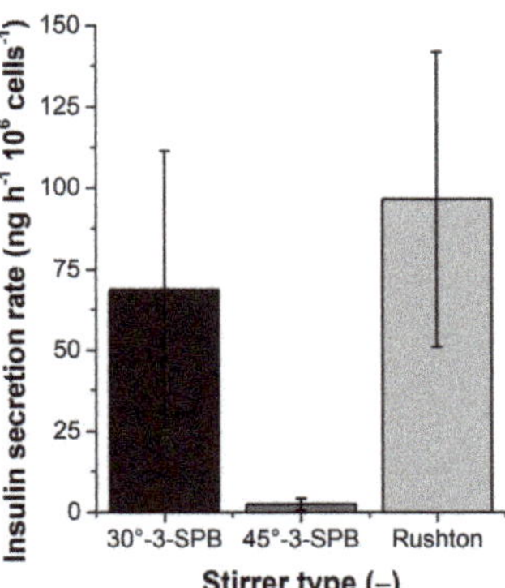

Figure 7. The functionality of β-cell spheroids determined using a glucose tolerance test. Left panel: the significant (*** <0.001, Student's *t*-test) increased secretion of insulin during the acute phase, compared with basal secretion, showed that β-cells produced with a 30°-3-SPB stirrer and a Rushton turbine responded to varying glucose concentrations. The insulin profiles of the β-cell spheroids generated with a 45°-3-SPB are only presented for completeness, but were not considered for discussion. Middle panel: the insulin secretion rate per spheroid. Right panel: we used a conversion factor to approximate the number of β-cells contained in a spheroid, to compensate for spheroid size differences. For each stirrer type (*n* = 2), three technical replicates of the glucose tolerance test were performed (error bars = STDV).

The reduced functionality of the static cultured spheroids was attributed to a limited mass transport within the spheroids and, consequently, low viabilities. Thereby, the STR-based production of β-cell spheroids restored the insulin secretion of the INS-1 from 2D cultures, while providing highly viable spheroids. The insulin profiles in static cultures only serve as a trend, which also applies to the insulin profiles of INS-1 cells in the literature, where static INS-1 spheroids showed a basal secretion of 20 μg L^{-1}, and an acute secretion of 40 μg L^{-1}, SI = 2 [10]. Static and dynamic spheroid formation vary widely, which may result in the differential compaction of the spheroids (cells per volume), and static spheroids often show necrotic cores, whereas those from dynamic cultures are often completely viable due to a better mass transfer and, thus, a more efficient transport of insulin and glucose. The handling of spheroids also differs: whereas the glucose-stimulated insulin secretion assay can be applied to static cultures with spheroids alone and maybe only to one spheroid per well, the preparation of samples from dynamic cultures incorporates multiple spheroids, varying in size, along with single cells. This indicates the complexity of testing, and highlights the need for further investigation to determine the functionality of β-cells cultured under dynamic conditions.

Our STR-based process produced functional INS-1-cell spheroids with yields of $18 \times 10^6 \pm 2 \times 10^6$ spheroids L^{-1} for the 30°-3-SPB stirrer, $14 \times 10^6 \pm 4 \times 10^6$ spheroids L^{-1} for the 45°-3-SPB stirrer, and $22 \times 10^6 \pm 0.6 \times 10^6$ spheroids L^{-1} for the Rushton turbine (Table 3). By converting the spheroid mass to islet equivalents (IEQ), based on a standard islet of Langerhans' diameter of 150 μm and standard procedures for the transplantation of whole pancreatic islets [20], we calculated the IEQ counts for the complete STR volume (1 L) of $1.4 \times 10^6 \pm 0.6 \times 10^6$ for the 30°-3-SPB stirrer, $0.3 \times 10^6 \pm 0.2 \times 10^6$ for the 45°-3-SPB stirrer, and $1.1 \times 10^6 \pm 0.4 \times 10^6$ for the Rushton turbine (Table 3). Accordingly, one STR can produce sufficient spheroids for β-cell replacement therapies, given that the Edmonton protocol for islet transplantation requires $0.5–1 \times 10^6$ IEQs from up to three donors to achieve a post-transplantation insulin independence [22,63], and exceeds the yield of other culture systems (compare Table 1). Based on the high spheroid formation efficiency, the narrow spheroid distribution in combination with low single-cell counts, high viability and insulin secretion, and sufficient spheroid/IEQ yield, we recommend the 30°-3-SPB for the production of β-cell spheroids on a larger scale. In an earlier study, more than 1000 spheroids in the size range of 100–250 μm were produced from the β-cell line MIN6 [22]. The authors used a clinostat to simulate microgravity, and induced the self-assembly of spheroids, whose size was semi-regulated by adapting the seeding density (higher cell concentrations produced smaller spheroids). The MIN6-derived spheroids secreted more insulin than monolayers, and expressed functionally relevant genes, such as *insulin-2*, *glucokinase*, *SETD1A*, and *Kir6.2* at higher levels. The spheroids were also much more therapeutically effective than single cells, following a transplantation in a streptozotocin-induced diabetic mouse model. It is likely that spheroids perform better than single cells, not only because they are functionally superior, but also because single cells injected into the portal vein are too small for retention in the liver and most cells are lost, with their ultimate fate unclear. In contrast, spheroids larger than ~40 μm are more likely to be retained in the liver vessels. Islets < 100 μm in diameter performed better [21] than large spheroids (>250 μm), which showed lower functionality [22], probably reflecting the limited diffusive supply of nutrients, leading to the formation of necrotic cells. This shows the importance of the manufacturing of β-cell spheroids within a defined size range, which can be achieved using our STR setup.

4. Conclusions

We successfully used shaking flasks as a screening platform to determine the key process parameters (e.g., seeding density, spheroid strength) for the spheroid formation, using the β-cell line INS-1 as model. We then transferred the dynamic spheroid formation to a fully-controlled and monitored STR. We determined the mean energy dissipation as a transfer criterion to regulate the size of the spheroids formed under dynamic conditions, and investigated three different stirrer types to evaluate the effect of the stirrer design (and the associated forces). Using the same mean energy dissipation, the 30°-3-SPB stirrer was slightly better than the Rushton turbine in terms of spheroid size distribution, also reducing the dead single-cell count and increasing spheroid formation efficiency and biological performance (higher μ_{Vol} and viability), while maintaining similar insulin secretion properties. In contrast, the 45°-3-SPB stirrer achieved poor results in all these aspects. We developed an image-based protocol to determine spheroid size and viability, and implemented the inline monitoring of biomass. The resulting spheroids achieved a similar glucose-dependent insulin secretion as standard 2D cultures of INS-1 cells. The large-scale cultivation of INS-1-cells achieved spheroid counts of up to $22 \times 10^6 \pm 0.6 \times 10^6$, and the corresponding IEQ counts were sufficient for β-cell replacement therapy. Although primary β-cells do not proliferate in culture, the development of human β-cell lines and β-cell-like cells derived from iPSCs is progressing, and robust large-scale production processes are needed for both. Other cell types, such as human mesenchymal stem cells (MSCs), can also benefit from shear-guided spheroid formation in an STR-based manufacturing process. Our STR-based spheroid formation process offers the scalability, process monitoring, and full control required for manufacturing, and allows the regulation of spheroid size while maintaining β-cell functionality.

Supplementary Materials: The following supporting information can be downloaded at: https://www.mdpi.com/article/10.3390/pr10050861/s1, Figure S1: Schematic description of (A): the mechanical power input of the stirrer, the resulting eddy cascade and energy dissipation, followed by the two spheroid stress concepts involving (B) tensile forces, (C) surface erosion, and finally (D) spheroid agglomeration, until the hydrodynamic forces are in balance with the adhesion force Fad; Table S1: Summary of stirrer properties and the resulting bioreactor geometry including the ratios for swept volume VS and working volume V_L, stirrer height h_S and stirrer diameter d_S or tank diameter D_T, as well as the bottom clearance $C_B = h_{S,bottom}/d_S$ ($h_{s,bottom}$ = installation height of stirrer from the bottom). References [11,41–46,53,55,56,64–67] are cited in the supplementary materials.

Author Contributions: F.P. conceived, designed and performed the experiments and wrote the paper. D.S. helped to draft and revise the manuscript, and supervised the research. All authors have read and agreed to the published version of the manuscript.

Funding: The work was funded by the Forschungscampus Mittelhessen (FCHM).

Institutional Review Board Statement: Not applicable.

Informed Consent Statement: Not applicable.

Data Availability Statement: All relevant data are given within the manuscript. The raw data presented in this study are available on request from the corresponding author.

Acknowledgments: We would like to thank Peter Czermak for the opportunity to work at his institute, to use his laboratories, additional funding, all the fruitful scientific discussions, and his input into this work. The authors would like to thank Richard M. Twyman and Catharine Meckel-Oschmann for language editing.

Conflicts of Interest: The authors declare no conflict of interest.

References

1. International Diabetes Federation. IDF Diabetes Atlas Ninth Edition 2019. 2020. Available online: www.diabetesatlas.org (accessed on 12 April 2021).
2. Mir-Coll, J.; Moede, T.; Paschen, M.; Neelakandhan, A.; Valladolid-Acebes, I.; Leibiger, B.; Biernath, A.; Ämmälä, C.; Leibiger, I.B.; Yesildag, B.; et al. Human Islet Microtissues as an In Vitro and an In Vivo Model System for Diabetes. *Int. J. Mol. Sci.* **2021**, *22*, 1813. [CrossRef] [PubMed]
3. Islam, M.S. *The Islets of Langerhans*; Springer Netherlands: Dordrecht, The Netherlands, 2010; ISBN 978-90-481-3270-6.
4. Benthuysen, J.R.; Carrano, A.C.; Sander, M. Advances in β cell replacement and regeneration strategies for treating diabetes. *J. Clin. Investig.* **2016**, *126*, 3651–3660. [CrossRef] [PubMed]
5. Scharfmann, R.; Didiesheim, M.; Richards, P.; Chandra, V.; Oshima, M.; Albagli, O. Mass production of functional human pancreatic β-cells: Why and how? *Diabetes Obes. Metab.* **2016**, *18* (Suppl. 1), 128–136. [CrossRef] [PubMed]

6.   Rezania, A.; Bruin, J.E.; Arora, P.; Rubin, A.; Batushansky, I.; Asadi, A.; O'Dwyer, S.; Quiskamp, N.; Mojibian, M.; Albrecht, T.; et al. Reversal of diabetes with insulin-producing cells derived in vitro from human pluripotent stem cells. *Nat. Biotechnol.* **2014**, *32*, 1121–1133. [CrossRef] [PubMed]

7.   McCluskey, J.T.; Hamid, M.; Guo-Parke, H.; McClenaghan, N.H.; Gomis, R.; Flatt, P.R. Development and functional characterization of insulin-releasing human pancreatic beta cell lines produced by electrofusion. *J. Biol. Chem.* **2011**, *286*, 21982–21992. [CrossRef] [PubMed]

8.   Ravassard, P.; Hazhouz, Y.; Pechberty, S.; Bricout-Neveu, E.; Armanet, M.; Czernichow, P.; Scharfmann, R. A genetically engineered human pancreatic β cell line exhibiting glucose-inducible insulin secretion. *J. Clin. Investig.* **2011**, *121*, 3589–3597. [CrossRef]

9.   Scharfmann, R.; Pechberty, S.; Hazhouz, Y.; von Bülow, M.; Bricout-Neveu, E.; Grenier-Godard, M.; Guez, F.; Rachdi, L.; Lohmann, M.; Czernichow, P.; et al. Development of a conditionally immortalized human pancreatic β cell line. *J. Clin. Investig.* **2014**, *124*, 2087–2098. [CrossRef]

10.   Ntamo, Y.; Samodien, E.; Burger, J.; Muller, N.; Muller, C.J.F.; Chellan, N. In vitro Characterization of Insulin-Producing β-Cell Spheroids. *Front. Cell Dev. Biol.* **2020**, *8*, 623889. [CrossRef]

11.   Petry, F.; Weidner, T.; Czermak, P.; Salzig, D. Three-Dimensional Bioreactor Technologies for the Cocultivation of Human Mesenchymal Stem/Stromal Cells and Beta Cells. *Stem Cells Int.* **2018**, *2018*, 2547098. [CrossRef]

12.   Bernard, A.B.; Lin, C.-C.; Anseth, K.S. A microwell cell culture platform for the aggregation of pancreatic β-cells. *Tissue Eng. Part C Methods* **2012**, *18*, 583–592. [CrossRef]

13.   Bhang, S.H.; Jung, M.J.; Shin, J.-Y.; La, W.-G.; Hwang, Y.H.; Kim, M.J.; Kim, B.-S.; Lee, D.Y. Mutual effect of subcutaneously transplanted human adipose-derived stem cells and pancreatic islets within fibrin gel. *Biomaterials* **2013**, *34*, 7247–7256. [CrossRef] [PubMed]

14.   Achilli, T.-M.; Meyer, J.; Morgan, J.R. Advances in the formation, use and understanding of multi-cellular spheroids. *Expert Opin. Biol. Ther.* **2012**, *12*, 1347–1360. [CrossRef] [PubMed]

15.   Lammert, E.; Thorn, P. The Role of the Islet Niche on Beta Cell Structure and Function. *J. Mol. Biol.* **2020**, *432*, 1407–1418. [CrossRef] [PubMed]

16.   Petrenko, Y.; Syková, E.; Kubinová, Š. The therapeutic potential of three-dimensional multipotent mesenchymal stromal cell spheroids. *Stem Cell Res. Ther.* **2017**, *8*, 94. [CrossRef]

17.   Holt, R.I.G.; Cockram, C.S.; Flyvbjerg, A.; Goldstein, B.J. *Textbook of Diabetes*; Wiley-Blackwell: Oxford, UK, 2010; ISBN 9781444324808.

18.   Friberg, A.S.; Brandhorst, H.; Buchwald, P.; Goto, M.; Ricordi, C.; Brandhorst, D.; Korsgren, O. Quantification of the islet product: Presentation of a standardized current good manufacturing practices compliant system with minimal variability. *Transplantation* **2011**, *91*, 677–683. [CrossRef]

19.   Gmyr, V.; Bonner, C.; Lukowiak, B.; Pawlowski, V.; Dellaleau, N.; Belaich, S.; Aluka, I.; Moermann, E.; Thevenet, J.; Ezzouaoui, R.; et al. Automated digital image analysis of islet cell mass using Nikon's inverted eclipse Ti microscope and software to improve engraftment may help to advance the therapeutic efficacy and accessibility of islet transplantation across centers. *Cell Transplant.* **2015**, *24*, 1–9. [CrossRef]

20.   Buchwald, P.; Bernal, A.; Echeverri, F.; Tamayo-Garcia, A.; Linetsky, E.; Ricordi, C. Fully Automated Islet Cell Counter (ICC) for the Assessment of Islet Mass, Purity, and Size Distribution by Digital Image Analysis. *Cell Transplant.* **2016**, *25*, 1747–1761. [CrossRef]

21.   Huang, H.-H.; Harrington, S.; Stehno-Bittel, L. The Flaws and Future of Islet Volume Measurements. *Cell Transplant.* **2018**, *27*, 1017–1026. [CrossRef]

22.   Tanaka, H.; Tanaka, S.; Sekine, K.; Kita, S.; Okamura, A.; Takebe, T.; Zheng, Y.-W.; Ueno, Y.; Tanaka, J.; Taniguchi, H. The generation of pancreatic β-cell spheroids in a simulated microgravity culture system. *Biomaterials* **2013**, *34*, 5785–5791. [CrossRef]

23.   Zuellig, R.A.; Cavallari, G.; Gerber, P.; Tschopp, O.; Spinas, G.A.; Moritz, W.; Lehmann, R. Improved physiological properties of gravity-enforced reassembled rat and human pancreatic pseudo-islets. *J. Tissue Eng. Regen. Med.* **2017**, *11*, 109–120. [CrossRef]

24.   Abbasalizadeh, S.; Larijani, M.R.; Samadian, A.; Baharvand, H. Bioprocess development for mass production of size-controlled human pluripotent stem cell aggregates in stirred suspension bioreactor. *Tissue Eng. Part C Methods* **2012**, *18*, 831–851. [CrossRef]

25.   Chawla, M.; Bodnar, C.A.; Sen, A.; Kallos, M.S.; Behie, L.A. Production of islet-like structures from neonatal porcine pancreatic tissue in suspension bioreactors. *Biotechnol. Prog.* **2006**, *22*, 561–567. [CrossRef]

26.   Lock, L.T.; Laychock, S.G.; Tzanakakis, E.S. Pseudoislets in stirred-suspension culture exhibit enhanced cell survival, propagation and insulin secretion. *J. Biotechnol.* **2011**, *151*, 278–286. [CrossRef]

27.   Santo, V.E.; Estrada, M.F.; Rebelo, S.P.; Abreu, S.; Silva, I.; Pinto, C.; Veloso, S.C.; Serra, A.T.; Boghaert, E.; Alves, P.M.; et al. Adaptable stirred-tank culture strategies for large scale production of multicellular spheroid-based tumor cell models. *J. Biotechnol.* **2016**, *221*, 118–129. [CrossRef]

28.   Murray, H.E.; Paget, M.B.; Downing, R. Preservation of glucose responsiveness in human islets maintained in a rotational cell culture system. *Mol. Cell. Endocrinol.* **2005**, *238*, 39–49. [CrossRef]

29.   Chmiel, H. (Ed.) *Bioprozesstechnik*; Spektrum Akademischer Verl.: Heidelberg, Germany, 2011; ISBN 9783827424761.

30.   Petry, F.; Salzig, D. Impact of Bioreactor Geometry on Mesenchymal Stem Cell Production in Stirred-Tank Bioreactors. *Chem. Ing. Tech.* **2021**, *93*, 1537–1554. [CrossRef]

31. Lv, D.; Hu, Z.; Lu, L.; Lu, H.; Xu, X. Three-dimensional cell culture: A powerful tool in tumor research and drug discovery. *Oncol. Lett.* **2017**, *14*, 6999–7010. [CrossRef]

32. Schulz, T.C.; Young, H.Y.; Agulnick, A.D.; Babin, M.J.; Baetge, E.E.; Bang, A.G.; Bhoumik, A.; Cepa, I.; Cesario, R.M.; Haakmeester, C.; et al. A scalable system for production of functional pancreatic progenitors from human embryonic stem cells. *PLoS ONE* **2012**, *7*, e37004. [CrossRef]

33. Olmer, R.; Lange, A.; Selzer, S.; Kasper, C.; Haverich, A.; Martin, U.; Zweigerdt, R. Suspension culture of human pluripotent stem cells in controlled, stirred bioreactors. *Tissue Eng. Part C Methods* **2012**, *18*, 772–784. [CrossRef]

34. Manstein, F.; Ullmann, K.; Kropp, C.; Halloin, C.; Triebert, W.; Franke, A.; Farr, C.-M.; Sahabian, A.; Haase, A.; Breitkreuz, Y.; et al. High density bioprocessing of human pluripotent stem cells by metabolic control and in silico modeling. *Stem Cells Transl. Med.* **2021**, *10*, 1063–1080. [CrossRef]

35. Peter, C.P. Auslegung Geschüttelter Bioreaktoren für Hochviskose und Hydromechanisch Empfindliche Fermentationssysteme. Ph.D. Thesis, Rheinisch-Westfälischen Technischen Hochschule Aachen, Aachen, Germany, 2007.

36. Büchs, J.; Maier, U.; Milbradt, C.; Zoels, B. Power consumption in shaking flasks on rotary shaking machines: I. Power consumption measurement in unbaffled flasks at low liquid viscosity. *Biotechnol. Bioeng.* **2000**, *68*, 589–593. [CrossRef]

37. Büchs, J.; Maier, U.; Milbradt, C.; Zoels, B. Power consumption in shaking flasks on rotary shaking machines: II. Nondimensional description of specific power consumption and flow regimes in unbaffled flasks at elevated liquid viscosity. *Biotechnol. Bioeng.* **2000**, *68*, 594–601. [CrossRef]

38. Joo, D.J.; Kim, J.Y.; Lee, J.I.; Jeong, J.H.; Cho, Y.; Ju, M.K.; Huh, K.H.; Kim, M.S.; Kim, Y.S. Manufacturing of insulin-secreting spheroids with the RIN-5F cell line using a shaking culture method. *Transplant. Proc.* **2010**, *42*, 4225–4227. [CrossRef] [PubMed]

39. Cunha, B.; Aguiar, T.; Carvalho, S.B.; Silva, M.M.; Gomes, R.A.; Carrondo, M.J.T.; Gomes-Alves, P.; Peixoto, C.; Serra, M.; Alves, P.M. Bioprocess integration for human mesenchymal stem cells: From up to downstream processing scale-up to cell proteome characterization. *J. Biotechnol.* **2017**, *248*, 87–98. [CrossRef]

40. Serra, M.; Brito, C.; Sousa, M.F.Q.; Jensen, J.; Tostões, R.; Clemente, J.; Strehl, R.; Hyllner, J.; Carrondo, M.J.T.; Alves, P.M. Improving expansion of pluripotent human embryonic stem cells in perfused bioreactors through oxygen control. *J. Biotechnol.* **2010**, *148*, 208–215. [CrossRef]

41. Jarvis, P.; Jefferson, B.; Gregory, J.; Parsons, S.A. A review of floc strength and breakage. *Water Res.* **2005**, *39*, 3121–3137. [CrossRef]

42. KOBAYASHI, M.; Adachi, Y.; OOI, S. Effect of Particle Size on Breakup of Flocs in a Turbulent Flow. *Proc. Hydraul. Eng.* **2001**, *45*, 1249–1253. [CrossRef]

43. Mühle, K.; Domasch, K. Stability of particle aggregates in flocculation with polymers. *Chem. Eng. Processing Process Intensif.* **1991**, *29*, 1–8. [CrossRef]

44. Oyegbile, B.; Ay, P.; Narra, S. Flocculation kinetics and hydrodynamic interactions in natural and engineered flow systems: A review. *Environ. Eng. Res.* **2016**, *21*, 1–14. [CrossRef]

45. Schubert, H.; Mühle, K. The role of turbulence in unit operations of particle technology. *Adv. Powder Technol.* **1991**, *2*, 295–306. [CrossRef]

46. Wengeler, R.; Nirschl, H. Turbulent hydrodynamic stress induced dispersion and fragmentation of nanoscale agglomerates. *J. Colloid Interface Sci.* **2007**, *306*, 262–273. [CrossRef] [PubMed]

47. Moreira, J.; Cruz, P.E.; Santana, P.C.; Aunins, J.G.; Carrondo, M.J. Formation and disruption of animal cell aggregates in stirred vessels: Mechanisms and kinetic studies. *Chem. Eng. Sci.* **1995**, *50*, 2747–2764. [CrossRef]

48. Yeung, A.; Gibbs, A.; Pelton, R. Effect of Shear on the Strength of Polymer-Induced Flocs. *J. Colloid Interface Sci.* **1997**, *196*, 113–115. [CrossRef]

49. Sart, S.; Tsai, A.-C.; Li, Y.; Ma, T. Three-dimensional aggregates of mesenchymal stem cells: Cellular mechanisms, biological properties, and applications. *Tissue Eng. Part B Rev.* **2014**, *20*, 365–380. [CrossRef] [PubMed]

50. Parker, D. Floc Breakup in Turbulent Flocculation Processes. *J. Sanit. Eng.* **1972**, *98*, 79–99. [CrossRef]

51. David, B. Use of hydrodynamic shear stress to analyze cell adhesion. In *Principles of Cellular Engineering: Understanding the Biomolecular Interface*; King, M.R., Ed.; Elsevier Academic Press: Amsterdam, Boston, 2006; pp. 51–80; ISBN 9780123693921.

52. Böhm, L.; Hohl, L.; Bliatsiou, C.; Kraume, M. Multiphase Stirred Tank Bioreactors—New Geometrical Concepts and Scale-up Approaches. *Chem. Ing. Tech.* **2019**, *91*, 1724–1746. [CrossRef]

53. Henzler, H.J. Particle stress in bioreactors. *Adv. Biochem. Eng. Biotechnol.* **2000**, *67*, 35–82. [CrossRef]

54. Jüsten, P.; Paul, G.C.; Nienow, A.W.; Thomas, C.R. Dependence of mycelial morphology on impeller type and agitation intensity. *Biotechnol. Bioeng.* **1996**, *52*, 672–684. [CrossRef]

55. Langer, G.; Deppe, A. Zum Verständnis der hydrodynamischen Beanspruchung von Partikeln in turbulenten Rührerströmungen. *Chem. Ing. Tech.* **2000**, *72*, 31–41. [CrossRef]

56. Wollny, S. Experimentelle und Numerische Untersuchungen zur Partikelbeanspruchung in Gerührten (Bio-)Reaktoren. Doctoral Thesis, Technische Universität Berlin, Berlin, Germany, 2010.

57. Collignon, M.-L.; Delafosse, A.; Crine, M.; Toye, D. Axial impeller selection for anchorage dependent animal cell culture in stirred bioreactors: Methodology based on the impeller comparison at just-suspended speed of rotation. *Chem. Eng. Sci.* **2010**, *65*, 5929–5941. [CrossRef]

58. Jirout, T.; Rieger, F. Impeller design for mixing of suspensions. *Chem. Eng. Res. Des.* **2011**, *89*, 1144–1151. [CrossRef]

59. Torizal, F.G.; Kim, S.M.; Horiguchi, I.; Inamura, K.; Suzuki, I.; Morimura, T.; Nishikawa, M.; Sakai, Y. Production of homogenous size-controlled human induced pluripotent stem cell aggregates using ring-shaped culture vessel. *J. Tissue Eng. Regen. Med.* **2022**, *16*, 254–266. [CrossRef] [PubMed]
60. Kahn-Krell, A.; Pretorius, D.; Ou, J.; Fast, V.G.; Litovsky, S.; Berry, J.; Liu, X.M.; Zhang, J. Bioreactor Suspension Culture: Differentiation and Production of Cardiomyocyte Spheroids From Human Induced Pluripotent Stem Cells. *Front. Bioeng. Biotechnol.* **2021**, *9*, 674260. [CrossRef]
61. Isidro, I.A.; Vicente, P.; Pais, D.A.M.; Almeida, J.I.; Domingues, M.; Abecasis, B.; Zapata-Linares, N.; Rodriguez-Madoz, J.R.; Prosper, F.; Aspegren, A.; et al. Online monitoring of hiPSC expansion and hepatic differentiation in 3D culture by dielectric spectroscopy. *Biotechnol. Bioeng.* **2021**, *118*, 3610–3617. [CrossRef] [PubMed]
62. Huang, H.-H.; Ramachandran, K.; Stehno-Bittel, L. A replacement for islet equivalents with improved reliability and validity. *Acta Diabetol.* **2013**, *50*, 687–696. [CrossRef] [PubMed]
63. Shapiro, A.M.J.; Ricordi, C.; Hering, B.J.; Auchincloss, H.; Lindblad, R.; Robertson, R.P.; Secchi, A.; Brendel, M.D.; Berney, T.; Brennan, D.C.; et al. International trial of the Edmonton protocol for islet transplantation. *N. Engl. J. Med.* **2006**, *355*, 1318–1330. [CrossRef] [PubMed]
64. Hewitt, C.J.; Lee, K.; Nienow, A.W.; Thomas, R.J.; Smith, M.; Thomas, C.R. Expansion of human mesenchymal stem cells on microcarriers. *Biotechnol. Lett.* **2011**, *33*, 2325–2335. [CrossRef]
65. Pattappa, G.; Heywood, H.K.; de Bruijn, J.D.; Lee, D.A. The metabolism of human mesenchymal stem cells during proliferation and differentiation. *J. Cell. Physiol.* **2011**, *226*, 2562–2570. [CrossRef]
66. Heyter, A.; Wollny, S. Einfluss verschiedener Stromstörerausführungen auf die Bewehrung eines mehrstufigen Rührbehälters. *Chem. Ing. Tech.* **2017**, *89*, 416–423. [CrossRef]
67. Kraume, M. *Mischen und Rühren: Grundlagen und Moderne Verfahren*; Wiley-VCH: Weinheim, Germany, 2003; ISBN 3527307095.

MDPI

Review

Modern Sensor Tools and Techniques for Monitoring, Controlling, and Improving Cell Culture Processes

Sebastian Juan Reyes [1,2], Yves Durocher [2,3], Phuong Lan Pham [2] and Olivier Henry [1,*]

1   Department of Chemical Engineering, Polytechnique Montreal, Montreal, QC H3T 1J4, Canada; juan-sebastian.reyes@polymtl.ca
2   Human Health Therapeutics Research Center, National Research Council of Canada, Montreal, QC H4P 2R2, Canada; Yves.Durocher@cnrc-nrc.gc.ca (Y.D.); PhuongLan.Pham@cnrc-nrc.gc.ca (P.L.P.)
3   Department of Biochemistry and Molecular Medicine, University of Montreal, Montreal, QC H3T 1J4, Canada
*   Correspondence: olivier.henry@polymtl.ca

Abstract: The growing biopharmaceutical industry has reached a level of maturity that allows for the monitoring of numerous key variables for both process characterization and outcome predictions. Sensors were historically used in order to maintain an optimal environment within the reactor to optimize process performance. However, technological innovation has pushed towards on-line in situ continuous monitoring of quality attributes that could previously only be estimated off-line. These new sensing technologies when coupled with software models have shown promise for unique fingerprinting, smart process control, outcome improvement, and prediction. All this can be done without requiring invasive sampling or intervention on the system. In this paper, the state-of-the-art sensing technologies and their applications in the context of cell culture monitoring are reviewed with emphasis on the coming push towards industry 4.0 and smart manufacturing within the biopharmaceutical sector. Additionally, perspectives as to how this can be leveraged to improve both understanding and outcomes of cell culture processes are discussed.

Keywords: sensors; cell culture; spectroscopy; PAT; smart biomanufacturing; bioprocess; monitoring; soft-sensor

Citation: Reyes, S.J.; Durocher, Y.; Pham, P.L.; Henry, O. Modern Sensor Tools and Techniques for Monitoring, Controlling, and Improving Cell Culture Processes. *Processes* **2022**, *10*, 189. https://doi.org/10.3390/pr10020189

Academic Editors: Ralf Pörtner and Johannes Möller

Received: 12 November 2021
Accepted: 11 January 2022
Published: 18 January 2022

Publisher's Note: MDPI stays neutral with regard to jurisdictional claims in published maps and institutional affiliations.

1. Introduction

The global biotechnology market was valued at 752 million USD in 2020 with a significant portion of the market size belonging to the biopharmaceutical industry [1,2]. This key sector is expected to be valued at 526 million USD by 2025 [3], with a compounded annual growth rate of 13.8% [3]. The relevant value-added products include monoclonal antibodies, interferons, hormones, growth and coagulation factors, vaccines, and others. Monoclonal antibodies have dominated the global biopharmaceuticals market, due to their use in the treatment of chronic diseases, such as cancer [3]. An increase in R&D with respect to oncology drug development is also expected to increase the growth of monoclonal antibody production and market size [3]. It is important to note that as the biopharmaceutical industry matures, older patents of approved biologics expire. Thus, non-brand companies can begin to manufacture generic versions of the biotherapeutic. Within the biopharmaceutical industry, these non-brand drugs are called biosimilars and are analogous to generic drugs in the pharmaceutical industry [4,5]. This market is expected to grow at high compounded annual growth rates (24–34%) by 2025 [4,5]. Importantly, the COVID-19 pandemic has acted as a catalyst for biopharmaceutical growth given that numerous biological compounds have been produced with the purpose of tackling the virus and reducing its strain on healthcare systems. This includes the development of monoclonal antibody treatments and novel vaccine platforms [6,7]. Biopharmaceutical production needs large manufacturing capacities that must be designed with cost and time efficiency in mind because antibody therapies require long periods of time to be effective [6]. Mammalian cells, which are employed in the manufacturing of antibodies, have historically been associated with low yield and manufacturing complexity since the cells are shear sensitive and require specialized media additives to be able to grow properly [6]. However, recent advances in process optimization and cell line engineering have allowed antibody production to generate high yields of up to 10–20 g/L in fed-batch mode [6]. As a consequence of increased manufacturing capability, 570 therapeutic monoclonal antibodies (mAbs) have been approved for clinical trials by biopharma companies [7]. Of those tested, 79 mAbs have been approved by the Food and Drug Administration (FDA) for commercial use. A substantial majority of the approved mAbs are used in treatments of cancer and autoimmune disorders [7]. Considering that between 2008 and 2021, 48 of the currently approved 79 antibodies have been developed, it is possible to assert that increased understanding in mammalian cell platforms have allowed for such increase in antibody manufacturing capability. This is especially true bearing in mind that the first antibody approved for commercial use was murine IgG2a CD3 in 1986 [7].

To guide the biomanufacturing sector towards better production efficiencies while still ensuring maximum process safety within a timely manner, the Quality by Design (QbD) and Process Analytical Technologies (PAT) initiatives were established. Lot-to-lot variation indicated that the established processes were not as robust as imagined [8–12]. Since these inefficiencies also caused fewer products to be commercialized, drug manufacturing became more costly. Additionally, due to globalization, quality control guidelines became fragmented, making it difficult for pharmaceutical companies to meet all regulatory requirements [11]. Due to this regulatory difficulty, the FDA created an initiative, denominated current Good Manufacturing Practices (cGMP) for the 21st century [11]. This placed emphasis in a Quality by Design approach rather than relying on post quality control batch testing. QbD is a scientific, risk-based holistic approach that relies on defining and identifying the Critical Quality Attributes (CQA) of a product as well as defining an appropriate design space [11]. By designing and formulating production processes and product formulations around these CQA, the pharmaceutical company can continually monitor and update its manufacturing platform to assure consistent product quality [12]. These CQAs are generally defined thanks to in vitro and animal studies that help characterize the pharmaceutical compound. Once the CQA have been defined, developing a manufacturing process that will yield the desired product with the appropriate attributes is needed [12]. Because of this, the design space is developed early during each study. For example, during cell culture development, study ranges for temperature, pH, and feed timing are characterized [12]. With the help of design of experiments, characterization is done to evaluate the impact of multiple variables (Critical Process Parameters, CPPs) and how changes in these variables can affect the product quality or lack thereof. This allows the manufacturer to define the acceptable operating conditions in which the product maintains regulatory-approved quality [12]. However, it must be stated that a biopharmaceutical production platform deploys multiple steps that may be serial or parallel in nature. Because of this, the development of the design space must be evaluated in a big picture manner that takes into account various possible process conditions [8–12].

Since characterizing the design space and controlling the CQA are the fundamental pillars of QbD, Process Analytical Technologies have become important tools [12]. PAT analyze the CQA during various stages of biomanufacturing. These analyses are often conducted on-line to yield large amounts of data that can then be analyzed in order to make real-time adjustments to the process parameters [10–12]. Ideally, this would be employed at every stage of the manufacturing process, from the cell culture to the final purification and formulation steps [11,12]. Once the design space is established, the regulatory filing includes the acceptable ranges for the CQA. These parameters are then monitored to ensure that the process is performing within the specified design space [11,12]. This entails that an appropriately defined, expanded design space allows for a more flexible approach by regulatory agencies. Thus, process changes within the design space do not require additional regulatory filing and approval. This is in stark contrast to operating outside the design space where changes in the process or raw materials require formal filings and approval from regulatory agencies [12]. This flexibility is incredibly advantageous since process improvements can take place during the production cycle and, as such, the operating space can be revised within the design space without needing approval from regulators. In this way, a historically conservative industry is encouraged to innovate and improve its production platforms by adopting new technologies as they emerge to enhance process monitoring without additional regulatory burden. This concept can be visualized by Figure 1. Here, the knowledge space is a non-design space that requires regulatory approval before being ready for human use. The design space is the approval process by the FDA while the control space is the process configuration of the biomanufacturing process. Approval of a design space is key since it gives the manufacturer the flexibility of changing certain process parameters without additional regulatory requirements.

In this review, we first present an overview of biotherapeutic production modes. We then discuss why key metabolite accumulation and substrate consumption need to be routinely monitored to generate the appropriate environment within the bioreactors to maximize protein yield. The main technological tools used for bioprocess monitoring are then presented and we also describe recent advances on how data driven, mechanistic, or hybrid models can be used in tandem with technological tools to indirectly estimate additional parameters.

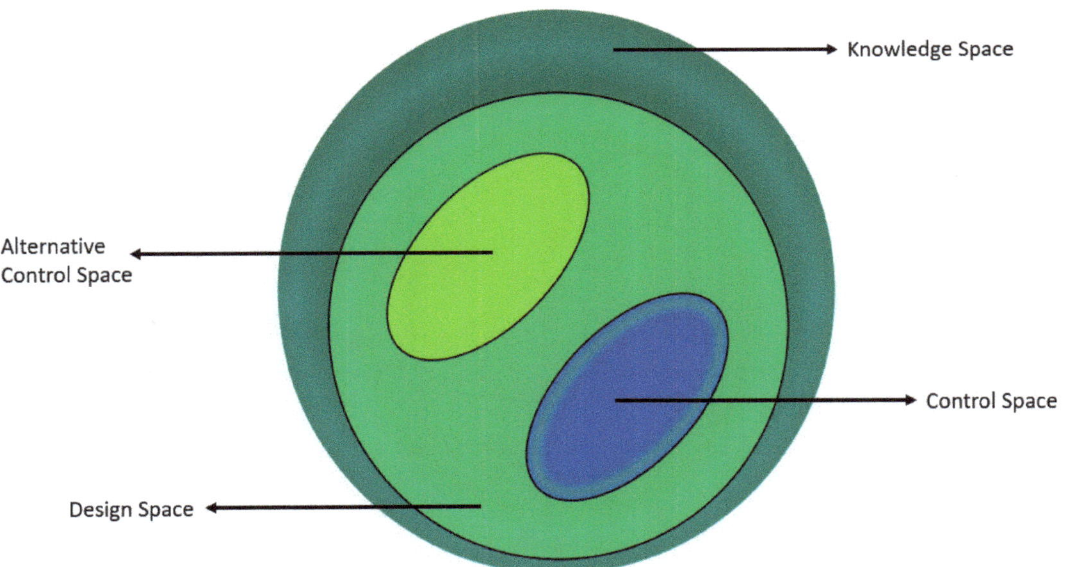

Figure 1. Knowledge, design, and control spaces. The knowledge space is a non-design space that requires regulatory approval. The design space is the pre-approved process by the regulatory body, while the control space is the process configuration of the biomanufacturing process.

2. Bioreactor Modes of Operation

In the production of biologics, there are several types of cell culture modes such as batch, fed-batch, concentrated fed-batch, and continuous (perfusion or chemostat). Batch production refers to the culturing of cells within a vessel that are grown with an initial known concentration of feed source and medium [13,14]. No further nutrition addition or removal is performed on the system. As the biomass within the vessel grows, the initial feed source begins to be depleted while metabolic waste and product are accumulated in the medium [13]. Even though this form of cultivation has drawbacks, such as limiting the maximum cellular density that can be achieved and limiting the culture run time due to accumulation of metabolic waste, it is a relatively simple arrangement that does not require complex control loops in order to manage subsequent feeding or removal of waste [15]. In a fed-batch operation, nutrients are fed continuously or periodically (bolus) to the system to supplement reactor contents and control overall substrate concentrations [13,15–19]. Constant measurement of relevant metabolic products and feed source concentrations is needed to have knowledge of the relevant feed additives that are critical for the cell culture. This mode is widely used in the industry since it is excellent for the production of non-growth-associated products as well as providing a strong alternative to complex continuous feeding regimens [13,15–19]. Additionally, it is used to control substrate concentrations since high levels can be inhibitory or can cause shifts in the metabolic pathways [14]. It is important that space is allowed in the system to permit medium addition. Two main methods exist for fed-batch: the constant feeding strategy and the constant substrate concentration strategy [14]. With the constant substrate concentration strategy, a constant growth rate can be initially maintained and the number of cells in the bioreactor will, thus, increase exponentially as a function of time. However, it also means that the system must be supplied exponentially with substrates. Even with concentrated feed, this can cause significant volume changes in the systems. Because of this, maintaining the feed rate constant is used as a viable alternative even though the system rapidly may become substrate limited [14]. Practically, industry often uses periodic feeding (bolus) due to its simplicity and high efficiency [14–17]. The batch and fed-batch modes are depicted in Figure 2.

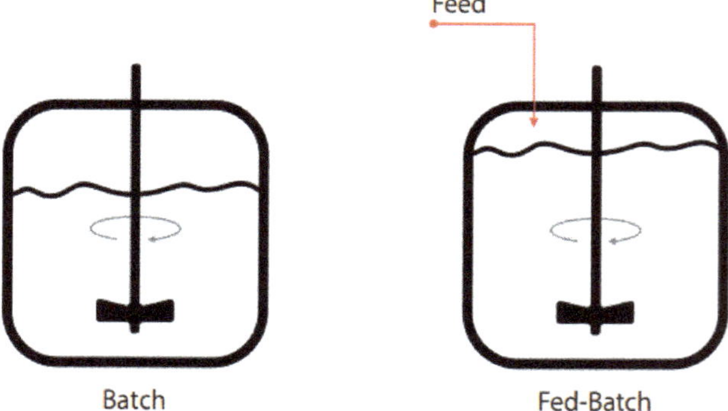

Figure 2. Batch and fed-batch bioreactor operations. During fed-batch operation, nutrients are fed continuously or periodically (bolus) to the system to control overall substrate concentration. Conversely, during a batch cultivation, no feed is added to the system nor is any medium extracted.

Given that the culture environment changes throughout the bioreactor run, because of cellular growth, substrate consumption and metabolite and product formation, continuous processes that replenish fresh nutrient medium while at the same time withdrawing the spent media from the system have been developed [14,15,18]. With these continuous processes, cell growth and product formation may be prolonged for longer periods, when compared to batch and fed-batch modes. For example, in a chemostat, the balance of feed addition and removal can be controlled so as to attain a steady state where nutrient, product, and cell concentrations are held constant [13,14,18]. An important characteristic of the chemostat is that a time-invariant growth environment is created and, thus, the net growth rate is equal to the dilution rate, which is determined by the flow rate into the vessel [14,18]. Consequently, the growth rate can be directly manipulated, making this a valuable tool to conduct kinetic studies. However, product dilution, resulting in large purification volumes, makes it generally unattractive for industrial biomanufacturing. In order to limit biomass loss in the outflow, cell recycling mechanisms can be employed [13,14,18]. The retention of cells can be achieved through the use of membranes, screens, or centrifuges. Perfusion systems have the advantage of removing toxic or inhibitory metabolic by-products that can be detrimental to either cellular growth or product formation [13,14,18]. Additionally, the protein of interest has a hydraulic residence time much shorter than the cells, reducing its exposition to varying culture conditions such as pH fluctuations or proteolytic enzymes. Cell concentrations in the range of 50–100 million cells/mL, which are comparatively much higher than batch or fed-batch modes, can be achieved [18]. Thus, high per unit volumetric productivity can be attained [14]. Importantly, given the continuous addition of nutrients and removal of toxic metabolic waste, perfusion cultures can maintain biomanufacturing operations for longer periods of time [13,14,18]. However, given the additional complexity of cell retention and recycling, as well as the constant addition of fresh medium, the increase in volumetric productivity may not be enough to merit the increase in operational costs and its implementation is largely dependent on the economic feasibility of the process [13,14]. It must also be noted that this mode of operation also adds burden on downstream processing given the large volumes of continuous fluids that must be handled. Recently, the concentrated fed-batch mode was developed; it is a hybrid system between perfusion and fed-batch [20]. Here, cell recycling is also used. However, an ultrafiltration module is employed and, thus, the protein product and the cells are recycled back into the reactor while still removing the spent media and waste by-products. Concentrated fed-batch systems can achieve high densities, above 100+ million cells/mL, and product yields of 25–30 g/L [20]. An important characteristic of this system is that the active protein of interest is retained within the reactor and, consequently, harvest day signifies the end of the fed-batch culture [20]. Because of this simplicity in harvesting and increases in both product yield and cell concentration, the concentrated fed-batch is being increasingly used within the biopharmaceutical industry, especially in companies that have well-established fed-batch facilities as it serves as a good nexus point to be able to begin implementing perfusion style manufacturing [20]. In Figure 3, a schematic of concentrated fed-batch and perfusion systems can be visualized. Such arrangements are the most prominent continuous systems in biomanufacturing.

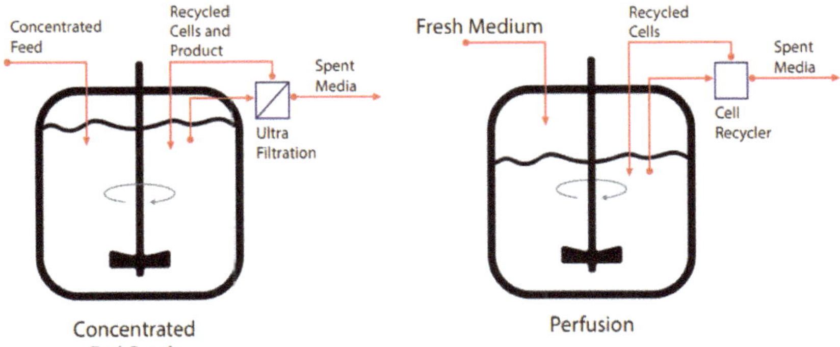

Figure 3. Concentrated fed-batch and perfusion modes of operation. Perfusion systems continuously replenish medium into the reactor while spent media are continually extracted. Cells are separated from the extracted media and recycled back into the stirred tank reactor. In a concentrated fed-batch system, concentrated feed formulations are added while extracting spent media. However, an ultrafiltration module is used to recycle back both cells and product into the system.

3. Mammalian Cell Metabolism in Culture

The metabolism of mammalian cells during biomanufacturing runs is known to vary depending on nutrient availability. Mammalian cells use substrates as carbon and nitrogen sources and their metabolism can be either mostly glycolytic or oxidative [21–23]. Through the glycolytic pathway, glucose is consumed at a high rate and only two adenosine triphosphate (ATP) are produced with lactate being generated as a by-product (Figure 4). Lactate is a key metabolite resulting from the conversion of pyruvate by the lactate dehydrogenase (LDH) enzyme even in the presence of oxygen. The conversion of pyruvate to lactate limits the full oxidation of glucose to carbon dioxide and water in aerobic conditions [21–23] and causes a carbon flux away from the tricarboxylic acid (TCA) cycle. The corresponding flux lessens energy production and instead allows the carbon backbones to be used for biomass formation. This is a phenomenon commonly observed in cancer cells and it is denominated the Warburg effect, aerobic glycolysis, or overflow metabolism [22,24,25]. In cancerous cells, the rate of glucose uptake in rapidly growing cells is generally many orders of magnitude larger when compared to cells that make up non-growing differentiated tissue [24]. Given that mammalian cells used in bioprocess are derived from immortalized cell lines, it is logical to see metabolic similitudes with cancerous cells.

The two ATPs formed by this glycolytic pathway are in stark contrast to the 36 ATPs that are generated through the oxidative pathway [21–23]. Given the low energy efficiency of the glycolytic pathway, mammalian cells in bioprocesses are also known to use oxidative phosphorylation for the production of ATP and, thus, for their energy requirements. In the TCA cycle, pyruvate, which is generated at the end of the glycolytic pathway, is used as the primary substrate. Amino acid catabolism is another substrate source for the TCA cycle, for example, glutamine is readily catabolized as a source of energy to form glutamate and ammonia [21–23]. Changes from the glycolytic pathway to oxidative pathway can vary throughout culture run and can even be controlled through mediating process conditions. For instance, cells grown in low-glucose environments are able to upregulate the oxidative pathway and, thus, maximize ATP synthesis [22].

The metabolism of cells during the production process is regarded as inefficient and suboptimal because the nutrients supplied in the media and/or feeds at given concentrations can lead to accumulation of toxic by-products, intermediates, and metabolites [21–23]. Substrates are not fully used for production of recombinant proteins or biomass. For instance, 35 to 70 % of the glucose consumed can be diverted into the formation of by-products [22]. This hints at the existence of metabolic bottlenecks in relevant pathways as well as inefficient flux distribution. These metabolic inefficiencies that lead to compound accumulation can cause decreases in cell growth and product titer, as well as alter the product glycosylation profile [21–23]. Lactate is one of the main toxic metabolites that is accumulated in cell culture processes [21,22,26,27]. It has been reported to inhibit cell growth and to induce apoptosis, as well as to reduce the productivity of recombinant protein production because of osmolality increase and changes in the pH [21]. Interestingly, within a Chinese Hamster Ovary (CHO) cell culture process, two distinct phases regarding lactate metabolism have been described: (1) lactate production at the start of the culture as a consequence of glucose uptake through the glycolytic pathway and (2) lactate consumption following rapid cell growth [21,22,26,27]. It is worth noting that concomitant consumption of glucose and lactate has also been observed [22]. There are two important LDH genes: LDHA and LDHB. The LDHB gene encodes the LDH-H protein while the LDHA gene encodes the LDH-M protein; together they make the important subunits of the LDH enzyme [21]. Given that the LDH enzyme is tetrameric, it can be found in five different isoforms, which differ in the ratio of the subunits (LDH-H and LDH-M). The LDH-M has a higher affinity for pyruvate; thus, isoenzymes with a majority of this subunit will catalyze the reaction of pyruvate to lactate [21]. Conversely, the LDH-H protein has a higher affinity for lactate and isoenzymes with the majority of this subunit catalyzing the reaction of lactate to pyruvate. A link has been established between lactate consumption and increased recombinant protein productivity [22,28]. The lactate consumption phase is observed in cells that are in a stationary phase of their growth. In this way, the consumption of lactate reduces its own accumulation and thus limits the negative effects on cell behavior [21,22,26,27]. It is thought that the lactate shift is originated by an upregulation, which causes lactate to be converted into pyruvate; the latter is then incorporated into either the TCA cycle or to monocarboxylate transporters (MCT) through which lactate can enter and exit the cell in co-transport with H+ ions [21].

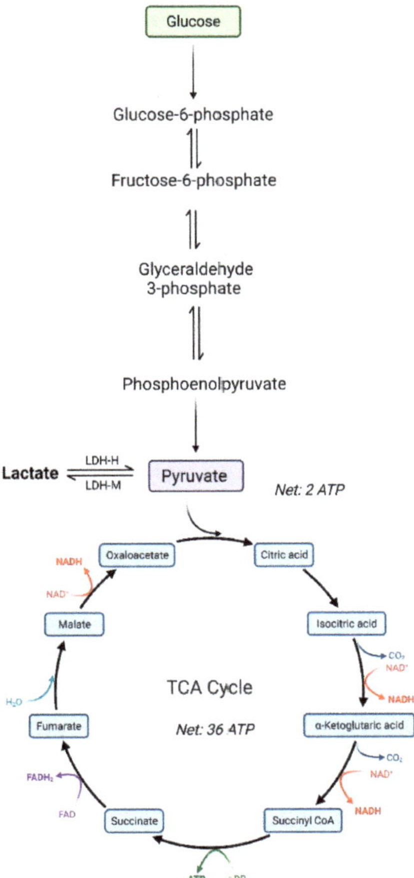

Figure 4. Schematic of the glycolysis pathway and TCA cycle. Two ATPs are formed in the glycolytic pathway while 36 ATPs are generated through the oxidative pathway.

Important metabolic intermediates that accumulate during the bioprocess include citrate, succinate, fumarate, and malate [22]. This particularly happens during oxidative metabolism and has been observed to occur after the addition of feed, which indicates the existences of bottlenecks [21–23]. Amino acids also play an important role in mammalian cell culture and are either supplied in the medium and feed or some can be produced through biosynthetic pathways by the cells. The amino acids support cell growth and are used for the synthesis of protein. Through the catabolism of amino acids, the formation of TCA cycle intermediates can be used to support energy production [21–23]. However, if the amino acids are supplied in excess, accumulation of the TCA cycle intermediates can lead to the formation of ammonium thanks to a series of transamination and deamination reactions that terminate in the release of an ammonium ion [22]. Ammonia is mainly formed as a result of glutamine breakdown, but other amino acids such as serine and threonine can also produce ammonia through direct deamination [22]. Accumulation of ammonia can negatively impact cell growth and recombinant protein productivity [27,29,30]. One hypothesis for this is that, as the ammonia concentration increases, it alters the electrochemical gradient and acidifies the intracellular compartments. Thus, normal enzymatic activity is impaired and apoptosis is induced [22,27,31]. Defined amino acid concentrations at different phases of the bioprocess have been correlated to cellular inhibition and apoptosis. Asparagine depletion has been determined to have a negative effect on cell growth while the production of alanine has been determined to inhibit the TCA cycle as well as contributing towards ammonium accumulation [32]. Alanine has been shown to accumulate during the first days of culture and then will either continue to build up or be consumed in a similar way as observed with lactate (our internal data not shown). Excess lysine has been observed to be associated with cell death. The catabolism of phenylalanine, tyrosine, tryptophan, methionine, leucine, serine, threonine, and glycine has been determined to produce intermediates that inhibit cellular growth [22,33,34]. The accumulation of intermediates is a direct consequence of non-optimally regulated pathways and, as a result, formulation of defined amino acid concentrations in the media and feed can improve cell growth and recombinant protein yields [22,34]. During the production phase, the metabolism is shifted towards the TCA cycle and, thus, the cells are subjected to higher levels of oxidative stress [22]. To counteract this, glutathione is biosynthesized de novo and interacts with reactive oxygen species, which in turn limits the toxic effect of the oxidative stress [22]. Glutathione has also been determined to function as a marker for productivity given that its presence indicates adequate recombinant protein production [35]. Recently, it has been determined that the accumulation of phenylalanine-tyrosine by-products, which are caused by secondary branching pathways when the key enzymes in the main catabolic pathway are under expressed, can also lead to growth inhibition [36].

Cellular respiration is another important component of metabolism given that oxygen requirements and carbon dioxide production are closely linked to cell growth. Thus, oxygen in the culture can be understood as a substrate while carbon dioxide can be conceptualized as a metabolic by-product. Oxygen must be monitored and routinely controlled as it is a crucial nutrient for aerobic mammalian cell survival. Additionally, it has been observed that dissolved oxygen can impact glycosylation profiles, which are key in determining the protein pharmacodynamics [37]. Very high dissolved oxygen (DO) levels can cause the formation of superoxides or peroxides, which have a detrimental effect on the cell membrane; thus, finding the optimal operating range for dissolved oxygen (DO) is instrumental [37]. Conversely, dissolved CO_2 is also quickly becoming recognized as a critical process parameter (CPP) [38,39]. This is partially because high CO_2 concentrations have been found to impact glycosylation profiles. Cell growth and protein productivity can also be significantly reduced if dissolved CO_2 values exceed a certain threshold (68 mmHg at bench scale to 179 mmHg at pilot scale) [38]. This is thought to be due to detrimental effects on internal pH and cellular metabolism [38,39]. This effect can be counteracted with the addition of base into the medium to maintain a constant pH. However, this can generate its own set of problems as progressive increases in osmolality can also negatively impact cell culture performance [38,39]. It has been observed that at higher dissolved carbon dioxide concentrations, in insect cells cultures, reduced glutamine consumption occurs [40]. This is interesting because even though low glutamine consumption was observed, glucose consumption remained unchanged [40]. This could suggest that ammonia and lactate were probably produced from the metabolism of non-glutamine amino acids because the TCA cycle was not efficient [40]. Thus, dissolved carbon dioxide measurement can be used to directly correlate with other metabolic fluxes and consequently we can gain a deeper understanding of the cell culture process. Given that low pH by itself does not have the same impact on cell metabolism, CO_2 is an important parameter to control on its own and not only in conjunction with pH. The latter is generally accomplished through double-sided pH control loops [40].

4. Sensor Types and Characteristics

Sensors are commonly used to control and measure the aforementioned relevant metabolic parameters as well as to control pertinent process variables (Table 1). They can be used to directly measure the main metabolites (e.g., lactate and ammonia), detect changes in substrate (e.g., glucose), or measure metabolism indirectly through cellular respiration by detecting changes in gas composition (e.g., oxygen and carbon dioxide). Through the continuous measurement of these parameters, feeding strategies, process conditions, and scale-up procedures can be rationally established. This is important for maintaining an optimal environment in which the cells can grow, and it can also be used as a way to construct dynamic feeding strategies (feed on-demand), which are automatically triggered after it is determined through the sensor measurements that important nutrients are becoming limiting. Broadly speaking, sensors in the upstream monitoring of the bioprocess require the measurement of three different types of variables: (1) Physical variables such as temperature, stir speed, and foam level; (2) Biological variables such as cell count, product concentration, and cell metabolism [41]; and (3) Chemical variables such as nutrient concentration, pH, dissolved oxygen, and dissolved carbon dioxide concentrations [41].

Table 1. Measured process variables.

Variable Type	Bioprocess Parameter	Sensor Type
Physical	Temperature	Thermostat, thermistor
	Foam	Conductance
	Viscosity	Viscometer
	Pressure	Capacitance
	Stirring	Torque
Chemical	Oxygen	Optical, electrochemical
	pH	Electrochemical, optical
	Lactate	Spectroscopic, biochemical
	Glucose	Spectroscopic, biochemical
	Carbon Dioxide	Optical, electrochemical
Biological	Cell count (viable cell density, total cell density, viability, cell size, aggregation)	Microscopy, spectroscopic
	Protein	Spectroscopy
	Cellular morphology	Flow cytometry, spectroscopic
	Intermediate metabolites	Spectroscopy

These sensors vary in application depending on how they are connected to the bioprocess (Figure 5). If they share a direct interface with the culture, they are denominated in-line sensors [41–44]. These sensors are also referred to as in situ sensors and do not require any type of manual or autonomous sampling. If the sensor module lies in close proximity to the production process and manual or automatic sampling is required to analyze the predetermined variables, then the sensors are denominated at-line [41–44]. If the data are analyzed in a continuous fashion, the sensors are determined to be on-line sensors [41]. The continuity of the measurement depends on the response time of the signal and the flow rate of the sampling procedures, which must be small when compared to the dynamics of the process. Thus, if the data points generated from the sensor occur at spaced-out time intervals, the sensor is determined to be quasi on-line [41]. Alternatively, if a sample is required to be taken off the system and analyzed in the laboratory after proper pre-treatments (e.g., dilution, filtration, or digestion), the employed sensor is determined to be off-line [44]. These sampling events, whether at-line or off-line, need to maintain rigorous sterility standards to prevent contamination and protect the cell culture [44]. Given that the isolated sterile bioreactor compartment has to be opened and a sample has to be withdrawn, cell-free sampling is usually assured by sterile barriers, such as microporous filters [44]. For the purpose of process control, on-line measurements are desirable given the fact that the data can be readily used in feedback control loops that regulate the process [41,42]. For the sensor to be considered on-line, the sensed variable must be measured more frequently than it can change in the process. In the case of mammalian cell culture, metabolic rates are slower when compared to microbial cultures. Thus, it is expected that substrate concentrations or metabolic by-products are subjected to low variations within a 1-h period. Consequently, a sensor capable of detecting metabolite concentrations every 30 min can be considered to be on-line because enough time is given to the system to gather and act on the received data through the established control loop [42].

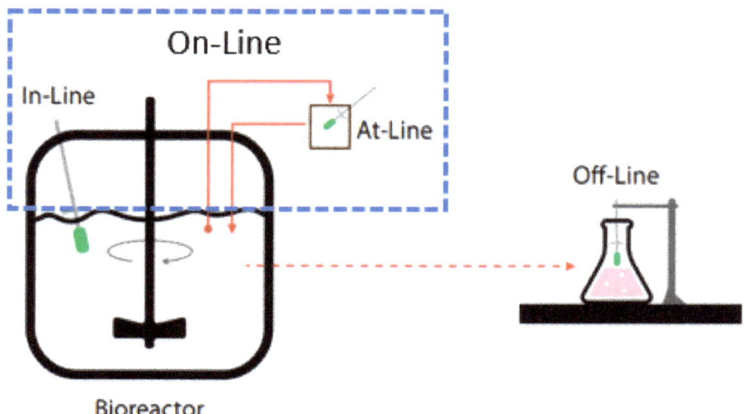

Figure 5. Different sensor types used in cell culture monitoring. In-line sensors are also referred to as in situ sensors because they are sterilized and placed within the reactor. At-line sensors require an alternative flow loop to realize measurements; this alternate flow is closed off from the environment to ensure sterility. Off-line sensors require sampling from the reactor and specialized lab equipment to analyze. Within this definition, both in-line and at-line sensors can be considered to be on-line.

These distinctions are important because, depending on the proximity of the sensor to the culture, varying degrees of sterility must be taken to protect the bioprocess as well as considerations as to the true representation of the data with respect to the bioprocess. For example, in-line sensors that are in direct contact with the bioprocess must be subjected to varying sterilization techniques. If steam heat sterilization is used, as in the case of glass or stainless steel vessels, the sensor must be resistant to high temperatures and varying pressures. Conversely, in the case of single-use reactors, gamma radiation is used and, thus, the sensor must remain in working condition after prolonged exposure [41–44]. Therefore, the sensor must be manufactured with the fore knowledge of the sterilization procedures that will be employed. Additionally, since the sensor forms an interface with the bioprocess, the sensor must be resistant to fouling and it should not interfere in any way with the medium components, the cells, or the product. Off-line or at-line sensors run the risk of not being completely representative of the process, given that a small volume is sampled [41–44]. Table 2 provides a list of critical characteristics that should guide sensor selection.

Table 2. Important sensor characteristics [41,45–47].

Characteristic	Definition	Remarks
Selectivity	Ability to detect analyte of interest or a group of analytes.	One example of selectivity in a biosensor is the interaction of an antigen with the antibody.
Reproducibility	Capacity of the sensor to generate identical responses in separate experimental runs.	This is usually characterized by measuring variance, standard deviation, or coefficient of variation. This is important in bioprocess, given that manufacturing runs depend on specific sensors that are reused as in the case of stainless-steel or glass bioreactors.
Accuracy	Ability of the sensor to determine a mean value similar to the true value when the analyte is measured more than once.	It is generally expressed as a percentage of full-range output. If the accuracy of the sensor is high, the difference between the measured analyte value and the real analyte value is small.
Stability	Capacity of the sensor to produce an identical output for a constant input over a certain period of time.	It represents the degree of susceptibility that the sensor has to environmental disturbances. Over compounded time, such disturbances can generate a drift in output signals.
Sensitivity	Magnitude of output signal per unit change in the variable of interest. It is the relationship between the input physical signal and the output electrical signal.	The sensitivity can also be described as the Limit Of Detection (LOD) of the sensor, which is the concentration at which the mean output signal value is equal to two standard deviations. If a sensor possesses both high selectivity and high sensitivity, it is able to detect and quantify small concentrations of the analyte of interest in the presence of various substances.
Resolution	Smallest change in variable that is sufficient to elicit a response from the sensor.	This is key in metabolite monitoring where concentrations within the cell culture broth can be very low and, thus, differentiating slight changes of small concentrations is critical.
Linearity	Accuracy of the output response with respect to a straight line.	Non-linearity is an indication of deviation of the measurements from the curve of ideal measurement.
Response time	Speed of change in an output signal relative to a stepwise change of the input variable.	Response time should be small relative to the measured process dynamics given that long response times complicate efficient control of the process.
Robustness	Durability of the sensor when subjected to varying environmental conditions.	This is key in sensors that undergo sterilization and sensors that will be used on-line for long periods.

Figure 6 summarizes different sensor techniques available on the market together with various process parameters monitored online and offline.

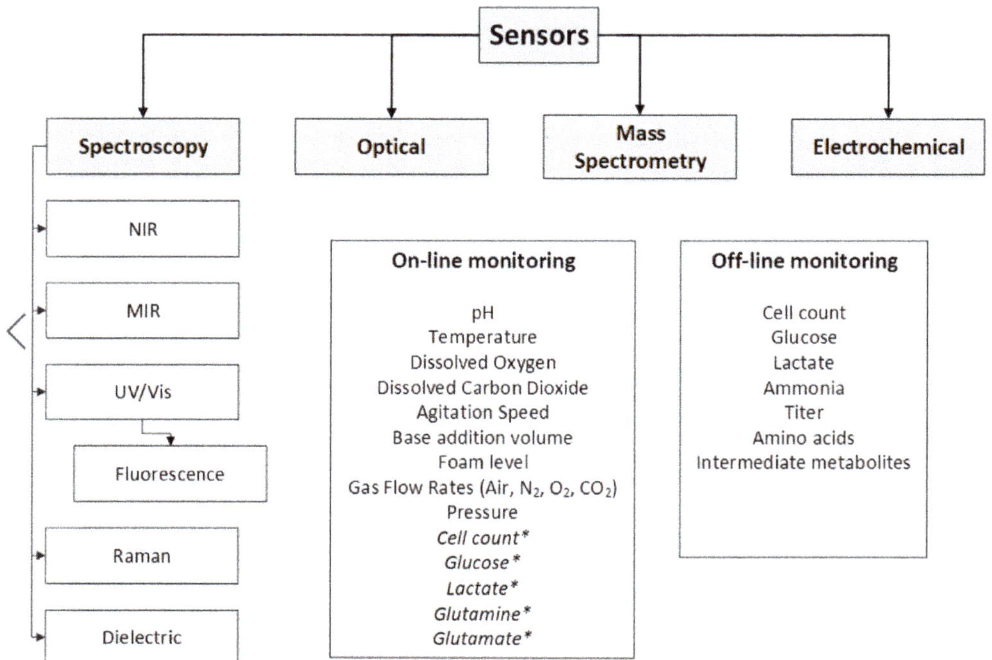

Figure 6. Sensor techniques and their application in process monitoring on-line and off-line. * Italic text shows the desirable online measurements that are currently estimated offline.

Table 3 provides a summary of selected studies demonstrating the use of the various sensing techniques that can provide direct or indirect assessment of culture performance. Single-use sensors have seen strong use with the surge of single-use bioreactors (SUB). These sensors are sterilized together with the SUB by γ-radiation [48,49]. These single-use sensors can be of a wide variety, such as electrochemical pH sensors, ion selective field effect transistors (ISFETs), optical CO_2, optical O_2, optical pH, and chemo/biosensors [48,49]. The last type of single-use sensors are of interest given that they can be used for the detection of metabolite concentrations [48,49].

Another interesting development is that of free floating wireless sensors, which can measure temperature, conductivity, pH, pressure, and turbidity [49]. Since the sensor is floating in the cultivation broth, its movement follows the fluid flow, thus giving a true representation of the concentration within the bioreactor [49].

Table 3. Overview of available sensors and techniques for process monitoring and control.

Technique	Sensing Attributes	References	Developer
NIR spectroscopy	Glucose, lactate, glutamine, and ammonia	[50–53]	Matrix F FT-NIR, Bruker, Billerica, MA, USA Fossanalytics, Hillerød, Denmark FossNIRSystems Inc., Silver Spring, MD, USA ABB Bomem FT-NIR spectrophotometer, Quebec City, QC, Canada Antaris II MX, Thermo Fisher Scientific, Madison, WI, USA Sartorius Stedim Biotech GmbH, Göttingen, Germany
MIR spectroscopy	Cell density, cell viability, lactate dehydrogenase (LDH), secreted antibody, glutamate, lactate, glucose, glutamine, ammonia.	[54,55]	Merck Millipore, Danvers, MA, USA MATRIX-MF, Bruker Optik GmbH, Ettlingen, Germany Mettler Toledo AutoChem, Inc., Columbia, SC, USA
UV-Vis spectroscopy	Cell density, viability, glutamine, glutamate, glucose, lactate.	[56–60]	J&M Analytik AG, Esslingen, Germany Thermo Fisher Scientific, Waltham, MA, USA
Fluorescence spectroscopy	Cell density, cell viability, recombinant protein, glucose, and ammonia concentrations.	[61–64]	J&M Analytik AG, Essingen, Germany LabX, Midland, ON, Canada Cary Eclipse CA, USA BioView, Delta Light and Optics, Denmark Horiba Jobin Yvon Fluoromax-4, Kyoto, Japan
Raman spectroscopy	Glycoprotein yield, Glucose, glutamine, lactate, ammonia, glutamate, cell density and viability	[65–74]	VALON Instruments Ltd., Belfast, United Kingdom Kaiser Optical Instruments, Ann Arbor, MI, USA. Perkin-Elmer, Waltham, MA, USA
Dielectric spectroscopy	Cell density and viability, viable cell volume	[75–82]	Aber instruments, Aberystwyth, United Kingdom Hamilton, NV, USA
* Optical sensors: O_2	Dissolved oxygen (DO)	[83–86]	Ocean Insight, Orlando, FL, USA Presens, Regensburg, Germany Mettler Toledo, Greifensee, Switzerland. Hamilton, NV, USA
* Optical sensors: pH	pH	[41,87]	Pyroscience, Aachen, Germany Presens, Regensburg, Germany Mettler Toledo, Greifensee, Switzerland Hamilton, NV, USA
* Optical sensors: CO_2	Dissolved CO_2	[38,88–90]	Presens, Regensburg, Germany Mettler Toledo, Greifensee, Switzerland Hamilton, NV, USA
Mass spectrometry	CO_2, O_2 Volatile organic compounds Aglycosylation, glycosylation, and glycation profiles	[91–103]	Q Exactive, Thermo Fisher Scientific, Winsford, UK Ionimed Analytik, Innsbruck, Austria Xevo G2-XS Q-TOF, Waters, Milford, CT, USA
Free-floating wireless sensors	Temperature, conductivity, pH, pressure, and turbidity	[49]	smartCAPS, smartINST, Lyon, France
Biosensors	Glucose, lactate, glutamate	[104,105]	C-CIT Sensors AG, Switzerland

* These technologies are available as single-use sensors.

Because the on-line monitoring technologies can generate a data point every few seconds, a systematic methodology for storing and handling the large quantity of generated data is key. While developing custom-made IT infrastructure that can handle and store the data is feasible, plug-and-play platforms already exist. For example, SynTQ® from Optimal Industrial Automation, SIPAT® from Siemens, BioPAT SIMCA® from Sartorius, and Unscrambler Process Pulse II® from Camo are some of the most popular platforms in the market [106]. These platforms do not only allow for the handling of data from analytical instruments, they can also realize aspects of data preprocessing, multivariate data analysis, and visualization [106].

5. Spectroscopy-Based Techniques

Sensing devices that are not in direct contact with the bioprocess interface are regarded as non-invasive sensors. One important example is spectroscopic sensors. Such sensors rely on the interaction of electromagnetic waves and the analyte of interest [107,108]. The electromagnetic waves interact through absorption, emission, or scattering. The wavelength that can be employed exists within a wide range such as ultraviolet–visible (UV/Vis), near-infrared (NIR), mid-infrared (MIR), far-infrared (FIR), Raman, terahertz, and nuclear magnetic resonance (NMR) [107,108].

5.1. Applications of Near-Infrared (NIR) and Mid-Infrared (MIR) Spectroscopy Techniques

The two types of infrared spectroscopy are often used differently. MIR is more sensitive when compared to NIR and can detect functional groups of molecules. However, NIR devices are known to be more stable against interference and are cheaper to implement. The IR light is able to incite specific vibrational modes in different molecules [41,107–110]. As such, each organic or inorganic compound has its own unique spectral signal. Most excitation of unique molecular vibrations exists within the MIR range while vibration combinations and overtones exist within the NIR region [41,107–110]. Because of the higher energy of the near-infrared region, the spectra are less defined and, thus, spectrometers with high signal-to-noise ratios are required. Conversely, MIR spectroscopy has high absorption capacity and well-defined peaks. The greater resolution of MIR spectroscopy allows it to be employed in the detection of components in aqueous solutions at low concentrations and, thus, has been applied to measure glucose, lactate, fructose, ammonia, acetic acid, and antibodies in bioprocess [41,107–110]. Given the large complexity of data gathered through NIR and MIR spectroscopy, multivariate analysis (MVA) is often used. For qualitative variance analysis, principal component analysis (PCA) is commonly employed [109,111]. Alternatively, quantitative analysis requires a reference dataset to calibrate a model that can be used to correlate signals with relevant process variables. For this purpose, artificial neural networks (ANN), partial least squares regression (PLS), and multiple least squares regression (MLS) are often used to correlate the absorption of NIR/MIR data to the analytical data [109,111].

NIR spectroscopy can be used for in situ measurement of bioprocess with an optical probe; the ex situ approach can also be realized by using a flow-through cell, by employing a reflectance probe on the glass wall of the reactor or by utilizing fiber-optic cables that allow the sensor to be used within the reactor. Given the much less defined spectra in NIR spectroscopy, it is more used as a qualitative monitoring of the bioprocess [41,107–110]. Importantly, it has been applied in the monitoring of glucose, lactate, ammonium, and biomass [109]. NIR has been found to be able to monitor seven different parameters in parallel and on-line, including osmolality, glucose concentration, product titer, packed cell volume (PCV), integrated viable packed cell volume (ivPCV), viable cell density (VCD), and integrated viable cell density (iVCD), by using PCA and PLS in order to relate off-line measurements with the spectral data that were acquired on-line [50]. When comparing NIR to MIR, it has additionally been determined that, although MIR has a higher accuracy regarding the prediction of single analytes, NIR is better at predicting concentration of multiple analytes [51]. This is because absorption coefficients and absorption bands are much lower and wider (respectively) in NIRs than they are in MIRs. Thus, analyte concentrations for glucose and lactate can be detected with higher accuracy and in lower concentrations with MIRs. However, due to the low penetration depth in MIR spectroscopy, total cell concentration cannot be measured directly while NIR can employ light-scattering effects to measure cell density and cellular viability. It was also determined that ammonia, glutamate, and glutamine could not be adequately detected by NIR or MIR techniques. It was concluded that NIR spectrometers are inherently more robust and better suited for production processes when compared to MIR spectrometers because they have the added benefit of predicting cellular viability parameters as well as the concentration of single analytes [51]. Low-cost MIR probes have been applied in the monitoring of cell viability, lactate dehydrogenase, secreted antibodies, and lactate and glutamate concentrations in an at-line arrangement [54]. Glutamate could be predicted with high accuracy, but antibody concentration could only be achieved with good results at concentrations above 0.4 mg/L. Additionally, lactate dehydrogenase (LDH) activity could not be accurately predicted in low-activity regimes. Lactate prediction was determined to be deficient while viability could be determined with an error of 8.8% at ranges between 20 and 95% [54]. Another application where MIR is gaining increased interest relates to the monitoring of product quality and impurity. This is commonly done with spectral acquisition (Bruker Matrix MF®, Mettler Toledo ReactIR®) [106]. Furthermore, NIR has been applied in raw material characterization so as to generate a qualitative analysis of cell culture media components through spectral fingerprinting. This is particularly useful in terms of diminishing lot-to-lot variability [106].

5.2. Applications of Ultraviolet-Visible (UV/Vis) Spectroscopy

Ultraviolet-Visible spectroscopy is a sensitive method that employs ultraviolet and visible light with wavelengths in the range of 200–780 nm [107,111,112]. The absorbance of UV/Vis light is restricted to molecular function groups, known as chromophores, whose electrons are excited. Thus, unique absorption spectra can be obtained for molecules with chromophore groups and the correlation between light absorption, light path of the sample, and concentration of the absorbing molecules can be realized through the Beer–Lambert law. With the absorption measurement, the concentration of the analyte of interest can be determined. Differentiation of proteins through UV spectra is difficult and quantification is realized after purification procedures [107,111,112]. However, a method has been reported that is capable of selective protein quantification in protein mixtures, which would bypass the need for chromatography or electrophoresis as purification steps, by using PCA for cluster analysis and based on spectral similarity [56]. Mid-UV (200–300 nm) absorption spectroscopy is used for protein quantification. This range offers an advantage given that there is low impact of water vibration at these wavelengths. Within this range, peptide bonds and a few amino acid residues are responsible for the absorption. The aromatic structures of phenylalanine, tyrosine, and tryptophan contribute to the mid-UV absorption [107,111,112]. Cysteine residues and peptide bonds also absorb mid-UV light, mainly below 260 nm. However, high-energy molecules such as saturated hydrocarbons and sugars cannot be detected through UV-Vis spectroscopy. UV spectroscopy is also used for biomass concentration estimation as a function of turbidity in the sample [113]. These applications correlate linearly at the start of cell culture where cell density is low. Correlation is poor in later stages of cell culture when optical density cannot be used to differentiate between viable and dead cells [112,114]. ANN have been used along with UV spectroscopy to develop off-line monitoring of glutamine, glutamate, and viable cell concentrations on the basis of spectra monitoring [57].

Fluorescence spectroscopy exists within the UV-Vis range and it is another relevant tool for bioprocess monitoring, given that a lot of biological components within the culture media have fluorescence properties including amino acids, enzymes, cofactors, and vitamins [115]. When a fluorescent compound absorbs a photon, it is transferred to a higher energy state. As the energy of the molecule drops to a lower energy state, it emits a photon at a different frequency than the one that was used for excitation. Thus, by analyzing the range of frequencies of emitted photons from the original excitation frequency, an emission spectrum can be developed. Historically, in situ fluorescence sensors were based on a single wavelength pair, thus limiting analysis to a single fluorophore [107,108]. Nicotinamide adenine dinucleotide (NAD) and nicotinamide adenine dinucleotide phosphate (NADP) are the most used fluorophores given that they are cofactors involved in several metabolic pathways. For instance, they can be employed as indirect measurements for biomass estimation [116]. However, several secondary effects disturb fluorescence analyses such as inner filter effects that occur when non-fluorescent compounds absorb the exciting radiation, cascade effects that occur when the emission of a fluorophore excites another, and quenching effects that cause a decrease in fluorescence intensity [115]. Recently, 2D fluorescence spectroscopy was developed, whereby several excitation and emission wavelengths are employed. More elements can be analyzed at the same time over the course of the bioprocess [116]. These spectrometers can be used in situ and non-invasively with fiber-optical probes just as NIR/MIR spectroscopic sensors [116]. When compared to MIR or NIR, fluorescence spectroscopy is better suited for monitoring bioprocesses that require the measurement of species at low concentrations [108].

In mammalian cells, none of the fluorescent amino acids (tryptophane, tyrosine, phenylalanine) is synthesized by the cells and must be added to the medium. During the cell culture, these amino acids are taken up by the cells and incorporated into proteins. Once this happens, a quenching effect is observed where the fluorescence is diminished by neighboring protonated acid groups such as aspartate and glutamate [117]. When a direct linear correlation is not possible, multivariate techniques may be employed. For example, 2D fluorometry has been applied in a Baby Hamster Kidney (BHK) cell culture for determination of viable cell count and recombinant protein production by using PCA and PLS regression [61]. A similar method was developed for CHO cell lines expressing glutamine synthetase (GS) where the authors captured data on multiple fluorophores present in animal cell culture bulks in a single scan [62,63]. Modelling of viable cell density and antibody titers was realized through PCA and PLS regression. The 2-D fluorescence spectrometry has been used along with multivariate data analysis to differentiate between viable, dead, and lysed cell populations in mammalian cell culture. This is of value given that, with standard methodologies, it can be problematic to differentiate between populations with high resolution and accuracy [64].

5.3. Applications of Raman Spectroscopy

Raman spectroscopy is centered on the detection of inelastic scattering of monochromatic light that occurs when incident light interacts with the molecules of a sample [65,107,111,112,118]. When the light interacts with the vibrational frequencies of the molecules, most of it scattered without a change in frequency. This is known as Rayleigh scattering. However, a small fraction of the scattered electromagnetic energy is shifted from its original wavelength, and this is known as Raman scattering. The wavelength shift between the original monochromatic light and the scattered light is linearly dependent on the chemical bonds that caused the Raman scattering in the first place [65,112]. Thus, the detected light can give information regarding vibrational and rotational characteristics of the molecule. Importantly, this can give both qualitative and quantitative information about the sample composition. Raman spectroscopy has been applied for in situ bioprocess monitoring through the use of fiber optics as the delivery and collection system. Raman spectra are not sensitive to water, which is advantageous in cell culture monitoring. One of the drawbacks for its implementation is that biological molecules fluoresce in the Raman spectra region, generating interference. Thus, the selection of an appropriate laser wavelength that can maximize Raman signal and minimize fluorescence is a critical parameter when implementing Raman spectroscopy for bioprocess monitoring [65,107,111,112,118].

This technology has been applied to real-time in-line monitoring of glucose, glutamine, glutamate, lactate, ammonia, and viable cell density by coupling the technique with PLS modelling [66]. This is a promising move towards process monitoring and control as, previously, Raman spectroscopy had been applied to off-line monitoring of nutrients/metabolites in supernatants [67]. Raman spectroscopy along with PLS was also used as an in situ monitoring technique of glucose, glutamine, lactate, ammonia, glutamate, and total cell density in a CHO cell fed-batch process [68]. This technique was proven to be transferable across scales as it was tested at 3-L and 15-L scales with similar results.

Raman spectra analysis has also been used in a study to test changes in developmental scales (3-L, 200-L) and clinical manufacturing scale (2000-L) [69]. It was determined that glucose, lactate, and osmolality could be adequately modelled regardless of scale, while viable cell density and total cell density could achieve accurate predictive models but some scale-dependent variations limited across-scale predictions. Raman spectroscopy has also been applied in the prediction of glycoprotein yield at every stage, from small scale up to the final 5000-L bioreactor of a CHO cell process, demonstrating accurate predictions with relative errors between 2.1% and 3.3% [70]. In a similar research endeavor, Raman spectroscopy spectra were employed to build generic PLS models capable of predicting glucose, lactate, glutamate, ammonia, viable cell concentration, and total cell concentration values in a CHO cell culture process at 5-L and 10-L scales [71]. These models were in accordance with the off-line measurement error. The built models were found to be independent of cell line, given that model calibration and model validation were done with different cell lines [71]. However, glutamate and product yield could not be properly monitored in the process and it is thought that more sensitive off-line methods and the inclusion of more data could improve the estimation [71]. Similar problems have also been found in other CHO cell lines where adequate glucose and lactate models are readily built but issues arise in the development of glutamine and product titer models that fail to be specific and accurate enough in validation tests [72]. Automated feeding strategies that automatically maintain glucose at a low set point in order to limit lactate accumulation have seen difficulties when developed with Raman spectroscopy given that the measurement error is generally around 0.3–0.5 g/L [73]. However, to get around this problem, closed loop control schemes that measured both lactate and glucose concentrations have been developed. With this strategy, when lactate concentrations in the culture exceeded a predetermined set point, glucose addition was stopped [73]. Conversely, glucose was automatically fed (up to a maximum desired set point) when lactate levels were beneath a predetermined set point. By successfully limiting lactate accumulation, this approach increased cell counts and viability when compared to historical fed-batch cultures with the same cell line. This led to an 84% increase in final titer, thus demonstrating that real-time monitoring of cultures with spectroscopic techniques, along with feedback control loops, can be utilized to improve the production process [73]. Given that some critical parameters, such as glucose concentration, do not need high-precision and accuracy measurements, maintaining glucose levels within a predefined range has been determined to be a good strategy. Consequently, applying Raman spectroscopy in the domain of feedback control in order to maintain the glucose concentration autonomously has been explored [74]. This strategy allowed for the production of a target protein in a glucose concentration range that was not possible to achieve under daily bolus feeding strategies. Interestingly, glycation profiles were observed to be diminished from 9% to 4%, demonstrating that product quality attributes could be controlled with appropriate feedback controls [74].

From the aforementioned studies, it is clear that spectroscopic techniques are not so much in competition with one another but must be viewed as complementary in nature. However, it is important to highlight that Raman spectroscopy has seen more applicability when compared to NIR, MIR, and UV-VIS, given its unique ability to measure various types of compounds of interest by analyzing its spectral signal.

5.4. Applications of Dielectric Spectroscopy

Dielectric spectroscopy is another widely used method to monitor relevant variables in bioprocesses. The technique centers around the measurement of the passive dielectric properties of cells within a conductive medium. This is done by detecting the permittivity, which is the measurement of polarization, and the conductance. The method can be used on-line with a sterilizable permittivity probe [75,119–121]. This is possible because the cell is encapsulated by a lipid layer that is not conductive, while the cytoplasm is a complex, highly conductive medium containing water, salts, proteins, nucleic acids, and organelles. When an electric field is applied to cells in a suspension medium, cellular polarization occurs because the intracellular and extracellular ions move towards the electrode with the opposite charge but are stopped by the lipid layer [75,119–121]. Consequently, cells that have an intact membrane exhibit capacitance behavior and, thus, permittivity values, while cells with compromised membrane structure are unable to polarize. Because of this, the technique can be employed to measure cell density in a culture. However, since not all cells have the same diameter, in reality, permittivity is a measurement of biovolume (or biomass). When this measurement is used to get an estimate of cell concentration, it is under the assumption that all cells have the same diameter. Alternatively, solids or fragments, which are not part of the biomass, are unable to polarize given that they lack a nonconductive lipid layer.

Interestingly, it was determined that changes in capacitance relative to frequency could be related to cellular morphology. Indeed, a PLS regression allowed estimating the median cell diameter with a measurement error of only 2% [76]. It was also found that nutrient availability can be monitored by the permittivity signal because, immediately after a feeding event, signal values were observed to suddenly increase, while declines in permittivity values were correlated to states of nutrient depletion [76]. When the cell radius remained constant, as is mostly the case in the exponential phase, good correlations between oxygen uptake rate (OUR) and permittivity values were obtained, indicating that metabolic activity could be at least partly assessed via dielectric spectroscopy [76]. It should be noted that linear models used in conjunction with dielectric spectroscopy data were only accurate in the early phase of cell culture, while they failed to predict viable cell density during the decline phase [75]. In contrast, multivariate approaches such as PLS or alternative modelling techniques like Cole–Cole models were shown to predict viable cell density with high accuracy throughout the whole culture process [75]. In insect cell cultures using the baculovirus expression system, dielectric spectroscopy was able to estimate cellular growth and determine an appropriate time of infection (TOI) for the production of β galactosidase [77]. Furthermore, dielectric spectroscopy allowed for intimate process tracking of the progress of infection. Given that cell infection is characterized by a significant increase in cell biovolume, the capacitance signal is able to capture this phenomenon. Additionally, it has also been proposed that on-line dielectric spectroscopy estimation of adherent cells in microcarriers was a more accurate alternative than using off-line protein estimation as it is usually performed [121].

Given the variation in cell size distribution during cell culture, biocapacitance data can decouple from viable cell density data towards the end of a fed-batch cycle [78]. However, by analyzing the complex scanning data with multivariate data analysis (MVA) and constructing non-linear models with PLS or OPLS (orthogonal partial least squares), prediction of viable cell density instead of viable cell volume can be accomplished [78]. An on-line multivariate model resulted in viable cell density estimations that were a better fit to the death phase of the fed-batch cycle. To test the robustness of the model for real-time estimation of viable cell density, the fed-batch culture was subjected to dilutions (30 vol%) of the culture broth at various stages in the process. It was determined that the dilution steps were detectable in the on-line signal and correlated with the off-line cell counts [78]. Work has been done towards the development of a single, universal model that can be easily transferred across scales or clones without negatively impacting viable cell density prediction, especially in the declining phase of the fed-batch culture [82]. This is important because PLS models of multivariate signals generally result in low errors (when compared to viable cell density measurements) when created off-line after the process is finalized [82]. However, the same model quickly becomes unreliable for estimating the viable cell concentration of a new process with different operating parameters or cell line [82].

On-line biocapacitance measurements can be used as a surrogate for cell growth estimation in order to dynamically adjust feed rates [80]. This is of great importance given that it simplifies its integration into feed algorithms. Thanks to cell growth estimation, constant recalculation of growth rates was done automatically and used as a parameter to determine bolus feeding frequency [80]. When compared to manual bolus feeding, the results demonstrated that the dynamic feeding strategy had equal or better performance in terms of maximum viable cell concentration and maximum titer and metabolite accumulation [80]. Dynamic feeding strategies that are coupled with feedback control can avoid overfeeding and underfeeding throughout a fed-batch process [80]. This technology has gained great interest within the biomanufacturing industry given that the signal is independent of scale, as proved by Biogen when comparing biocapacitance trends of cultures of production cell lines at 5-L, 200-L, 315-L, and 15,000-L scales [79]. Interestingly, given this consistency across scales, it was possible to automate seeding trains within the production floor [79]. Additionally, with the generated data, it is possible to develop dynamic feeding strategies that calculate and add a volume feed as needed, as opposed to calculating a feed volume daily and feeding that bolus. The resulting feed strategy was responsive to actual culture performance and this reduced the risk of process failures caused by underfeeding [79].

Additional Biogen studies found that integrated cumulative integral of cell growth can be directly related to integrated biocapacitance (IBC) [122]. A linear correlation between the IBC and total feed amount was found in the entire process and, thus, it was possible to directly use biocapacitance data to control feed addition [122]. Based on the hypothesis that biocapacitance-based auto feeding could mitigate underfeeding/overfeeding phenomena in fed-batch culture process, bioreactors were intentionally seeded at low seeding densities. The fixed feed strategy caused overfeeding, which consequently caused cell growth inhibition. Alternatively, the biocapacitance-based feeding strategy automatically reduced the feed accordingly to biomass need. This led to lower ammonium and lactate levels when compared to the fixed process that was intentionally seeded at lower cell densities. It also led to higher titers and higher product purities given that the fixed process affected glycation and trisulfide formation [122]. Consequently, it was determined that capacitance data can improve process robustness by providing consistency in both productivity and product quality regardless of process variation. This simplified application is encouraging as it does not require complex multivariate analysis models, which run the risk of being overfit for a certain process to develop dynamic feeding strategies in the manufacturing floor.

Alternative strategies include biocapacitance data and glucose measurements to determine glucose uptake rates, which can then be used to feed the cultures as needed. Thus, by setting up a strict, target-specific glucose consumption rate for the whole process runtime of a fed-batch, the culture can be fed optimally [81]. It was shown that feeding based on specific glucose consumption and not on glucose itself can improve lactic acid profiles.

6. Optical Sensing Techniques

Optical chemosensors, also known as optodes, work thanks to the interaction of an analyte and a matrix-embedded indicator that is immobilized at the sensor tip [41,87,118,123]. The indicator is illuminated by a diode through optical fiber, and a change in optical properties that are detected by a photodiode is correlated with the concentration of the analyte of interest. The change in optical properties that is directly correlated to the variable of interest can be photoluminescence intensity, absorption, or reflection. These sensor types can be used in situ in stainless-steel bioreactors through standard ports or in small-scale systems such as deep well plates and shake flasks through patches. This application is of interest in systems with low volume where in situ sensors are not possible or would directly impact the hydrodynamics of the system in question, such as the case of systems at the millimeter scale [118]. Optical sensors in patch form have also been applied extensively in single-use bioreactors to monitor dissolved oxygen, carbon dioxide, and pH, given their ease of use and disposability and because they can be readily sterilized by using gamma radiation [41,118]. They have several advantages over standard electrochemical sensors, namely, that no direct electrical contact between analyte and electronics is required [87]. Moreover, contrary to amperometry sensors, no analyte is consumed, thus there is no net change in the concentration of the variable of interest. One interesting example is the electrochemical oxygen sensors, which actively consume oxygen in the process of measurement, which may influence the measurement especially in miniaturized systems [124].

Dissolved oxygen (DO) is commonly measured using optical sensors through fluorescence quenching of an immobilized fluorescence dye by molecular oxygen [41,87,118]. The oxygen-sensitive indicator is an organometallic dye that is immobilized in an oxygen-permeable polymer matrix. These optical oxygen sensors can be readily miniaturized as opposed to Clark electrode sensors and, thus, measurements with high spatial resolution in small volumes are practicable [41,118]. A published study was able to compare optical oxygen sensors with respect to their electrochemical counterparts and determined that their Pearson correlation was of 98.7%, thus demonstrating that both sensors were in agreement with the measure values [125]. They also highlighted that the accuracy of the optical probes demonstrated an ability to detect parameters shifts that could impact cell growth, production kinetics, and protein quality in a significant way [125]. It has been postulated that the electrochemical oxygen sensors and optical oxygen sensors may be complementary rather than competitive, given that electrochemical sensors perform best at high oxygen concentrations, while optical sensors have an optimal sensitivity at low concentrations (<50% of air saturation) [124,126]. By employing the same mechanism as an optical dissolved oxygen sensor, disposable optical in-line glucose monitoring sensors have been developed [127]. Here, a PreSens oxygen sensor is coated with a cross-linked glucose oxidase layer. Thus, the oxygen partial pressure within the crosslinked enzyme layer is monitored non-invasively and its signal is inversely proportional to the glucose concentration within the sample [127]. The dynamic range of the biosensor could be tuned for specific purposes by covering the enzyme layer with different hydrophilic perforated membranes [127].

Optical carbon dioxide sensors work relatively like their electrochemical counterparts, whereby a change in pH of a bicarbonate buffer system due to carbon dioxide presence is measured [41,118,124]. In this buffer system, fluorometric or colorimetric pH sensitive indicators are added, and the system is isolated from the bioprocess broth by a carbon dioxide permeable membrane. From the pH bicarbonate system equilibrium, it is possible to calculate the carbon dioxide partial pressure from a change in pH in the medium. The response time is within the range of minutes given that it is diffusion dependent. Frequent buffer solution change is required because the optical measurement is dependent on ionic strength of the buffer [41,118,124]. These sensors can be used to study dissolved carbon dioxide concentrations within the media to estimate the carbon evolution rate (CER), which can serve as a proxy for cellular respiration estimation. Optical pH sensitive sensors function in a very similar way to carbon dioxide sensors given that a change of pH is the variable of interest. They can be constructed based on absorbance or on fluorescent dyes, which are covalently immobilized on cellulose matrixes [41,87,118]. Optical pH and DO sensors have become standard equipment for single-use vessels given that they can monitor non-invasively the bioreactor. Importantly, these sensors are connected to a readout unit through a reusable fiber optic [49].

7. Mass Spectrometry Techniques

Mass spectrometry can be a powerful analytical tool given its high specificity, selectivity, sensitivity, dynamic range, and resolution, as well as mass accuracy [91]. Thanks to this, on-line application of mass spectrometry can contribute to the analysis of numerous components as well as various attributes of heterogeneous biological compounds. The most important component in mass spectrometry is the generation of a high vacuum that is below 10^{-5} mbar [91]. The samples are commonly introduced to the system through thermospray, electrospray, or direct liquid inlet. The samples are consequently ionized through electron impact, chemical ionization, or desorption ionization. The data gathered by mass spectrometers are quantitative, can be assigned to specific compounds, and, when analyzed with multivariate statistical tools, they can be used to build models that predict variables of interest.

On-line mass spectrometry is readily applied for the analysis of gas phase samples from bioreactor exhaust gas. It has been used to detect oxygen uptake rate and carbon dioxide production rates from cell culture runs in 10-L bioreactors. It was possible to correlate oxygen uptake rate to viable cell growth throughout the course of the bioprocess [92]. Additionally, it was also determined that changes in the oxygen uptake rate during the cell culture indicated occurrences of limitation of nutrients within the medium. Thus, an on-line mass spectrometry gas analyzer can be useful in timing important events such as splits or harvest time [93]. Interestingly, mass spectrometry gas analysis has been realized for fed-batch cultivation of mammalian cell cultures in 5-L and 50-L vessels to measure its application in scale-up systems. Correlation between viable cell concentration and oxygen concentration of the inlet gas into the bioreactor was high, irrespective of scale for a CHO-GS cell line that expressed chimeric IgG4 monoclonal antibodies [94]. Additionally, the oxygen mass transfer coefficient ($k_L a$) could also be identified throughout the culture with the mass spectrometer, and an impact of antifoam on $k_L a$ was found [94]. The respiratory quotient (RQ), defined as carbon dioxide evolution rate (CER)/oxygen uptake rate (OUR), was also estimated and a distinct correlation between RQ and the metabolic state of the cell culture was established; when a cell culture was determined to be in the lactate production phase, the average RQ was above 1. Conversely, when the culture was determined to be in a lactate consumption phase, the average RQ was below 1 [94,95].

Liquid chromatography mass spectrometry (LC-MS) has been used for the monitoring of glycan profiles and charge variants as well as purity. Given its multi-attribute monitoring, it has the potential to replace current electrophoretic and chromatographic tools that are used in quality control [96]. Another important technique for biological mass spectrometry is matrix-assisted laser desorption/ionization time-of-flight mass spectrometry, which has been used to determine process consistency and suitability of the cell line used for production [92]. When mass spectrometry is coupled with at-line liquid chromatography, simultaneous analysis of glycosylation patterns, heavy chain/light chain dimers, and C-terminal lysine residues is possible at the time of harvest [92]. Liquid phase measurement can be realized through membrane-inlet mass spectrometry because it allows the analytes to be moved from a complex aqueous solution to the ionic source thanks to a semi-permeable membrane [91]. This is specially used for the monitoring of metabolites that are released by the cells during the culture process, particularly volatile or semi-volatile organic compounds. Volatile organic compounds, within the context of cell culture, participate in a variety of biological functions such as growth inhibiting/promoting agents or cellular communication between cells [97]. Proton transfer reaction mass spectrometry (PTR-MS) was recently applied to monitor a recombinant CHO cell culture process [98]. In total, eight volatile organic compounds that showed high relevance to cultivation conditions could be identified: Methanol; Acetaldehyde; Methanethiol; Ethanthiol; Isoprene; Ester organic acid; α, γ-Butyrolacton, pentanal, 2-pentanon, and 3-methyl-3-buten-1-on; and 4-Methylpentan-2-on and 3-methylpentan-2-on. Among these, methanethiol could be directly correlated to metabolic shifts or nutrient limitation. In conjunction with PLS modelling, this technique was able to predict physiological cell culture parameters such as specific glutamine uptake rate and viable cell density.

Because of its multi-utility, this tool has also gained popularity in bioprocess development given that it can be utilized as a platform quality control method during the establishment of a manufacturing process [99]. With LC-MS, it was possible to demonstrate correlation between specific process parameters (pH, DO, glucose, temperature, seeding density) and the levels of glycosylated and glycated species [100]. LC-MS could thus be incorporated into an automated platform capable of monitoring, in real time, quality attribute outcomes for feedback control [100]. A multi-attribute method using non-reduced peptide mapping with quadrupole Dalton (QDa) detection has been reported in the literature [101]. QDa is a single quadrupole mass detector with an electrospray ionization source [101]. As opposed to other mass spectrometers, it is cost effective and relatively easy to maintain. The method was tested in upstream and downstream process development in order to monitor fucosylation, deamidation, and glycosylation in the Fc region, as well as disulfide bond-related modifications, such as trisulfide, thioether, free thiols, and cysteinylation [101]. The monitoring of glycosylation profiles of monoclonal antibody in cell culture samples through microfluidic chips has also been reported in the literature [102]. The method characterizes the sample through charge-based separation using microfluidic capillary electrophoresis and high-resolution mass spectrometry. It is suggested that this method can detect undesired shifts in product quality and, thus, has potential for at-line and in-line cell culture monitoring [102]. Thus, even though spectroscopic techniques can determine concentration of analytes in a cell culture process, they lack the sensitivity and specificity to measure post-translational modifications such as deamination and oxidation [103]. Consequently, mass spectrometry has shown great promise in being able to measure increasingly important product quality attributes. This is clear when looking at therapeutic proteins' license applications between 2000–2015. Of 80 therapeutic proteins, 79 employed mass spectrometry in their workflows to determine protein and purity characterization [103].

8. Electrochemical Sensing Techniques

This family of sensors can detect variations in electrical properties or charged species through chemical reactions. These sensors can be classified as potentiometric, conductometric, voltametric, and amperometric. Conductometric sensors measure variations in conductance while potentiometric sensors measure differences in electrical potential with respect to a reference electrode. Voltametric sensors measure changes in charge transport when the applied potential is varied. Importantly, amperometric sensors rely on the same principle but measure changes in charge transport while the potential is kept constant [118,128]. The standard electrochemical sensors routinely used in bioprocessing are pH and dissolved oxygen probes. These parameters are important to measure and control to ensure that cells remain in a favorable environment. The pH must be controlled to a range that is optimal for a specific cell line while dissolved oxygen must be maintained, usually within a range of 20–60%, to allow for cellular respiration. Process-induced shifts such as temperature (growth arrest) or pH shifts (lactate consumption trigger) have been explored in optimizing the productivity of a bioprocess; thus, the reliance of accurate in-line sensors is important [21,129]. More sophisticated sensors that use electrochemical enzymatic arrangements have been developed to monitor glucose and lactate on-line [130]. These sensors are advantageous in that they diminish the need for sampling and increase the real-time knowledge within the reactors regarding substrate consumption and accumulation of lactate. The drawback, however, is that these sensors are usable for up to 21 days, meaning that they can only be used for the fed-batch mode and not in perfusion reactors where the process can be maintained for longer [130]. Electrochemical single-use biosensors for on-line measurement of glucose and glutamate are also available. These sensors employ enzymatic oxidation processes that direct electron transfer from the measured substrate to an electrode. Such sensors are delivered ready to use and can be integrated to shake flask or disposable bioreactors [49].

In the case of pH sensors, potentiometric sensors are generally used while, for dissolved oxygen sensors, Clark electrodes are considered the standard. Of importance, Clark electrodes are amperometric in nature [41]. However, the drawback is that oxygen is consumed during the measurement, which means that the utility of such probes is limited in situations where oxygen can be depleted, such as in small-scale models. Additionally, dissolved carbon dioxide can also be measured by using Severinghaus electrodes [118,128]. Given that dissolved CO_2 has been determined to have a significant impact on quantity and quality of final product titer, tight control of dissolved CO_2 has been hypothesized to benefit fed-batch cell culture productivity. In one investigation, two equivalent CHO fed-batch cultures were realized [88], one with CO_2 tightly controlled at 10% and one where the CO_2 was let freely to accumulate up to 20%. The tightly CO_2-controlled cell culture resulted in a longer productive phase and higher protein yield. Interestingly, the pH between both reactors was equivalent [88]. Monitoring and controlling CO_2 with dedicated feedback control loops that have optimized sparging/CO_2 stripping strategies could benefit the overall cell culture process. It has been reported that batch-to-batch reproducibility can also be increased if pH control is managed without base addition. Because of this, strategies towards pH regulation without base addition (only using gas sparging) has been tested across varying scales with great success [38].

Dissolved carbon dioxide probes have also proved to be extremely useful for the scale-up of bioreactors. For example, strategies focused around the k_La ratio of k_La (O_2)/k_La (CO_2) have been proposed [90]. This scale-up criterion has the benefit that CO_2 does not tend to accumulate in larger vessels. When this ratio is kept constant, it is highly likely that the dissolved CO_2 profile during cultivation will be similar across different bioreactor scales, hence helping to maintain product and process consistency. This strategy has been suggested to be especially useful for technology transfer of large stainless steel tanks or single-use reactors, where accurate knowledge regarding the geometry of the vessel may be lacking and, thus, the standard Power/Volume (P/V) scale-up strategy may be deficient [90]. During scale-up procedures, dissolved CO_2 probes are also useful in determining removal rates of dissolved carbon dioxide in large reactors. This can help to assess the impact of sparging and headspace purging strategies. For example, it was found that headspace air flow rates do not have a significant impact on carbon dioxide removal, while sparging and varying specific power inputs were determined to be more effective for stripping [89].

DO sensor can be employed in the estimation of oxygen uptake rate through the dynamic method, which is inherently discrete given that it requires the DO to fluctuate within a range of 70–20% after turning off aeration. The decrease in DO can then be directly related to the amount of oxygen that is being consumed by the cells. With this method, it has been observed that the oxygen uptake rate (OUR) is directly proportional to viable cell concentration during the exponential growth phase [83]. Importantly, the end of the exponential growth phase also correlated with OUR measurements. By estimating cell concentration in the bioreactor through the OUR, feeding of substrates could be predicted. This requires the assumption that, during the exponential growth phase, the specific oxygen consumption is constant [83]. This feeding control strategy resulted in cell growth maintained at the exponential phase. It was found that the glucose concentration in the bioreactor was kept between 0.9 and 1.2 mM, indicating that the control strategy could maintain a desired set point. The OUR estimation was able to detect the start of the cell death phase just before maximum cell concentration was achieved, which served as an early warning in order to avoid glucose accumulation from this moment on [83]. Alternatively, given the cumbersome nature of OUR estimation through the dynamic method, oxygen transfer rate (OTR) measurements can also be utilized to indirectly measure metabolic activity [84]. With this measurement, it was possible to relate cumulative glucose consumption with cumulative OTR. Based on this correlation, it was possible to generate an on-line prediction of glucose that can be incorporated into a control algorithm that manipulates the glucose feed rate [84]. It was then determined that the advanced process control strategy could adequately maintain glucose concentration at an adjustable set point. In a similar study, a DO control system was used to provide a measure of the culture gas phase partial pressure of oxygen to calculate OUR from an oxygen balance in the liquid phase and relating it to the head space with Henry's constant [85]. It was determined that the OUR could predict viable cell density in uninfected growing insect cell cultures. Interestingly, it was also determined that the OUR could follow the progress of baculovirus infection and that it could pinpoint the onset of the death phase of infected cell cultures [85]. This is key given that the best time for product harvest occurs within a relatively narrow interval imposed by cell lysis at the end of the infection cycle [85]. Consequently, determining the optimal time of product harvest could ensure better product yields [85]. Researchers have proposed adequate OUR estimation through OTR measurements by characterizing $k_L a$ values throughout the cell culture cycle [86]. Thanks to a two-segment linear model, it was determined that the OUR can be directly associated with viable biomass of the system [86]. This correlation was done directly with capacitance measurements given that capacitance data can be correlated with biomass. The segmented model was necessary due to a metabolic transition in which the specific consumption of oxygen changed [86]. The metabolic shifts could also be observed in the system when OUR and capacitance measurements were analyzed together given that changes in specific oxygen consumption could be observed [86].

It is clear that even the industry standard sensors, which are employed to control the environment within the bioreactor, can be used to gather more in-depth information about the metabolic activity or respiratory profile of the cells. This information can also be applied for scale-up purposes or feedback control algorithms, which in turn will optimize cell culture processes. Thus, sensors can be used beyond controlling the environment of the reactor to maintain physiologically relevant conditions in the interest of maximizing the production process.

9. Soft Sensors for Cell Culture Monitoring

A soft sensor is a term used to describe an approach that employs hardware devices and software-implemented models to gather new information about the process that would otherwise be impossible to derive exclusively with hardware sensor measurements [131–134]. In essence, these novel arrangements are employed with the purpose of using easily accessible on-line data to infer quantitative information about process variables that are difficult to measure directly or can only be measured at low sampling frequency [135]. Soft sensors can, thus, become useful for both monitoring and control applications in the bioprocessing industry if they are demonstrated to be robust and easy to implement [131–134]. In theory, a soft sensor should result in reduced need for extensive operational surveillance and educed maintenance work and should increase the interpretability of the results given the capacity of the models to relate various key variables. Because of this, soft sensors are perfect candidates for the PAT initiative and to contribute towards automated control [135]. Broadly speaking, soft sensors can be split into three global categories: data-driven sensors, model-driven sensors, and hybrid models [116].

9.1. Data-Driven Soft Sensors

Data-driven soft sensors employ common chemometric techniques such as PLS, PCA, and other complex non-linear regressions such as ANN and fuzzy logic [131–134]. ANN can be used as multi-input, single-output systems or multi-input, multi-output systems. Importantly, fuzzy logic sets are based on general rules that have also been shown to be capable of describing unknown state variables from known measurements [131–134]. This is particularly useful in mammalian cell cultures where a lot of the interactions between metabolism and process conditions remain unknown or highly cell line specific. Another widely used method is the PLS regression that is frequently applied in soft sensors [116]. This is notably the case with mass and optical spectral data, which are used as inputs to PLS or ANN models linked to outputs such as media analyte concentrations, cell count, cell viability, or expressed proteins [131–134]. Thus, it must be stressed that, when these methods are applied separately or in combination, they can predict critical process parameters that are not immediately available through the spectral signals or multi-sensor data but arise from the deconvolution of the datasets. Because of this, data-driven models do not provide further mechanistic understanding of the physical and biological processes and they require extensive calibration within operational ranges to make the correlation valid [116].

9.2. Model-Driven Soft Sensors

Model-driven sensors involve mechanistic models that are based on engineering principles and biological insights, such as mass or energy balances that provide an understanding of the transformation processes in the organism [116]. These models can incorporate culture conditions such as media composition and/or culture performance (cell growth, production yield) in order to build explicatory models. As such, these models exploit existing knowledge with kinetic equations to capture dynamic changes of important variables [131–134]. These types of soft sensors are generally built by incorporating reaction kinetics, transport phenomena, and thermodynamic constraints into the model [116]. Such models must be accompanied by parameter identification, uncertainty analysis, and sensitivity analysis to validate the model. Mechanistic models require extensive experimental data to be verified. However, if the model is reproducible and reliable, it can provide biologically interpretable information and simultaneously proven and increased understanding of the production process [116]. Model-driven sensors can be split into steady-state and dynamic models. The steady-state models are developed from mass and component balances or from mass and heat transfer laws, while dynamic models employ dynamic balances along with kinetic assumptions to describe rate expressions as functions of the state variables [136]. Flux balance analysis (FBA) and metabolic flux analysis (MFA) are two stoichiometric-based methods commonly employed to characterize cell metabolism and estimate intracellular fluxes by using extracellular metabolite consumption or production rates as constraints. Since quasi-steady state for intracellular metabolites is a critical assumption applied in stoichiometric-based models, these approaches are static in nature [136]. Kinetic models are generally expressed as a series of ordinary differential equations (ODEs) and consequently describe dynamic changes in metabolite concentrations, cell density, and product formation by describing its rate of change with respect to time during the cell culture process [131–134]. Thus, cell growth and death can be linked to critical nutrients and metabolic by-products, while the protein production is usually linked to cell growth and amino acid metabolism [136]. Kinetic/dynamic models can be structured with varying levels of complexity depending on the assumptions made regarding the culture system and intracellular processes. For example, a model can add complexity by considering the heterogeneity in a cell population or by taking into account the different cellular compartments. Models can also be simplified by lumping reactions to rate-limiting steps [136]. These model-driven sensors are complex to develop and, as such, in the biopharmaceutical industry, data-driven sensors, which rely on historic data or small-scale process development runs, tend to be more used [116].

9.3. Hybrid Models

Grey-box models are another important category of soft sensors, which can be considered as a combination of mechanistic models and data-driven models. They have the advantage of maximizing the benefits of each method while avoiding some of the disadvantages inherent to each approach [131–134]. To limit the shortcomings of black box models (lack of extrapolation capabilities within trained range) and white box models (large uncertainty over parameters estimation and susceptibility to noise), recursive state observers can be used to combine dynamic metabolic modelling and data-driven modelling by updating state estimates derived from noisy measurement and gradually reducing the estimation error covariance on the specific assumption of linear process and Gaussian distribution for the error terms. For this purpose, a Kalman filter can be used. Importantly, given that the process dynamics within a bioprocess are highly non-linear, the extended Kalman filter can be applied to non-linear systems, thanks to piecewise linearization of the process around the time trajectories of the variables through the estimation of Jacobian matrixes [137]. Another popular version of the Kalman filter for non-linear systems is the unscented Kalman filter, which uses a Taylor series expansion to linearize the model [138]. Since the accuracy of a hybrid soft sensor is significantly impacted by the accuracy of the mechanistic model, the latter must be extensively validated to ensure it can successfully represent the process [137].

In grey box/hybrid models, the biological system is described by a mechanistic framework but the cell-specific rates are defined by statistical expressions [131–134]. Thus, the material balance constrains the solution space for the model and the statistical cell-specific rate expressions can be automated [139]. For example, if multi-wavelength spectra are analyzed with PLS or an ANN and the resulting predictions are used as an input in a mechanistic model, this can be considered a grey box model. By mixing material balances with statistical models, direct links between data and physical bioprocess systems can be generated. This is because, within the hybrid model structure, the Kalman filter uses the prediction from the mechanistic model and the data gained from the data-driven model to recursively update the state estimators, thus synthesizing the information gained from both types of models [138,140]. In such a way, it is possible to imagine numerous applications where multivariate models generated from spectroscopic data or other on-line measurements and mechanistic models are used in tandem to develop models that use historical data while also describing the dynamics of the system. Some studies have demonstrated improved protein yield when employing hybrid models for adaptive feeding when compared to relying exclusively on data-driven models for adaptive feeding [137]. Kalman filtering can be used with both on-line and off-line data. For example, biocapacitance measurements used to estimate biomass on-line and infrequent sampling of ammonia and lactate can be coupled along with process dynamic equations to continuously estimate glucose and glutamine concentrations [141]. A similar study was realized with Raman measurements that were combined with dynamic metabolic models through adaptive, constrained, extended Kalman filters for the purpose of metabolite concentration tracking, which could then be applied in setpoint tracking of glucose [142].

9.4. Applications of Soft Sensors in Bioprocessing

It is clear from the aforementioned definitions that soft sensors are dependent on both the mathematical framework and the measurement device to be successfully merged into a single functional entity. The state-of-the-art techniques used to generate a soft sensor include many of the measurement devices that have been covered in the previous sections, ranging from NIR/MIR spectroscopy, fluorescence spectroscopy, dielectric spectroscopy, and Raman spectroscopy to mass spectrometry. Soft sensors can employ either in-line measurements or at-line measurements. They have been applied to non-invasive on-line spectroscopic methods such as NIR/MIR, 2D fluorescence, and Raman spectral data given the multidimensional complexity of the signal and the need of multivariate data analysis to relate the data to relevant process parameters [131–134]. While in situ sensors may be more attractive than at-line sensors given the higher sampling rates, on the other hand, precision, calibration, and stability generally often favor the at-line alternatives [131–134]. For example, a soft sensor capable of monitoring biomass subpopulations (viable cells, dead cells, and lysed cells) in a cell culture process was developed through the use of permittivity and turbidity sensors in conjunction with mechanistic models that describe the dynamics of the subpopulations [143]. This is of great value given that estimating lysed cell concentrations is difficult, generally requiring indirect methods such as the measurement of process-related impurities like DNA and host cell proteins in the supernatant.

By combining mechanistic metabolic modelling and multi-wavelength fluorescence spectroscopic data, it was proposed that the resulting soft sensor could filter noise in the data and produce estimates of culture variables in between fluorescent data samples. More precisely, a metabolic flux model capable of relating the main nutrients to by-products was combined with a PLS regression of fluorescent data. It was concluded that the dynamic model was capable of improving the accuracy of the data-driven fluorescence-based predictions [137]. Additionally, the extended Kalman filter model could generate accurate predictions of the temporal evolution of the culture variables in between sampling instances. This study is in addition to another publication by the same authors where a PLS model of fluorescent data was built [144]. Here, viable cell, glucose, recombinant protein, and ammonia concentrations were predicted accurately throughout the culture progress of CHO cells [144]. However, such soft sensor arrangements work as black box models and, thus, are inherently limited in their capacity to predict key process variables in discrete space or extrapolating the results for varying processes. As such, they are not able to accommodate the dynamic evolution of the variables in between measurement instances [144].

Recently, efforts have been made to develop data-driven models that aim to integrate varying 2-D fluorescence datasets into calibration phase, regardless of the process strategy-dependent diversity [145]. For this, special attention was directed towards the nutrient rich and, consequently, fluorescent feeding solutions that are suspected to hamper the generation of reliable chemometric models given that they alter the evolution of fluorescent components during cell culture cultivation [145]. It was determined that calibration of soft sensors was generally possible regardless of the process strategy. It was suggested that calibration of soft sensors with data that adequately incorporate process variations can facilitate the transfer of soft sensors from one production process to another [145]. This is an important improvement for data-driven applications within bioprocessing given that most developed models tend to be over fitted for their application, making model transferability difficult.

As mentioned previously, another spectroscopic technique that is readily applied to data-driven soft sensors is Raman spectroscopy. It has been proven to be a reliable way to attain in situ, real-time measurements of relevant process parameters while at the same time being translatable across different scales without the need of model recalibration [68]. A comparison of multivariate linear regression (MLR), principal component regression (PCR), and PLS regression (PLSR) algorithms for the creation of a data-driven model using Raman spectroscopic data has been performed. It was determined that PLSR, the most advanced algorithm of the three, delivered the lowest root mean square error prediction, indicating that it was the model that most accurately represented the empirical measurements [146]. This is doubly advantageous given that PLSR is the most available algorithm in commercial, multivariate data-analyzing software. Raman spectroscopic data have also been used to indirectly measure changes in pH [147]. This method was proposed as an in situ, real-time method for pH estimation given that standard electrochemical pH probes suffer from drifts that require correction by off-line measurement methods. Given that pH in a bioprocess is influenced by controller inputs (base and carbon dioxide additions) and cell culture metabolism (lactate, ammonia), a method capable of relating spectroscopic (X variables) to pH (Y variable) data in fed-batch mammalian cell cultures was developed. The model conveyed errors of 0.035 and 0.034 pH for two different CHO cell lines. This can be explained by the fact that the Raman spectroscopy model is not being directly correlated to pH; instead, it is correlated to bonds and molecules that influence pH within the bioprocess [147]. The advances in soft sensor modelling for data-driven sensors have been so vast that variables beyond routine measurements like glucose, lactate, and viable cell density have been suggested. Chemometric models for tyrosine, tryptophan, phenylalanine, and methionine have been developed with good correlation metrics. This, in turn, suggests that real-time monitoring and control of amino acids in cell culture is feasible and can help in the process of optimizing yields and product quality [148]. Historically, ultraviolet-visible spectroscopy has provided poor information of bioprocesses because of wide and unspecified bands, spectral interferences, and deficiency in detection of higher energy electronic levels in molecules. To circumvent this, ANN have been proposed to predict glutamine, glutamate, glucose, lactate, and viable cell concentrations [57].

In Figure 7, it is possible to observe how the different sensors covered can be used as varying sources of data streams to feed into a soft sensor. These soft sensors bifurcate in the methodology used (mechanistic mode, data-driven, and hybrids) but always have the goal of generating information that cannot be measured directly and individually with each hardware sensor. It is clear that soft sensors have an important role to play in the future of bioprocess monitoring. Black box-based sensors can combine information of newly applied spectroscopy data to gather information directly from the media without the need of repeated sampling, while mechanistic-based sensors can estimate information regarding the process dynamics and intracellular metabolic rates. This is of great value as well, as it can serve to develop control algorithms that take into account real-time data regarding the metabolic profile. Alternatively, hybrid-based sensors could be considered the best of both worlds given that they can integrate the large amount of data available with knowledge-based models.

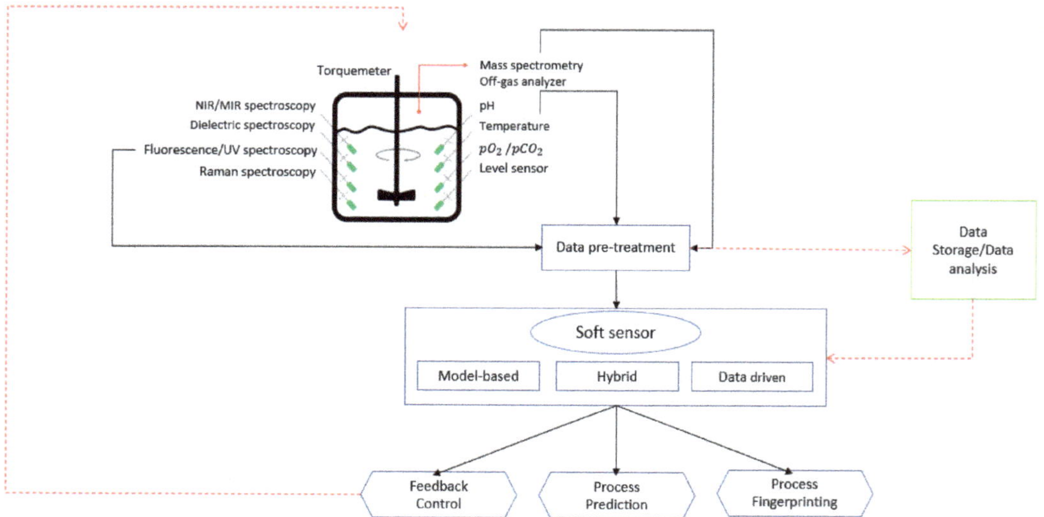

Figure 7. Outline of how different hardware sensor data are integrated into soft sensors and how historical data storage can further intensify cell culture processes.

9.5. Soft Sensor Implementation

To build the soft sensors, a variety of programming environments can be employed such as Matlab and Python [149]. These programming environments are popular for the development of soft sensors because they provide libraries and tool boxes for signal processing, data analysis, model calibration, and model validation [149]. Alternatively, in an industrial setting, commercial software systems for chemometrics are generally used, such as SIMCA (Sartorius AG) or Unscrambler (Aspen Technology Inc., Bedford, MA, USA) [150]. With these software systems, the flexibility of the programming environment offered by Python/Matlab is exchanged for streamlined platform. Vendors of on-line analytical tools also offer software modules for soft sensor development such as the OPUS suite by Bruker Corp., iC suite by Mettler Toledo Inc., Columbus, OH, USA, and GRAMS suite by Thermo Fisher Scientific Inc., Waltham, MA, USA, [150]. Finally, soft sensors can also be developed as internet of things such as Predix, MindSphere, and Sentience. Alternatively, cloud infrastructure can be used for soft sensor integration within a manufacturing plant by using Amazon web services, Microsoft Azure, Google cloud, or IBM's Watson IoT [151].

9.6. Soft Sensors for Bioprocess Control

Soft sensors are of increasing interest in order to develop non-linear control strategies. Control strategies in bioprocessing have the objective of supervising key parameters in a dynamically changing process in order to maintain the key variables within the desired design space. These control strategies are developed early in the process development cycle in order to work around unforeseen consequences of alternative control strategies.

Robust control strategies must be built on top of in-depth understandings of the process. Thus, mechanistic models, data-driven models, or hybrid models can be used to develop successful control strategies, which can be split up into open-loop strategies, closed-loop (or feedback) strategies, fuzzy control, and model predictive control [152,153].

Open loop strategies, for example, are used when applying predefined feed rates into the process, which are entirely dependent on the initial conditions and the predetermined process conditions. Such control strategy does not require any on-line monitoring. One large drawback of these strategies is that they require precomputed knowledge profiles of growth kinetics, which is difficult in non-linear systems with a dynamically changing metabolism, as in the case of mammalian cells. Additionally, open-loop control strategies are unable to perform corrective measures when the system has deviated from the designed space as a result of disturbances impacting the process [152,153].

Closed loop control systems were designed to overcome the largest disadvantage of open loop systems, namely, the inability to provide feedback on the process as regulatory control. These controllers are of standard use for pH control, temperature control, and dissolved oxygen control [153]. Cascade control involves two feedback controllers and is used to improve the dynamic response of the controllers by distributing the disturbance over a secondary loop where corrective measures are taken without affecting the primary loop. This type of controller has been successfully applied in bioprocessing, particularly to control dissolved oxygen [154].

More sophisticated techniques involve model predictive control, where the controller response is based on a process model, which can be mechanistic, hybrid, or data-driven in origin. The model is capable of forecasting process events given process conditions and measurements from various input sensors [133]. Thus, the model is key for successful prediction and accurate response to process variations. These control methods are more computationally expensive than standard control techniques, given that various inputs must be analyzed within the model function to generate an output.

Another promising control technique for the bioprocessing industry is fuzzy control [152,153,155]. This technique does not require a complex mathematical description of the process since fuzzy logic does not require initial knowledge of the system dynamics. Fuzzy control is centered on the transformation of quantitative data into qualitative parameters. This is done by converting numerical data into a membership function, which is a value between 0 and 1 that defines the degree to which a certain variable fits a given fuzzy set. The values in the 0–1 scale are dependent on a predetermined knowledge of the range of possible values. With this information, fuzzy rules can be enacted based on experience with the process by employing conditional statements [152,153,155]. This inherently incorporates process experience into the controller by combining user knowledge and trends with past data. However, this strategy differs from data-driven models given that they can only operate within the range of the specific datasets that were used to train the model [152,153,155].

10. Concluding Remarks

Given that bioreactor modes of operation, such as fed-batch, concentrated fed-batch, and perfusion, are gaining more application within the biopharmaceutical industry, the push towards sensors that are capable of monitoring the process beyond its immediate environment (temperature, pH, DO, stirring) is gathering strength. This is because most current research is also focused on understanding the complex metabolism of mammalian cells in order to gauge what additional parameters should be monitored in order to assess what is driving/hampering the protein production. Because of this, it has been understood that amino acid concentrations in the media must be monitored as to not cause metabolic bottlenecks. The trend for increased bioprocess monitoring applies both to small-scale and large-scale bioreactors since most of the technologies tend to be independent of scale. This is especially true for spectroscopic, biochemical, and optical sensors, which are now being used at all stages of process development and biomanufacturing. Importantly, the capacity to monitor metabolic activity routinely at different vessel sizes makes it easier to transfer processes across scales. Other methods such as off-gas analyzers generally work better at higher scales given their low resolution, but work is being done in order to reduce this limitation [156].

It has also been understood that dissolved carbon dioxide should be decoupled from standard pH control and optimized on its own. It is noteworthy that, given the importance of understanding the bioprocess, standard sensors have also been applied, with the help of mechanistic models, in the development of measuring in-depth parameters such as OUR and CER. This should allow better control of the extracellular space and, in turn, enable a favorable intracellular environment. The latter is what really determines the quality and quantity of the end product. Moreover, soft sensors can play an integral role in developing process models that are able to feed back into the system dynamics to maintain culture conditions within a desired design space. For this reason, process analytical technologies (PAT) in conjunction with multivariate data analysis (MVA) and mechanistic models such as the ones covered in this review can be used alone or in conjunction to gain understanding about the system. This will help develop outcome predictions and process optimization or support the development of new control strategies that are needed to automate certain critical steps in biomanufacturing. Another aspect in which PAT monitoring technologies in conjunction with soft sensors will play a key role in the future of biomanufacturing is in the development of digital twins. These digital twins can be thought of as the simulation of bioprocess in silico. This, in turn, means that the entire bioprocess could be simulated at different scales (metabolic models, genome-based models, first principles models), which can then be used to test out various predictive control techniques before implementing changes at the bench scale and before reaching the manufacturing floor. Thus, understanding of process variation can be obtained through digital twins, which could then allow for prediction of productivity, product quality, and process attributes and forecasting of costs of future physical experiments. This itself feeds into the wider push towards industry 4.0 where advanced autonomous manufacturing systems and infrastructure can improve product output. Within this framework, performance data can be stored and mined into *big data* structures that can consequently be readily analyzed with the help of both artificial intelligence algorithms and mechanistic models. This maximizes the use of both historical and real-time data in order to understand what is happening in the process and how it can be leveraged to improve the manufacturing process. Hence, given the trend in the biopharmaceutical industry towards biologization of manufacturing and at the same time a wider trend of digitalization of manufacturing, the canalization of monitoring technologies should be harnessed to improve both understanding and outcomes.

Author Contributions: S.J.R., P.L.P. and O.H. were responsible for conceptualizing and editing the manuscript. Y.D. provided mentorship and valuable comments on the manuscript. All authors have read and agreed to the published version of the manuscript.

Funding: This research received funding from NRC's Pandemic Response Challenge Program and the Natural Sciences and Engineering Research Council (RGPIN-2021-04048).

Informed Consent Statement: Not applicable.

Data Availability Statement: Not applicable.

Acknowledgments: Not applicable.

Conflicts of Interest: The authors declare no conflict of interest.

References

1. Research, G.V. Biotechnology Market Size, Share & Trends Analysis Report by Technology (DNA Sequencing, Nanobiotechnology), by Application (Health, Bioinformatics), by Region, and Segment Forecasts, 2021–2028. Available online: https://www.grandviewresearch.com/industry-analysis/biotechnology-market (accessed on 24 July 2021).
2. Intelligence, M. Biopharmaceuticals Market—Growth, Trends, COVID-19 Impact, and Forecasts (2021–2026). Available online: https://www.mordorintelligence.com/industry-reports/global-biopharmaceuticals-market-industry (accessed on 24 July 2021).

3.   Markets, R.A. Biopharmaceuticals Market by Type and Application—Global Opportunity Analysis and Industry Forecast, 2018–2025. Available online: https://www.researchandmarkets.com/reports/4612776/biopharmaceuticals-market-by-type-and-application (accessed on 24 July 2021).

4.   Research, G.V. Biosimilars Market Size, Share & Trends Analysis Report by Product, by Application (Oncology, Growth Hormone, Blood Disorders, Chronic & Autoimmune Disorders), by Region, and Segment Forecasts, 2018–2025. Available online: https://markets.businessinsider.com/news/stocks/biosimilars-market-size-worth-61-47-billion-by-2025-cagr-34-2-grand-view-research-inc-1027561345 (accessed on 24 July 2021).

5.   Intelligence, M. Biosimilars Market—Growth, Trends, COVID-19 Impact, and Forecasts (2021–2026). Available online: https://www.mordorintelligence.com/industry-reports/global-biosimilars-market-industry (accessed on 24 July 2021).

6.   Li, F.; Vijayasankaran, N.; Shen, A.Y.; Kiss, R.; Amanullah, A. Cell culture processes for monoclonal antibody production. *mAbs* **2010**, *2*, 466–479. [CrossRef] [PubMed]

7.   Lu, R.M.; Hwang, Y.C.; Liu, I.J.; Lee, C.C.; Tsai, H.Z.; Li, H.J.; Wu, H.C. Development of therapeutic antibodies for the treatment of diseases. *J. Biomed. Sci.* **2020**, *27*, 1. [CrossRef] [PubMed]

8.   Lustri, J. How to Implement Process Analytical Technology in Pharmaceutical Manufacturing. Available online: https://blog.isa.org/how-to-implement-process-analytical-technology-in-pharmaceutical-manufacturing (accessed on 27 July 2021).

9.   Uwe Kirschner, R.E.C.; Vangenechten, R.; François, K. Process Analytical Technology: An Industry Perspective. Available online: https://www.europeanpharmaceuticalreview.com/article/3643/process-analytical-technology-pharma-industry/ (accessed on 29 July 2021).

10.   John, D.O.; George, L.; Reid, I. An Introduction To Process Analytical Technology. Available online: https://www.pharmaceuticalonline.com/doc/an-introduction-to-process-analytical-technology-0001 (accessed on 28 July 2021).

11.   Mhatre, R.; Rathore, A.S. Quality by Design: An Overview of the Basic Concepts. *Qual. By Des. Biopharm.* **2009**, 1–8.

12.   Rathore, A.S.; Winkle, H. Quality by design for biopharmaceuticals. *Nat. Biotechnol.* **2009**, *27*, 26–34. [CrossRef] [PubMed]

13.   Doran, P.M. *Bioprocess Engineering Principles*; Academic Press: London, UK, 1995.

14.   Shuler, M.L.; Kargı, F.; DeLisa, M. *Bioprocess Engineering: Basic Concepts*; Prentice Hall: Hoboken, NJ, USA, 2017.

15.   O'Flaherty, R.; Bergin, A.; Flampouri, E.; Mota, L.M.; Obaidi, I.; Quigley, A.; Xie, Y.; Butler, M. Mammalian cell culture for production of recombinant proteins: A review of the critical steps in their biomanufacturing. *Biotechnol. Adv.* **2020**, *43*, 107552. [CrossRef]

16.   Wiegmann, V.; Giaka, M.; Martinez, C.B.; Baganz, F. Towards the development of automated fed-batch cell culture processes at microscale. *BioTechniques* **2019**, *67*, 238–241. [CrossRef] [PubMed]

17.   Chen, C.; Wong, H.E.; Goudar, C.T. Upstream process intensification and continuous manufacturing. *Curr. Opin. Chem. Eng.* **2018**, *22*, 191–198. [CrossRef]

18.   Bielser, J.-M.; Wolf, M.; Souquet, J.; Broly, H.; Morbidelli, M. Perfusion mammalian cell culture for recombinant protein manufacturing—A critical review. *Biotechnol. Adv.* **2018**, *36*, 1328–1340. [CrossRef]

19.   Wlaschin, K.F.; Hu, W.S. Fedbatch culture and dynamic nutrient feeding. *Adv. Biochem. Eng. Biotechnol.* **2006**, *101*, 43–74. [CrossRef]

20.   Yang, W.C.; Minkler, D.F.; Kshirsagar, R.; Ryll, T.; Huang, Y.M. Concentrated fed-batch cell culture increases manufacturing capacity without additional volumetric capacity. *J. Biotechnol.* **2016**, *217*, 1–11. [CrossRef] [PubMed]

21.   Hartley, F.; Walker, T.; Chung, V.; Morten, K. Mechanisms driving the lactate switch in Chinese hamster ovary cells. *Biotechnol. Bioeng.* **2018**, *115*, 1890–1903. [CrossRef] [PubMed]

22.   Pereira, S.; Kildegaard, H.F.; Andersen, M.R. Impact of CHO Metabolism on Cell Growth and Protein Production: An Overview of Toxic and Inhibiting Metabolites and Nutrients. *Biotechnol. J.* **2018**, *13*, 1700499. [CrossRef] [PubMed]

23.   Galleguillos, S.N.; Ruckerbauer, D.; Gerstl, M.P.; Borth, N.; Hanscho, M.; Zanghellini, J. What can mathematical modelling say about CHO metabolism and protein glycosylation? *Comput. Struct. Biotechnol. J.* **2017**, *15*, 212–221. [CrossRef] [PubMed]

24.   Locasale, J.W.; Cantley, L.C. Metabolic flux and the regulation of mammalian cell growth. *Cell Metab.* **2011**, *14*, 443–451. [CrossRef]

25.   Mulukutla, B.C.; Khan, S.; Lange, A.; Hu, W.-S. Glucose metabolism in mammalian cell culture: New insights for tweaking vintage pathways. *Trends Biotechnol.* **2010**, *28*, 476–484. [CrossRef]

26.   Mulukutla, B.C.; Gramer, M.; Hu, W.-S. On metabolic shift to lactate consumption in fed-batch culture of mammalian cells. *Metab. Eng.* **2012**, *14*, 138–149. [CrossRef]

27.   Ozturk, S.S.; Riley, M.R.; Palsson, B.O. Effects of ammonia and lactate on hybridoma growth, metabolism, and antibody production. *Biotechnol. Bioeng.* **1992**, *39*, 418–431. [CrossRef] [PubMed]

28.   Zagari, F.; Jordan, M.; Stettler, M.; Broly, H.; Wurm, F.M. Lactate metabolism shift in CHO cell culture: The role of mitochondrial oxidative activity. *New Biotechnol.* **2013**, *30*, 238–245. [CrossRef]

29.   Fan, Y.; Jimenez del Val, I.; Müller, C.; Wagtberg Sen, J.; Rasmussen, S.K.; Kontoravdi, C.; Weilguny, D.; Andersen, M.R. Amino acid and glucose metabolism in fed-batch CHO cell culture affects antibody production and glycosylation. *Biotechnol. Bioeng.* **2015**, *112*, 521–535. [CrossRef]

30.   Schneider, M.; Marison, I.W.; von Stockar, U. The importance of ammonia in mammalian cell culture. *J. Biotechnol.* **1996**, *46*, 161–185. [CrossRef]

31.   Martinelle, K.; Häggström, L. Effects of $NH4^+$ and K^+ on the energy metabolism in Sp2/0-Ag14 myeloma cells. *Cytotechnology* **1999**, *29*, 45–53. [CrossRef]

32. Duarte, T.M.; Carinhas, N.; Barreiro, L.C.; Carrondo, M.J.T.; Alves, P.M.; Teixeira, A.P. Metabolic responses of CHO cells to limitation of key amino acids. *Biotechnol. Bioeng.* **2014**, *111*, 2095–2106. [CrossRef]

33. Mohmad-Saberi, S.E.; Hashim, Y.Z.; Mel, M.; Amid, A.; Ahmad-Raus, R.; Packeer-Mohamed, V. Metabolomics profiling of extracellular metabolites in CHO-K1 cells cultured in different types of growth media. *Cytotechnology* **2013**, *65*, 577–586. [CrossRef]

34. Mulukutla, B.C.; Kale, J.; Kalomeris, T.; Jacobs, M.; Hiller, G.W. Identification and control of novel growth inhibitors in fed-batch cultures of Chinese hamster ovary cells. *Biotechnol. Bioeng.* **2017**, *114*, 1779–1790. [CrossRef] [PubMed]

35. Templeton, N.; Dean, J.; Reddy, P.; Young, J.D. Peak antibody production is associated with increased oxidative metabolism in an industrially relevant fed-batch CHO cell culture. *Biotechnol. Bioeng.* **2013**, *110*, 2013–2024. [CrossRef]

36. Mulukutla, B.C.; Mitchell, J.; Geoffroy, P.; Harrington, C.; Krishnan, M.; Kalomeris, T.; Morris, C.; Zhang, L.; Pegman, P.; Hiller, G.W. Metabolic engineering of Chinese hamster ovary cells towards reduced biosynthesis and accumulation of novel growth inhibitors in fed-batch cultures. *Metab. Eng.* **2019**, *54*, 54–68. [CrossRef]

37. O'Mara, P.; Farrell, A.; Bones, J.; Twomey, K. Staying alive! Sensors used for monitoring cell health in bioreactors. *Talanta* **2018**, *176*, 130–139. [CrossRef] [PubMed]

38. Hoshan, L.; Jiang, R.; Moroney, J.; Bui, A.; Zhang, X.; Hang, T.-C.; Xu, S. Effective bioreactor pH control using only sparging gases. *Biotechnol. Prog.* **2019**, *35*, e2743. [CrossRef] [PubMed]

39. Xing, Z.; Lewis, A.M.; Borys, M.C.; Li, Z.J. A carbon dioxide stripping model for mammalian cell culture in manufacturing scale bioreactors. *Biotechnol. Bioeng.* **2017**, *114*, 1184–1194. [CrossRef]

40. Meghrous, J.; Khramtsov, N.; Buckland, B.C.; Cox, M.M.J.; Palomares, L.A.; Srivastava, I.K. Dissolved carbon dioxide determines the productivity of a recombinant hemagglutinin component of an influenza vaccine produced by insect cells. *Biotechnol. Bioeng.* **2015**, *112*, 2267–2275. [CrossRef]

41. Biechele, P.; Busse, C.; Solle, D.; Scheper, T.; Reardon, K. Sensor systems for bioprocess monitoring. *Eng. Life Sci.* **2015**, *15*, 469–488. [CrossRef]

42. Zhao, L.; Fu, H.Y.; Zhou, W.; Hu, W.S. Advances in process monitoring tools for cell culture bioprocesses. *Eng. Life Sci.* **2015**, *15*, 459–468. [CrossRef]

43. Company, H. Bioprocess Monitoring and Control (Off-Line, At-Line, On-Line, In-Line/In-Situ). Available online: https://www.hamiltoncompany.com/process-analytics/process-analytical-technology/bioprocess-monitoring-and-control (accessed on 27 July 2021).

44. Gargalo, C.L.; Udugama, I.; Pontius, K.; Lopez, P.C.; Nielsen, R.F.; Hasanzadeh, A.; Mansouri, S.S.; Bayer, C.; Junicke, H.; Gernaey, K.V. Towards smart biomanufacturing: A perspective on recent developments in industrial measurement and monitoring technologies for bio-based production processes. *J. Ind. Microbiol. Biotechnol.* **2020**, *47*, 947–964. [CrossRef]

45. Hub, E. Introduction to Sensors and Transducers. Available online: https://www.electronicshub.org/sensors-and-transducers-introduction/ (accessed on 30 July 2021).

46. Bhalla, N.; Jolly, P.; Formisano, N.; Estrela, P. Introduction to biosensors. *Essays Biochem.* **2016**, *60*, 1–8. [CrossRef]

47. Kisaalita, W.S. Biosensor standards requirements. *Biosens. Bioelectron.* **1992**, *7*, 613–620. [CrossRef]

48. Steinwedel, T.; Dahlmann, K.; Solle, D.; Scheper, T.; Reardon, K.F.; Lammers, F. Sensors for disposable bioreactor systems. In *Single-Use Technology in Biopharmaceutical Manufacture*; John Wiley & Sons Inc.: Hoboken, NJ, USA, 2019; pp. 69–82.

49. Busse, C.; Biechele, P.; de Vries, I.; Reardon, K.F.; Solle, D.; Scheper, T. Sensors for disposable bioreactors. *Eng. Life Sci.* **2017**, *17*, 940–952. [CrossRef] [PubMed]

50. Clavaud, M.; Roggo, Y.; von Daeniken, R.; Liebler, A.; Schwabe, J.O. Chemometrics and in-line near infrared spectroscopic monitoring of a biopharmaceutical Chinese hamster ovary cell culture: Prediction of multiple cultivation variables. *Talanta* **2013**, *111*, 28–38. [CrossRef]

51. Sandor, M.; Rüdinger, F.; Bienert, R.; Grimm, C.; Solle, D.; Scheper, T. Comparative study of non-invasive monitoring via infrared spectroscopy for mammalian cell cultivations. *J. Biotechnol.* **2013**, *168*, 636–645. [CrossRef] [PubMed]

52. Riley, M.R.; Crider, H.M.; Nite, M.E.; Garcia, R.A.; Woo, J.; Wegge, R.M. Simultaneous measurement of 19 components in serum-containing animal cell culture media by fourier transform near-infrared spectroscopy. *Biotechnol. Prog.* **2001**, *17*, 376–378. [CrossRef]

53. Hakemeyer, C.; Strauss, U.; Werz, S.; Jose, G.E.; Folque, F.; Menezes, J.C. At-line NIR spectroscopy as effective PAT monitoring technique in Mab cultivations during process development and manufacturing. *Talanta* **2012**, *90*, 12–21. [CrossRef]

54. Capito, F.; Zimmer, A.; Skudas, R. Mid-infrared spectroscopy-based analysis of mammalian cell culture parameters. *Biotechnol. Prog.* **2015**, *31*, 578–584. [CrossRef]

55. Rosa, F.O.P.; Cunha, B.; Carmelo, J.G.; Fernandes-Platzgummer, A.; da Silva, C.L.; Calado, C.R.C. Mid-infrared spectroscopy: A groundbreaking tool for monitoring mammalian cells processes. In Proceedings of the 2017 IEEE 5th Portuguese Meeting on Bioengineering (ENBENG), Coimbra, Portugal, 16–18 February 2017; pp. 1–6.

56. Hansen, S.K.; Jamali, B.; Hubbuch, J. Selective high throughput protein quantification based on UV absorption spectra. *Biotechnol. Bioeng.* **2013**, *110*, 448–460. [CrossRef] [PubMed]

57. Takahashi, M.B.; Leme, J.; Caricati, C.P.; Tonso, A.; Fernández Núñez, E.G.; Rocha, J.C. Artificial neural network associated to UV/Vis spectroscopy for monitoring bioreactions in biopharmaceutical processes. *Bioprocess Biosyst. Eng.* **2015**, *38*, 1045–1054. [CrossRef] [PubMed]

58. Drieschner, T.; Ostertag, E.; Boldrini, B.; Lorenz, A.; Brecht, M.; Rebner, K. Direct optical detection of cell density and viability of mammalian cells by means of UV/VIS spectroscopy. *Anal. Bioanal. Chem.* **2020**, *412*, 3359–3371. [CrossRef]
59. Leme, J.; Fernández Núñez, E.G.; de Almeida Parizotto, L.; Chagas, W.A.; dos Santos, E.S.; Tojeira Prestia Caricati, A.; de Rezende, A.G.; da Costa, B.L.V.; Ventini Monteiro, D.C.; Lopes Boldorini, V.L.; et al. A multivariate calibration procedure for UV/VIS spectrometric monitoring of BHK-21 cell metabolism and growth. *Biotechnol. Prog.* **2014**, *30*, 241–248. [CrossRef] [PubMed]
60. Li, M.; Ebel, B.; Chauchard, F.; Guédon, E.; Marc, A. Parallel comparison of in situ Raman and NIR spectroscopies to simultaneously measure multiple variables toward real-time monitoring of CHO cell bioreactor cultures. *Biochem. Eng. J.* **2018**, *137*, 205–213. [CrossRef]
61. Teixeira, A.P.; Portugal, C.A.; Carinhas, N.; Dias, J.M.; Crespo, J.P.; Alves, P.M.; Carrondo, M.J.; Oliveira, R. In situ 2D fluorometry and chemometric monitoring of mammalian cell cultures. *Biotechnol. Bioeng.* **2009**, *102*, 1098–1106. [CrossRef] [PubMed]
62. Teixeira, A.P.; Duarte, T.M.; Oliveira, R.; Carrondo, M.J.T.; Alves, P.M. High-throughput analysis of animal cell cultures using two-dimensional fluorometry. *J. Biotechnol.* **2011**, *151*, 255–260. [CrossRef]
63. Teixeira, A.P.; Duarte, T.M.; Carrondo, M.J.T.; Alves, P.M. Synchronous fluorescence spectroscopy as a novel tool to enable PAT applications in bioprocesses. *Biotechnol. Bioeng.* **2011**, *108*, 1852–1861. [CrossRef] [PubMed]
64. Claßen, J.; Graf, A.; Aupert, F.; Solle, D.; Höhse, M.; Scheper, T. A novel LED-based 2D-fluorescence spectroscopy system for in-line bioprocess monitoring of Chinese hamster ovary cell cultivations—Part II. *Eng. Life Sci.* **2019**, *19*, 341–351. [CrossRef]
65. Esmonde-White, K.A.; Cuellar, M.; Uerpmann, C.; Lenain, B.; Lewis, I.R. Raman spectroscopy as a process analytical technology for pharmaceutical manufacturing and bioprocessing. *Anal. Bioanal. Chem.* **2017**, *409*, 637–649. [CrossRef]
66. Abu-Absi, N.R.; Kenty, B.M.; Cuellar, M.E.; Borys, M.C.; Sakhamuri, S.; Strachan, D.J.; Hausladen, M.C.; Li, Z.J. Real time monitoring of multiple parameters in mammalian cell culture bioreactors using an in-line Raman spectroscopy probe. *Biotechnol. Bioeng.* **2011**, *108*, 1215–1221. [CrossRef] [PubMed]
67. Xu, Y.; Ford, J.; Mann, C.; Vickers, T.; Brackett, J.; Cousineau, K.; Robey, W. *Raman Measurement of Glucose in Bioreactor Materials*; SPIE: Bellingham, WA, USA, 1997; Volume 2976.
68. Whelan, J.; Craven, S.; Glennon, B. In situ Raman spectroscopy for simultaneous monitoring of multiple process parameters in mammalian cell culture bioreactors. *Biotechnol. Prog.* **2012**, *28*, 1355–1362. [CrossRef]
69. Berry, B.; Moretto, J.; Matthews, T.; Smelko, J.; Wiltberger, K. Cross-scale predictive modeling of CHO cell culture growth and metabolites using Raman spectroscopy and multivariate analysis. *Biotechnol. Prog.* **2015**, *31*, 566–577. [CrossRef] [PubMed]
70. Li, B.; Ray, B.H.; Leister, K.J.; Ryder, A.G. Performance monitoring of a mammalian cell based bioprocess using Raman spectroscopy. *Anal. Chim. Acta* **2013**, *796*, 84–91. [CrossRef] [PubMed]
71. Webster, T.A.; Hadley, B.C.; Hilliard, W.; Jaques, C.; Mason, C. Development of generic Raman models for a GS-KOTM CHO platform process. *Biotechnol. Prog.* **2018**, *34*, 730–737. [CrossRef]
72. Santos, R.M.; Kessler, J.M.; Salou, P.; Menezes, J.C.; Peinado, A. Monitoring mAb cultivations with in-situ raman spectroscopy: The influence of spectral selectivity on calibration models and industrial use as reliable PAT tool. *Biotechnol. Prog.* **2018**, *34*, 659–670. [CrossRef] [PubMed]
73. Matthews, T.E.; Berry, B.N.; Smelko, J.; Moretto, J.; Moore, B.; Wiltberger, K. Closed loop control of lactate concentration in mammalian cell culture by Raman spectroscopy leads to improved cell density, viability, and biopharmaceutical protein production. *Biotechnol. Bioeng.* **2016**, *113*, 2416–2424. [CrossRef]
74. Berry, B.N.; Dobrowsky, T.M.; Timson, R.C.; Kshirsagar, R.; Ryll, T.; Wiltberger, K. Quick generation of Raman spectroscopy based in-process glucose control to influence biopharmaceutical protein product quality during mammalian cell culture. *Biotechnol. Prog.* **2016**, *32*, 224–234. [CrossRef]
75. Párta, L.; Zalai, D.; Borbély, S.; Putics, Á. Application of dielectric spectroscopy for monitoring high cell density in monoclonal antibody producing CHO cell cultivations. *Bioprocess Biosyst. Eng.* **2014**, *37*, 311–323. [CrossRef]
76. Justice, C.; Brix, A.; Freimark, D.; Kraume, M.; Pfromm, P.; Eichenmueller, B.; Czermak, P. Process control in cell culture technology using dielectric spectroscopy. *Biotechnol. Adv.* **2011**, *29*, 391–401. [CrossRef]
77. Zeiser, A.; Bédard, C.; Voyer, R.; Jardin, B.; Tom, R.; Kamen, A.A. On-line monitoring of the progress of infection in Sf-9 insect cell cultures using relative permittivity measurements. *Biotechnol. Bioeng.* **1999**, *63*, 122–126. [CrossRef]
78. Metze, S.; Blioch, S.; Matuszczyk, J.; Greller, G.; Grimm, C.; Scholz, J.; Hoehse, M. Multivariate data analysis of capacitance frequency scanning for online monitoring of viable cell concentrations in small-scale bioreactors. *Anal. Bioanal. Chem.* **2020**, *412*, 2089–2102. [CrossRef] [PubMed]
79. Moore, B.; Sanford, R.; Zhang, A. Case study: The characterization and implementation of dielectric spectroscopy (biocapacitance) for process control in a commercial GMP CHO manufacturing process. *Biotechnol. Prog.* **2019**, *35*, e2782. [CrossRef]
80. Lu, F.; Toh, P.C.; Burnett, I.; Li, F.; Hudson, T.; Amanullah, A.; Li, J. Automated dynamic fed-batch process and media optimization for high productivity cell culture process development. *Biotechnol. Bioeng.* **2013**, *110*, 191–205. [CrossRef]
81. Konakovsky, V.; Clemens, C.; Müller, M.M.; Bechmann, J.; Berger, M.; Schlatter, S.; Herwig, C. Metabolic Control in Mammalian Fed-Batch Cell Cultures for Reduced Lactic Acid Accumulation and Improved Process Robustness. *Bioengineering* **2016**, *3*, 5. [CrossRef]
82. Konakovsky, V.; Yagtu, A.C.; Clemens, C.; Müller, M.M.; Berger, M.; Schlatter, S.; Herwig, C. Universal Capacitance Model for Real-Time Biomass in Cell Culture. *Sensors* **2015**, *15*, 22128–22150. [CrossRef] [PubMed]

83. Casablancas, A.; Gámez, X.; Lecina, M.; Solà, C.; Cairó, J.J.; Gòdia, F. Comparison of control strategies for fed-batch culture of hybridoma cells based on on-line monitoring of oxygen uptake rate, optical cell density and glucose concentration. *J. Chem. Technol. Biotechnol.* **2013**, *88*, 1680–1689. [CrossRef]

84. Goldrick, S.; Lee, K.; Spencer, C.; Holmes, W.; Kuiper, M.; Turner, R.; Farid, S.S. On-Line Control of Glucose Concentration in High-Yielding Mammalian Cell Cultures Enabled Through Oxygen Transfer Rate Measurements. *Biotechnol. J.* **2018**, *13*, e1700607. [CrossRef] [PubMed]

85. Kamen, A.A.; Bédard, C.; Tom, R.; Perret, S.; Jardin, B. On-line monitoring of respiration in recombinant-baculovirus infected and uninfected insect cell bioreactor cultures. *Biotechnol. Bioeng.* **1996**, *50*, 36–48. [CrossRef]

86. Pappenreiter, M.; Sissolak, B.; Sommeregger, W.; Striedner, G. Oxygen Uptake Rate Soft-Sensing via Dynamic kLa Computation: Cell Volume and Metabolic Transition Prediction in Mammalian Bioprocesses. *Front. Bioeng. Biotechnol.* **2019**, *7*. [CrossRef]

87. Lam, H.; Kostov, Y. Optical instrumentation for bioprocess monitoring. *Adv. Biochem. Eng. Biotechnol.* **2009**, *116*, 125–142. [CrossRef]

88. Becker, M.; Junghans, L.; Teleki, A.; Bechmann, J.; Takors, R. The Less the Better: How Suppressed Base Addition Boosts Production of Monoclonal Antibodies with Chinese Hamster Ovary Cells. *Front. Bioeng. Biotechnol.* **2019**, *7*, 76. [CrossRef] [PubMed]

89. Xing, Z.; Kenty, B.M.; Li, Z.J.; Lee, S.S. Scale-up analysis for a CHO cell culture process in large-scale bioreactors. *Biotechnol. Bioeng.* **2009**, *103*, 733–746. [CrossRef]

90. Doi, T.; Kajihara, H.; Chuman, Y.; Kuwae, S.; Kaminagayoshi, T.; Omasa, T. Development of a scale-up strategy for Chinese hamster ovary cell culture processes using the kLa ratio as a direct indicator of gas stripping conditions. *Biotechnol. Prog.* **2020**, *36*, e3000. [CrossRef]

91. Heinzle, E. Present and potential applications of mass spectrometry for bioprocess research and control. *J. Biotechnol.* **1992**, *25*, 81–114. [CrossRef]

92. Lyubarskaya, Y.; Kobayashi, K.; Swann, P. Application of mass spectrometry to facilitate advanced process controls of biopharmaceutical manufacture. *Pharm. Bioprocess.* **2015**, *3*, 313–321. [CrossRef]

93. Behrendt, U.; Koch, S.; Gooch, D.D.; Steegmans, U.; Comer, M.J. Mass spectrometry: A tool for on-line monitoring of animal cell cultures. *Cytotechnology* **1994**, *14*, 157–165. [CrossRef] [PubMed]

94. Goh, H.-Y.; Sulu, M.; Alosert, H.; Lewis, G.L.; Josland, G.D.; Merriman, D.E. Applications of off-gas mass spectrometry in fed-batch mammalian cell culture. *Bioprocess Biosyst. Eng.* **2020**, *43*, 483–493. [CrossRef] [PubMed]

95. Floris, P.; Dorival-García, N.; Lewis, G.; Josland, G.; Merriman, D.; Bones, J. Real-time characterization of mammalian cell culture bioprocesses by magnetic sector MS. *Anal. Methods* **2020**, *12*, 5601–5612. [CrossRef]

96. Rogers, R.S.; Nightlinger, N.S.; Livingston, B.; Campbell, P.; Bailey, R.; Balland, A. Development of a quantitative mass spectrometry multi-attribute method for characterization, quality control testing and disposition of biologics. *mAbs* **2015**, *7*, 881–890. [CrossRef] [PubMed]

97. Luchner, M.; Gutmann, R.; Bayer, K.; Dunkl, J.; Hansel, A.; Herbig, J.; Singer, W.; Strobl, F.; Winkler, K.; Striedner, G. Implementation of proton transfer reaction-mass spectrometry (PTR-MS) for advanced bioprocess monitoring. *Biotechnol. Bioeng.* **2012**, *109*, 3059–3069. [CrossRef] [PubMed]

98. Schmidberger, T.; Gutmann, R.; Bayer, K.; Kronthaler, J.; Huber, R. Advanced online monitoring of cell culture off-gas using proton transfer reaction mass spectrometry. *Biotechnol. Prog.* **2014**, *30*, 496–504. [CrossRef]

99. Rogers, R.S.; Abernathy, M.; Richardson, D.D.; Rouse, J.C.; Sperry, J.B.; Swann, P.; Wypych, J.; Yu, C.; Zang, L.; Deshpande, R. A View on the Importance of "Multi-Attribute Method" for Measuring Purity of Biopharmaceuticals and Improving Overall Control Strategy. *AAPS J.* **2017**, *20*, 7. [CrossRef] [PubMed]

100. Dong, J.; Migliore, N.; Mehrman, S.J.; Cunningham, J.; Lewis, M.J.; Hu, P. High-Throughput, Automated Protein A Purification Platform with Multiattribute LC–MS Analysis for Advanced Cell Culture Process Monitoring. *Anal. Chem.* **2016**, *88*, 8673–8679. [CrossRef] [PubMed]

101. Xu, W.; Jimenez, R.B.; Mowery, R.; Luo, H.; Cao, M.; Agarwal, N.; Ramos, I.; Wang, X.; Wang, J. A Quadrupole Dalton-based multi-attribute method for product characterization, process development, and quality control of therapeutic proteins. *mAbs* **2017**, *9*, 1186–1196. [CrossRef] [PubMed]

102. Wang, Y.; Feng, P.; Sosic, Z.; Zang, L. Monitoring glycosylation profile and protein titer in cell culture samples using ZipChip CE-MS. *J. Anal. Bioanal. Tech.* **2017**, *8*, 2. [CrossRef]

103. Rogstad, S.; Faustino, A.; Ruth, A.; Keire, D.; Boyne, M.; Park, J. A Retrospective Evaluation of the Use of Mass Spectrometry in FDA Biologics License Applications. *J. Am. Soc. Mass Spectrom.* **2017**, *28*, 786–794. [CrossRef]

104. Bauer, I.; Poggendorf, I.; Spichiger, S.; Spichiger-Keller, U.E.; John, G. Novel single-use sensors for online measurement of glucose. *BioProcess Int.* **2012**, *10*, 56–60.

105. Spichiger, S.; Spichiger-Keller, U.E. New single-use sensors for online measurement of glucose and lactate: The answer to the PAT Initiative. In *Single-Use Technology in Biopharmaceutical Manufacture*; Eibl, R., Eibl, D., Eds.; Wiley: Hoboken, NJ, USA, 2010; pp. 295–299.

106. Wasalathanthri, D.P.; Rehmann, M.S.; Song, Y.; Gu, Y.; Mi, L.; Shao, C.; Chemmalil, L.; Lee, J.; Ghose, S.; Borys, M.C.; et al. Technology outlook for real-time quality attribute and process parameter monitoring in biopharmaceutical development—A review. *Biotechnol. Bioeng.* **2020**, *117*, 3182–3198. [CrossRef] [PubMed]

107. Claßen, J.; Aupert, F.; Reardon, K.F.; Solle, D.; Scheper, T. Spectroscopic sensors for in-line bioprocess monitoring in research and pharmaceutical industrial application. *Anal. Bioanal. Chem.* **2017**, *409*, 651–666. [CrossRef]

108. Kara, S.; Mueller, J.J.; Liese, A. Online analysis methods for monitoring of bioprocesses. *Chim. Oggi-Chem. Today* **2011**, *29*, 38–41.

109. Landgrebe, D.; Haake, C.; Höpfner, T.; Beutel, S.; Hitzmann, B.; Scheper, T.; Rhiel, M.; Reardon, K.F. On-line infrared spectroscopy for bioprocess monitoring. *Appl. Microbiol. Biotechnol.* **2010**, *88*, 11–22. [CrossRef] [PubMed]

110. Scarff, M.; Arnold, S.A.; Harvey, L.M.; McNeil, B. Near infrared spectroscopy for bioprocess monitoring and control: Current status and future trends. *Crit. Rev. Biotechnol.* **2006**, *26*, 17–39. [CrossRef] [PubMed]

111. Rolinger, L.; Rüdt, M.; Hubbuch, J. A critical review of recent trends, and a future perspective of optical spectroscopy as PAT in biopharmaceutical downstream processing. *Anal. Bioanal. Chem.* **2020**, *412*, 2047–2064. [CrossRef]

112. Lourenço, N.D.; Lopes, J.A.; Almeida, C.F.; Sarraguça, M.C.; Pinheiro, H.M. Bioreactor monitoring with spectroscopy and chemometrics: A review. *Anal. Bioanal. Chem.* **2012**, *404*, 1211–1237. [CrossRef]

113. Ude, C.; Schmidt-Hager, J.; Findeis, M.; John, G.T.; Scheper, T.; Beutel, S. Application of an online-biomass sensor in an optical multisensory platform prototype for growth monitoring of biotechnical relevant microorganism and cell lines in single-use shake flasks. *Sensors* **2014**, *14*, 17390–17405. [CrossRef]

114. Edlich, A.; Magdanz, V.; Rasch, D.; Demming, S.; Aliasghar Zadeh, S.; Segura, R.; Kähler, C.; Radespiel, R.; Büttgenbach, S.; Franco-Lara, E.; et al. Microfluidic reactor for continuous cultivation of Saccharomyces cerevisiae. *Biotechnol. Prog.* **2010**, *26*, 1259–1270. [CrossRef]

115. Teixeira, A.P.; Oliveira, R.; Alves, P.M.; Carrondo, M.J.T. Advances in on-line monitoring and control of mammalian cell cultures: Supporting the PAT initiative. *Biotechnol. Adv.* **2009**, *27*, 726–732. [CrossRef] [PubMed]

116. Randek, J.; Mandenius, C.-F. On-line soft sensing in upstream bioprocessing. *Crit. Rev. Biotechnol.* **2018**, *38*, 106–121. [CrossRef]

117. Ghisaidoobe, A.B.T.; Chung, S.J. Intrinsic tryptophan fluorescence in the detection and analysis of proteins: A focus on Förster resonance energy transfer techniques. *Int. J. Mol. Sci.* **2014**, *15*, 22518–22538. [CrossRef] [PubMed]

118. Beutel, S.; Henkel, S. In situ sensor techniques in modern bioprocess monitoring. *Appl. Microbiol. Biotechnol.* **2011**, *91*, 1493–1505. [CrossRef]

119. Flores-Cosío, G.; Herrera-López, E.J.; Arellano-Plaza, M.; Gschaedler-Mathis, A.; Kirchmayr, M.; Amaya-Delgado, L. Application of dielectric spectroscopy to unravel the physiological state of microorganisms: Current state, prospects and limits. *Appl. Microbiol. Biotechnol.* **2020**, *104*, 6101–6113. [CrossRef] [PubMed]

120. Cannizzaro, C.; Gügerli, R.; Marison, I.; von Stockar, U. On-line biomass monitoring of CHO perfusion culture with scanning dielectric spectroscopy. *Biotechnol. Bioeng.* **2003**, *84*, 597–610. [CrossRef]

121. Yardley, J.E.; Kell, D.B.; Barrett, J.; Davey, C.L. On-Line, Real-Time Measurements of Cellular Biomass using Dielectric Spectroscopy. *Biotechnol. Genet. Eng. Rev.* **2000**, *17*, 3–36. [CrossRef] [PubMed]

122. Zhang, A.; Tsang, V.L.; Moore, B.; Shen, V.; Huang, Y.M.; Kshirsagar, R.; Ryll, T. Advanced process monitoring and feedback control to enhance cell culture process production and robustness. *Biotechnol. Bioeng.* **2015**, *112*, 2495–2504. [CrossRef]

123. Ulber, R.; Frerichs, J.-G.; Beutel, S. Optical sensor systems for bioprocess monitoring. *Anal. Bioanal. Chem.* **2003**, *376*, 342–348. [CrossRef] [PubMed]

124. Harms, P.; Kostov, Y.; Rao, G. Bioprocess monitoring. *Curr. Opin. Biotechnol.* **2002**, *13*, 124–127. [CrossRef]

125. Hanson, M.A.; Ge, X.; Kostov, Y.; Brorson, K.A.; Moreira, A.R.; Rao, G. Comparisons of optical pH and dissolved oxygen sensors with traditional electrochemical probes during mammalian cell culture. *Biotechnol. Bioeng.* **2007**, *97*, 833–841. [CrossRef]

126. Stine, J.M.; Beardslee, L.A.; Sathyam, R.M.; Bentley, W.E.; Ghodssi, R. Electrochemical Dissolved Oxygen Sensor-Integrated Platform for Wireless In Situ Bioprocess Monitoring. *Sens. Actuators B Chem.* **2020**, *320*, 128381. [CrossRef]

127. Tric, M.; Lederle, M.; Neuner, L.; Dolgowjasow, I.; Wiedemann, P.; Wölfl, S.; Werner, T. Optical biosensor optimized for continuous in-line glucose monitoring in animal cell culture. *Anal. Bioanal. Chem.* **2017**, *409*, 5711–5721. [CrossRef] [PubMed]

128. Vojinović, V.; Cabral, J.M.S.; Fonseca, L.P. Real-time bioprocess monitoring: Part I: In situ sensors. *Sens. Actuators B Chem.* **2006**, *114*, 1083–1091. [CrossRef]

129. Xu, J.; Tang, P.; Yongky, A.; Drew, B.; Borys, M.C.; Liu, S.; Li, Z.J. Systematic development of temperature shift strategies for Chinese hamster ovary cells based on short duration cultures and kinetic modeling. *mAbs* **2019**, *11*, 191–204. [CrossRef] [PubMed]

130. KDBIO. Flow Cell Biosensor for Glucose and/or Lactate. Available online: https://www.kdbio.com/products/glucose-lactate-flow-cell-biosensor/ (accessed on 22 August 2021).

131. Luttmann, R.; Bracewell, D.G.; Cornelissen, G.; Gernaey, K.V.; Glassey, J.; Hass, V.C.; Kaiser, C.; Preusse, C.; Striedner, G.; Mandenius, C.-F. Soft sensors in bioprocessing: A status report and recommendations. *Biotechnol. J.* **2012**, *7*, 1040–1048. [CrossRef]

132. Carrondo, M.J.T.; Alves, P.M.; Carinhas, N.; Glassey, J.; Hesse, F.; Merten, O.-W.; Micheletti, M.; Noll, T.; Oliveira, R.; Reichl, U.; et al. How can measurement, monitoring, modeling and control advance cell culture in industrial biotechnology? *Biotechnol. J.* **2012**, *7*, 1522–1529. [CrossRef]

133. Sommeregger, W.; Sissolak, B.; Kandra, K.; von Stosch, M.; Mayer, M.; Striedner, G. Quality by control: Towards model predictive control of mammalian cell culture bioprocesses. *Biotechnol. J.* **2017**, *12*, 1600546. [CrossRef]

134. Luo, Y.; Kurian, V.; Ogunnaike, B.A. Bioprocess systems analysis, modeling, estimation, and control. *Curr. Opin. Chem. Eng.* **2021**, *33*, 100705. [CrossRef]

135. Mandenius, C.-F.; Gustavsson, R. Mini-review: Soft sensors as means for PAT in the manufacture of bio-therapeutics. *J. Chem. Technol. Biotechnol.* **2015**, *90*, 215–227. [CrossRef]

136. Sha, S.; Huang, Z.; Wang, Z.; Yoon, S. Mechanistic modeling and applications for CHO cell culture development and production. *Curr. Opin. Chem. Eng.* **2018**, *22*, 54–61. [CrossRef]
137. Ohadi, K.; Legge, R.L.; Budman, H.M. Development of a soft-sensor based on multi-wavelength fluorescence spectroscopy and a dynamic metabolic model for monitoring mammalian cell cultures. *Biotechnol. Bioeng.* **2015**, *112*, 197–208. [CrossRef]
138. Narayanan, H.; Behle, L.; Luna, M.F.; Sokolov, M.; Guillén-Gosálbez, G.; Morbidelli, M.; Butté, A. Hybrid-EKF: Hybrid model coupled with extended Kalman filter for real-time monitoring and control of mammalian cell culture. *Biotechnol. Bioeng.* **2020**, *117*, 2703–2714. [CrossRef]
139. Tsopanoglou, A.; Jiménez del Val, I. Moving towards an era of hybrid modelling: Advantages and challenges of coupling mechanistic and data-driven models for upstream pharmaceutical bioprocesses. *Curr. Opin. Chem. Eng.* **2021**, *32*, 100691. [CrossRef]
140. Simutis, R.; Lübbert, A. Hybrid Approach to State Estimation for Bioprocess Control. *Bioengineering* **2017**, *4*, 21. [CrossRef]
141. Amribt, Z.; Dewasme, L.; Wouwer, A.V.; Bogaerts, P. Parameter Identification for State Estimation: Design of an Extended Kalman Filter for Hybridoma Cell Fed-Batch Cultures. *IFAC Proc. Vol.* **2014**, *47*, 1170–1175. [CrossRef]
142. Hille, R.; Brandt, H.; Colditz, V.; Classen, J.; Hebing, L.; Langer, M.; Kreye, S.; Neymann, T.; Krämer, S.; Tränkle, J.; et al. Application of Model-based Online Monitoring and Robust Optimizing Control to Fed-Batch Bioprocesses. *IFAC-Pap.* **2020**, *53*, 16846–16851. [CrossRef]
143. Kroll, P.; Stelzer, I.V.; Herwig, C. Soft sensor for monitoring biomass subpopulations in mammalian cell culture processes. *Biotechnol. Lett.* **2017**, *39*, 1667–1673. [CrossRef] [PubMed]
144. Ohadi, K.; Aghamohseni, H.; Legge, R.L.; Budman, H.M. Fluorescence-based soft sensor for at situ monitoring of chinese hamster ovary cell cultures. *Biotechnol. Bioeng.* **2014**, *111*, 1577–1586. [CrossRef] [PubMed]
145. Schwab, K.; Amann, T.; Schmid, J.; Handrick, R.; Hesse, F. Exploring the capabilities of fluorometric online monitoring on chinese hamster ovary cell cultivations producing a monoclonal antibody. *Biotechnol. Prog.* **2016**, *32*, 1592–1600. [CrossRef] [PubMed]
146. Kozma, B.; Salgó, A.; Gergely, S. Comparison of multivariate data analysis techniques to improve glucose concentration prediction in mammalian cell cultivations by Raman spectroscopy. *J. Pharm. Biomed. Anal.* **2018**, *158*, 269–279. [CrossRef]
147. Rafferty, C.; O'Mahony, J.; Burgoyne, B.; Rea, R.; Balss, K.M.; Latshaw II, D.C. Raman spectroscopy as a method to replace off-line pH during mammalian cell culture processes. *Biotechnol. Bioeng.* **2020**, *117*, 146–156. [CrossRef]
148. Bhatia, H.; Mehdizadeh, H.; Drapeau, D.; Yoon, S. In line monitoring of amino acids in mammalian cell cultures using Raman spectroscopy and multivariate chemometrics models. *Eng. Life Sci.* **2018**, *18*, 55–61. [CrossRef] [PubMed]
149. Chen, Y.; Yang, O.; Sampat, C.; Bhalode, P.; Ramachandran, R.; Ierapetritou, M. Digital Twins in Pharmaceutical and Biopharmaceutical Manufacturing: A Literature Review. *Processes* **2020**, *8*, 1088. [CrossRef]
150. Brunner, V.; Siegl, M.; Geier, D.; Becker, T. Challenges in the Development of Soft Sensors for Bioprocesses: A Critical Review. *Front. Bioeng. Biotechnol.* **2021**, *9*, 730. [CrossRef] [PubMed]
151. Kabugo, J.C.; Jämsä-Jounela, S.-L.; Schiemann, R.; Binder, C. Industry 4.0 based process data analytics platform: A waste-to-energy plant case study. *Int. J. Electr. Power Energy Syst.* **2020**, *115*, 105508. [CrossRef]
152. Mears, L.; Stocks, S.M.; Sin, G.; Gernaey, K.V. A review of control strategies for manipulating the feed rate in fed-batch fermentation processes. *J. Biotechnol.* **2017**, *245*, 34–46. [CrossRef]
153. Rathore, A.S.; Mishra, S.; Nikita, S.; Priyanka, P. Bioprocess Control: Current Progress and Future Perspectives. *Life* **2021**, *11*, 557. [CrossRef]
154. Velez-Suberbie, M.L.; Betts, J.P.J.; Walker, K.L.; Robinson, C.; Zoro, B.; Keshavarz-Moore, E. High throughput automated microbial bioreactor system used for clone selection and rapid scale-down process optimization. *Biotechnol. Prog.* **2018**, *34*, 58–68. [CrossRef]
155. Honda, H.; Kobayashi, T. Fuzzy control of bioprocess. *J. Biosci. Bioeng.* **2000**, *89*, 401–408. [CrossRef]
156. Anderlei, T.; Schulte, A.; Laidlaw, D. Kuhner TOM for off-gas analysis in shake flasks. In Proceedings of the Recent Advances in Fermentation Technology (RAFT 13), Bonita Springs, FL, USA, 27–30 October 2019.

processes

MDPI

Review

On the Use of Surface Plasmon Resonance-Based Biosensors for Advanced Bioprocess Monitoring

Jimmy Gaudreault [1], Catherine Forest-Nault [1,2], Gregory De Crescenzo [1], Yves Durocher [2,3] and Olivier Henry [1,*]

[1] Department of Chemical Engineering, Polytechnique Montreal, Montreal, QC H3T 1J4, Canada; jimmy.gaudreault@polymtl.ca (J.G.); catherine.forest-nault@polymtl.ca (C.F.-N.); gregory.decrescenzo@polymtl.ca (G.D.C.)

[2] Human Health Therapeutics Research Centre, National Research Council of Canada, Montreal, QC H4P 2R2, Canada; yves.durocher@cnrc-nrc.gc.ca

[3] Department of Biochemistry and Molecular Medicine, University of Montreal, Montreal, QC H3T 1J4, Canada

* Correspondence: olivier.henry@polymtl.ca

Citation: Gaudreault, J.; Forest-Nault, C.; De Crescenzo, G.; Durocher, Y.; Henry, O. On the Use of Surface Plasmon Resonance-Based Biosensors for Advanced Bioprocess Monitoring. *Processes* **2021**, *9*, 1996. https://doi.org/10.3390/pr9111996

Academic Editors: Ralf Pörtner and Johannes Möller

Received: 18 October 2021
Accepted: 2 November 2021
Published: 9 November 2021

Publisher's Note: MDPI stays neutral with regard to jurisdictional claims in published maps and institutional affiliations.

Abstract: Biomanufacturers are being incited by regulatory agencies to transition from a quality by testing framework, where they extensively test their product after their production, to more of a quality by design or even quality by control framework. This requires powerful analytical tools and sensors enabling measurements of key process variables and/or product quality attributes during production, preferably in an online manner. As such, the demand for monitoring technologies is rapidly growing. In this context, we believe surface plasmon resonance (SPR)-based biosensors can play a role in enabling the development of improved bioprocess monitoring and control strategies. The SPR technique has been profusely used to probe the binding behavior of a solution species with a sensor surface-immobilized partner in an investigative context, but its ability to detect binding in real-time and without a label has been exploited for monitoring purposes and is promising for the near future. In this review, we examine applications of SPR that are or could be related to bioprocess monitoring in three spheres: biotherapeutics production monitoring, vaccine monitoring, and bacteria and contaminant detection. These applications mainly exploit SPR's ability to measure solution species concentrations, but performing kinetic analyses is also possible and could prove useful for product quality assessments. We follow with a discussion on the limitations of SPR in a monitoring role and how recent advances in hardware and SPR response modeling could counter them. Mainly, throughput limitations can be addressed by multi-detection spot instruments, and nonspecific binding effects can be alleviated by new antifouling materials. A plethora of methods are available for cell growth and metabolism monitoring, but product monitoring is performed mainly a posteriori. SPR-based biosensors exhibit potential as product monitoring tools from early production to the end of downstream processing, paving the way for more efficient production control. However, more work needs to be done to facilitate or eliminate the need for sample preprocessing and to optimize the experimental protocols.

Keywords: surface plasmon resonance (SPR); bioprocess; monitoring; biosensor; quality by design (QbD); process analytical technology (PAT); biotherapeutics production; vaccines production

1. Introduction

Biomanufacturers are subject to strict directives imposed by regulatory agencies, such as the Food and Drug Administration (FDA) in the United States and the European Medicine Agency (EMA) in Europe, to ensure their products are fit to the highest standards of quality, efficacy, and safety. The biotherapeutic market of today is rapidly growing and evolving, with notable contributions attributed to the rise of therapeutic monoclonal antibodies [1] and expiring patents allowing the creation of more and more biosimilars [2].

Amid increasing demands, biomanufacturers aim to increase throughputs while retaining the quality and safety of their products. Products are typically extensively tested after production with fixed process parameters in what is commonly called a quality by testing framework. During the 2000s, the FDA introduced the concept of Quality by Design (QbD) to guide quality control [3]. QbD aims to build quality within the product at the design stage rather than test it after its production. To accomplish that, the critical quality attributes (CQA) of a product must be identified to construct a quality target product profile (QTPP), which links CQAs to the critical process parameters (CPP). Such relations can be known a priori or investigated using concepts of design of experiments and a statistical analysis. Ensuring repeatability by aiming to maintain CQAs constant from batch to batch, rather than the process parameters, enables more process flexibility [4,5]. Doing so requires a thorough understanding of which attributes of a biotherapeutic are critical to its efficacy and safety and of their dependency on the process variables. Risk assessment and continuous improvement are also key concepts of QbD [6]. Knowing the link between CQAs and CPPs, quality by control (QbC) becomes possible by manipulating CPPs to regulate CQAs. QbC necessitates measuring various process variables during production that are either CQAs themselves or that allow the prediction of CQAs via predictive modeling techniques [7,8].

Process analytical technology (PAT) is another initiative from the FDA from the middle of the 2000s [9]. PAT encompasses the development of sensors allowing monitoring and process control methods taking advantage of the measurements provided by these sensors [10]. Techniques and devices aiming to measure the critical process variables linked to CQAs in a timely manner compared to the process dynamics are very much a part of the PAT framework [11]. As such, there is now a vast interest in techniques allowing process variable measurements online or at-line of the production vessel. Such tools enable the monitoring of bioprocesses either during upstream or downstream processing. Techniques that allow the monitoring of cell growth or cell metabolism may not be sufficient to perform QbC, introducing a need for sensors allowing measurements on the product itself. Such sensors would not only be useful for production at the industrial scale but also during process development, when multiple conditions need to be tested. More efficient process development leads to a lower time to market, which is highly beneficial. To this day, adoption of the PAT framework is much more prominent at the R&D stage than it is at the production stage, but we believe powerful new tools and ingenious data analysis methods could change that in the future. Figure 1 illustrates how real-time quality assessment tools can lead to a greater production efficiency by allowing decision-making during the production rather than a posteriori. Such process analytical technologies would enable the implementation of a quality by control framework in which production issues may be solved in real time by adjusting the production parameters. This would lead to a greater efficiency by ensuring more production batches are of acceptable quality. In this mindset, this review investigates the potential of surface plasmon resonance (SPR)-based biosensors as a monitoring tool that will allow real-time quality assessments and/or quantitation.

The surface plasmon resonance phenomenon was established by the pioneering work of Otto, Kretschmann and Raether in the late 1960s [12–14]. It was first commercialized in the form of a biosensor capable of detecting interactions between an immobilized species and its solubilized binding partner by Pharmacia (subsequently Biacore, now commercialized by Cytiva, formerly GE Healthcare) in the 1990s. Advances in the liquid handling and control systems have greatly improved the precision and sensitivity of SPR-based biosensors in the last three decades [15–21]. On top of this, robust protocols [22] and data analysis works [23–31] have helped establish SPR as a premiere technique in the field of biomolecule interaction analyses. Its main advantage lies in its ability to detect interactions without a label, which simplifies the assay design. As the interaction depends on a biological event between the immobilized and solubilized species, the technique only detects bioactive compounds rather than all compounds harboring a given label. Another significant benefit of SPR is its ability to track the interaction in real time. This enables a kinetic analysis of the interaction on top of affinity measurements [15]. A concentration analysis is also possible [32–35], and as such, SPR represents an interesting alternative to enzyme-linked immunosorbent assays (ELISA) for quantitation purposes [36]. SPR technology has been used in multiple fields, such as drug screening [16–18,21]; drug potency assessments [37]; vaccine quantitation [38–42]; food safety (the detection of bacteria, pathogens, and other impurities) [43–50]; biotherapeutics characterization and quality control [51–57]; biotherapeutic safety [58]; medical diagnostics [20,59–63]; environmental monitoring [64]; and bioproduction monitoring [55,65–68]. Although little work has been done for the aim of using an SPR biosensor online or at-line from a bioreactor in an automatic fashion [67,68], a growing body of works showing accurate concentration measurements in complex media such as cell lysate [55,65,67,68] and human serum [69] on top of advancements in antifouling technologies [70,71] pave the way for using surface plasmon resonance-based biosensors in a bioprocess monitoring framework.

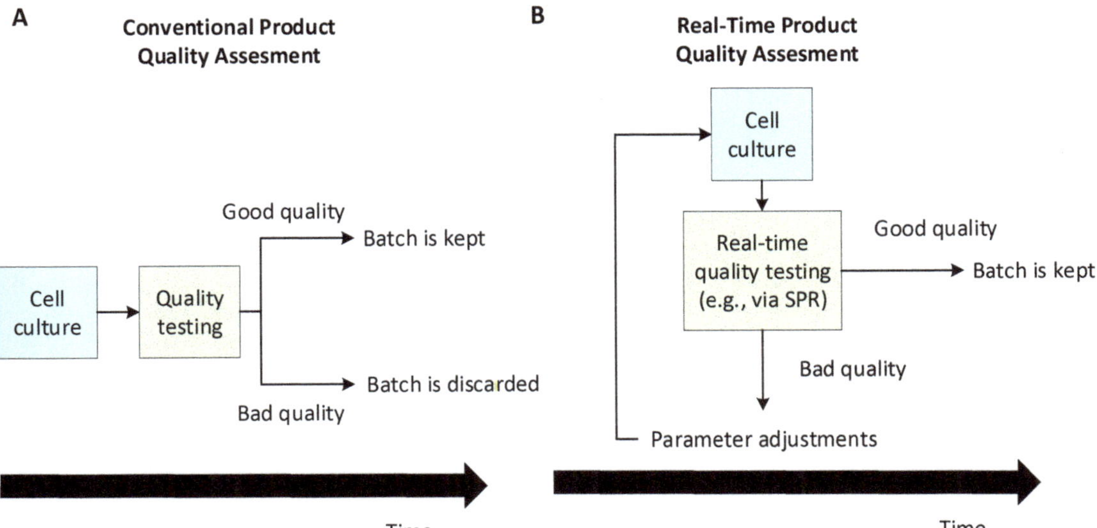

Figure 1. Conceptual differences between a conventional a posteriori quality assessment (**A**) and real-time quality assessment (**B**) of a cell culture-derived product. The conventional framework allows the detection of production issues only after the cell culture is finished, which may cause significant losses if a production batch is found to be inadequate and needs to be discarded. On the other hand, in a real-time quality assessment framework, a batch can be preemptively stopped if production problems start being detected during the cell culture. Eventually, a quality by control framework becomes possible, where one adjusts the production parameters to solve quality problems during the production. This can help discard fewer batches and, thus, greatly increase the efficiency.

First, the SPR phenomenon will be presented with the main considerations for a robust SPR assay, as well as general directions for kinetic and concentration analyses using SPR. The applications of SPR in monitoring biotherapeutics, vaccine antigens, viral particles, and bacteria will then be presented with some details on the assay designs found in the literature. The limitations of SPR in a monitoring context will emerge from this literature review, and they will be discussed next. We will then examine the recent biosensor and modeling headways. These are mainly aimed at increasing the throughput or treating more complex samples. Some compelling developments have not yet been applied to monitoring but could prove powerful tools in the near future. We will conclude with some perspectives on which key technological advancements are necessary to establish SPR as a predominant monitoring tool in the future.

2. Surface Plasmon Resonance: Basic Principles and Methods

Surface plasmon resonance-based biosensors rely on the excitation of an electron cloud, called a plasmon, at the interface between a noble metal—typically gold—and a dielectric medium following the projection of a polarized monochromatic light on the metal. This results in the propagation of an evanescent wave. When the near-infrared incident light projected by a laser hits the metal surface such that the component of the incident wave light that is parallel to the surface (k_x) perfectly aligns with the surface plasmon waves (k_{sp}), the oscillations of the surface plasmon are amplified; hence, there is resonance. A glass prism is typically used to ensure the total internal reflection of the light (see Figure 2A). This corresponds to the geometry suggested by Kretschmann and Raether [12,13]. Therefore, when resonance occurs, energy that would otherwise be used in reflecting the light instead goes to amplifying the oscillations of the electrons, resulting in a drop in the reflected light intensity, which can be measured in real time. The light incident angle for which resonance occurs, called the SPR angle, depends on the refractive index of the dielectric media in the evanescent wave propagation zone. As the refractive index depends on the concentration of material near the surface, SPR can be used to track the accumulation of proteins at the surface in real time by rapidly changing the incident angle of the projected light and monitoring the changes in the SPR angle.

Some SPR biosensors vary the wavelength of the incident light and use a spectrophotometer as a detector. It is also possible to keep the angle and wavelength constant while monitoring the intensity shift. This technique facilitates measurements at multiple detection spots simultaneously using SPR imaging instruments (SPRi; see Section 7). However, configurations that directly measure the SPR angle or wavelength are preferred, as the intensity shift is only a derivative measurement of the SPR angle/wavelength, which are linked more directly to the quantity of accumulated material near the surface [72]. Shifts in the phase of the light when SPR occurs can also be monitored. Such setups have not been broadly commercialized, as they require complex instrumentation, but they allow better sensitivity as the phase shift of the light is more abrupt than the variations of its intensity [73,74] (see Section 7). For more details on the SPR phenomenon and SPR biosensor configurations, the reader is referred to various reviews on SPR biosensing [15–21,74]. A non-exhaustive list of SPR instrument manufacturers is given in Table 1.

SPR biosensors allow real-time and label-free measurements of the interaction between a ligand immobilized on a biosensor metallic surface and an analyte introduced near the surface, for example, via a microfluidic channel. The biological interaction between the analyte and the ligand results in the accumulation of analytes near the biosensor, causing a change in the refractive index and SPR angle (see Figure 2B), which can be monitored to obtain a SPR sensorgram (see Figure 2C).

Ligand immobilization can be performed via multiple chemical approaches by using readily available SPR sensor chips with a tethered carboxymethylated dextran layer, as offered by Cytiva [15]. Oriented strategies are preferred when available, as they minimize the heterogeneity in the ligand interaction and steric hindrance, resulting in sensorgrams that are simpler to analyze.

Following ligand immobilization, multiple SPR cycles can be performed on the same sensor surface. A SPR cycle, or sensorgram, can be separated into three main phases (Figure 2D): first, a buffer injection phase (to obtain a baseline signal), an analyte injection phase during which analyte-ligand complexes are formed (association phase), and a second buffer injection phase to dissociate analyte-ligand complexes (dissociation phase). The association and dissociation durations, injection flow rate, and temperature can be set by the user for most high-grade SPR biosensors. The SPR signal is measured in RU (resonance units), with 1 RU being roughly equivalent to 1 pg/mm^2 of immobilized protein [75].

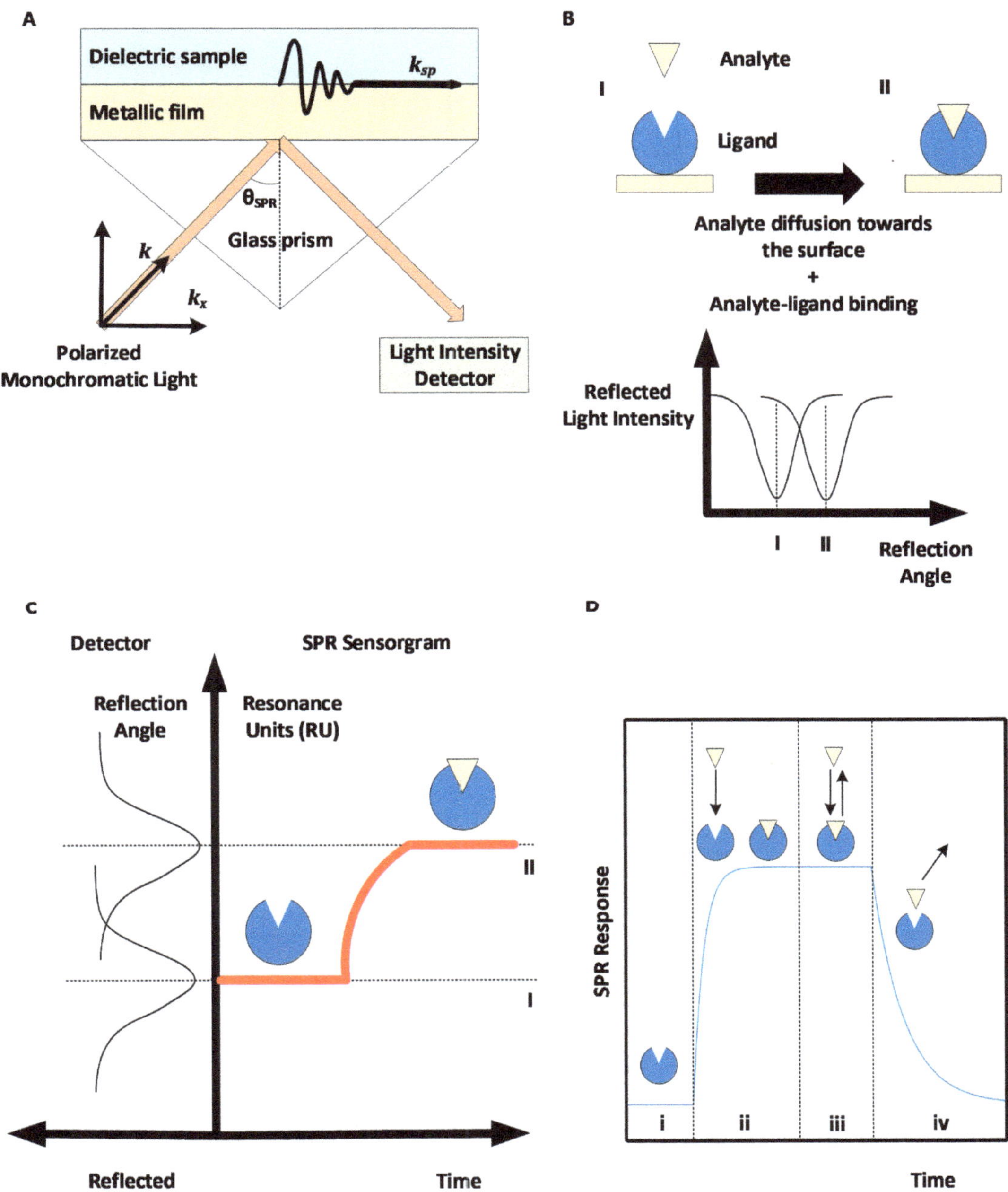

Figure 2. Surface plasmon resonance (SPR)-based biosensing principle. (**A**) Kretschmann-Raether biosensor configuration. Projecting near-infrared light at a specific angle (SPR angle) results in accrued oscillations of electron clouds (plasmon) at the interface between a dielectric sample and a metallic surface, hence the resonance of the surface plasmon. (**B**) Accumulation of materials (analyte molecules) on the metallic SPR surface creates a change in the SPR angle. (**C**) The change in the SPR angle can be monitored in real time to obtain a SPR sensorgram. (**D**) Phases of a SPR sensorgram: (i) baseline, (ii) association phase, (iii) equilibrium, and (iv) dissociation phase.

Table 1. Surface plasmon resonance biosensor manufacturers. SPR instruments can either vary the angle of the incident light or its wavelength. Some manufacturers utilize variations on the traditional SPR technique to increase the throughput or to increase the applicability to complex samples.

Manufacturer	SPR Method	Detection	Reference
Affinité Instruments	SPR	Wavelength	[76]
Biacore (Cytiva)	SPR	Angle	[77]
Bionavis	Multi Parametric SPR (MP-SPR)	Angle	[78]
Biosensing Instrument	SPR	Angle	[79]
Carterra	SPR Imaging (SPRi)	Angle	[80]
Reichert Technologies	SPR	Angle	[81]
Sierra Sensors (Bruker)	SPR Imaging (SPRi)	Angle	[82]

To obtain repeatable signals, it is essential that the amount of available, biologically active ligand molecules remains constant from one sensorgram to the other. In order to ensure this, all analyte molecules should be removed from the surface by the end of the dissociation phase. For slowly dissociating systems, a regeneration step may be necessary during which a harder solution (typically in terms of pH or salt concentration) is injected. Selecting an appropriate regeneration solution may be challenging, as regeneration needs to remove all analyte molecules without removing or damaging the ligand molecules [15,83,84].

SPR sensorgrams can be biased by nonspecific interactions between the analyte or other components of the injected sample and the SPR surface. Signal artefacts can also be observed (caused by sharp refractive index variations when switching from one buffer to another, by electric perturbations due to the biosensor's moving parts, etc.). As such, a robust experimental protocol includes a second SPR surface (mock) to perform a reference. The mock surface is exposed to the same sequence of injections as the active surface, except that it does not harbor any ligand molecules. Thus, subtracting the signal recorded on the reference surface from the signal recorded on the active surface removes nonspecific contributions. A second reference may be performed by repeating the same sequence of injections with a null concentration of the analyte or a blank (i.e., injecting buffer instead of the analyte). This removes signal drifts, which can occur when the temperature near the SPR surface is not constant during the experiment, as the refractive index also depends on the temperature. Sensorgrams obtained by subtracting the reference surface signal and the referenced blank injection signal are said to be double-referenced [22].

The following subsections will describe the two main uses of SPR: analyzing the kinetic and equilibrium behaviors of an analyte-ligand interaction and quantifying the solution of an analyte. Although the latter can perhaps seem more appropriate for bioprocess monitoring purposes, one should also strive to monitor the critical quality attributes (CQAs) of the product, which may influence the binding kinetics to its biological partners. As an example, monoclonal antibody (mAb) N-glycosylation (a CQA of mAbs) is known to influence IgG-FcγR binding kinetics, as measured by SPR, and, hence, the efficacy and safety of therapeutic mAbs [51].

2.1. SPR to Measure Kinetics and Affinity

As SPR biosensors allow real-time measurements, the kinetics of an analyte-ligand interaction can be measured. A typical kinetic SPR assay includes multiple cycles with distinct known analyte concentrations, from which the rate constants are extracted by globally fitting an interaction model to multiple double-referenced sensorgrams via optimization methods [15].

The simplest and most commonly used interaction model, called a 1:1 Langmuir model, contains an association rate $k_a\ [=]\ \mathrm{M}^{-1}s^{-1}$ and a dissociation rate $k_d\ [=]\ s^{-1}$, which mediate a reversible pseudo-reaction between a free analyte molecule A and an immobilized ligand molecule L to form an analyte-ligand complex AL [15,25]. Both kinetic constants have an impact during the association phase, whereas only k_d influences the signal during the dissociation phase. For the model to be appropriate, the predicted signal (with fitted model parameters) should adequately describe the recorded signal. This can be validated by observing that the residues (differences between fitted and measured signals) are randomly distributed with a null mean [23,25]. An inadequate fit to an ideal 1:1 Langmuir model can typically be explained either by a poor SPR assay design or by the presence of a more complex interaction scheme. Such complexity can hail from the presence of heterogeneity in the system (either in the analyte [26,28–31] or the ligand [85–88] molecules), the stoichiometry of the analyte-ligand interaction [89–92], or a conformational change following analyte-ligand binding [93,94]. Forest-Nault et al. [51] provided a description of the modeling approaches that can be used to analyze such complexities in SPR data and offered insight on the modeling of IgG-Fcγ receptor (FcγR) interactions, which has long been studied using SPR and for which nonideal behaviors have repeatedly been observed and confirmed.

The system may reach an equilibrium state during the association phase, for which the binding and dissociation of the analyte-ligand complex occur at the same rate. Such an equilibrium is detected by the appearance of a plateau in the SPR signal. In the Langmuir model, the value of this plateau is characterized by the equilibrium constant $K_D = \frac{k_d}{k_a}$ $[=]$ M or by the affinity $K_A = \frac{k_a}{k_d}$ $[=]$ M^{-1}. Interestingly, many complex models behave similarly to the ideal model at equilibrium, enabling thermodynamic measurements even if the kinetics are not properly understood.

2.2. SPR to Measure Concentrations

Kinetic analyses, like those described in the previous section, are typically performed in conditions for which the analyte-ligand interactions are limited by the interaction kinetics. This is achieved by performing experiments with a relatively low density of immobilized ligands and a high flow rate. Conversely, the conditions of mass transport limitations enable the quantitation of bioactive analytes. Indeed, when using a SPR surface that is highly concentrated in ligand molecules and a low flow rate, the diffusion of the analyte molecules from the bulk solution toward the SPR surface is the limiting step in the analyte-ligand binding process. This is characterized by a linear signal during the association phase of the sensorgrams, with the slope of the signal being proportional to the analyte concentration of the injected solution [34,35]. Interestingly, SPR has been used to estimate analyte concentrations in a multitude of sample media [33], including cell culture broths [55,65,67,68] and human serum [69]. Examples of applications of quantitation via calibration will be discussed in the following sections for different relevant types of analytes.

Analyte concentrations can be measured either with or without calibration, depending on the availability of an analyte solution of a known concentration. The calibration refers to the construction of a standard graph of the initial slope of the SPR signal with respect to the analyte concentration. Under conditions of mass transport limitations, such a graph should be linear, with a null initial value [32,34], although a four-parameter logistic function has also been used [65,95], allowing a broader applicability, since the linearity range may be limited. Rather than the slope, some authors used the SPR signal obtained at the end of a set analyte injection time as the measurement, which may have a broader applicability outside the linearity range [58,65,66,69].

The CFCA method (for calibration-free concentration analysis) requires estimating the mass transport kinetic constant k_M $[=]\frac{m}{s}$. Briefly, k_M is a function of the dimensions of the microfluidic channel, the diffusion coefficient of the analyte in the buffer, and the cubic root of the injection flow rate. An estimate equation of k_M has long been available [34,96,97], but its use remains complex, mainly due to the need to estimate the diffusion coefficient. This can be done using the Einstein-Sutherland equation, which is based on Stoke's law, but it requires knowledge of the relative frictional ratio and the solvent viscosity [32]. The molar weight of the analyte is also necessary to convert k_M from units of concentration to RU [32,34]. To determine the analyte concentration via CFCA, conditions of at least partial mass transport limitations are necessary, and the analyte sample must be injected at two different flow rates, meaning two different k_M values [35]. For details of the CFCA procedure, and a description of the acceptable degree of mass transport limitation, the reader is referred to a relevant review on CFCA [33].

Uncertainties of the value of the analyte diffusion coefficient and the molar weight may bias the computation of k_M and, hence, the concentration estimated via CFCA. Comparing two samples to obtain the ratio of their respective concentration with CFCA may prove more efficient and accurate, as the uncertainties cancel each other when computing the ratio [32,33,37]. A reference sample—for example, a quality control—is then compared to a second sample on the basis of their bioactivity, as SPR only detects bioactive analyte molecules that can interact with the ligand. Matching the absorbance measurements at 280 nm normalizes the quantity of the proteins in both samples, enabling potency comparison [32]. Potency could be compared across different batches to ensure batch-to-batch variability in product folding. Potency could also be monitored before and after each purification step of a bioprocess to detect product denaturation caused by the purification process [32,98] or before and after exposure to a stress such as heat or pH to quantify the product stability [33,37].

Relative CFCA was used in guiding the purification process of interferon α-2a [32,98] by helping select a chromatography resin and the purification conditions, which maximized the purified product bioactivity. Notably, by using a mouse anti-interferon α antibody as the ligand, a relative concentration (normalized via absorbance) above 100% was found between the samples before and after purification, indicating that purification indeed removed denatured products. CFCA was also used to evaluate the potency of TNF-α before and after temperature stress, proving an alternative to a traditional EC_{50} analysis [37]. Absolute CFCA was recently used to quantify human myoglobin with good agreement with isotope dilution mass spectrometry measurements [99], and a mathematical framework was developed to quantify analytes prone to self-association [100]. CFCA, especially in a relative framework, has already been proven to be efficient and accurate, and we predict its use will increase in the near future.

3. SPR Applications to Biotherapeutics Production Monitoring

Biotherapeutics are mainly produced by cultivating mammalian cells in a bioreactor. These bioprocesses can be sensitive to various operating parameters, such as pH, temperature, culture media, feeding strategy, bioreactor operation mode, cell line, etc. [101]. These parameters may be dynamic, and the length of the culture may also influence the product quality. As such, there is an obvious interest for tools allowing monitoring during production. Surface plasmon resonance-based biosensors represent an interesting option in that regard, as they can be used to quantify the product (as a faster alternative to ELISA) and to probe binding to its biological partners, which can often be indicative of the product quality, efficacy, and/or safety. SPR has been used to analyze various types of molecules in this context, including antibodies [65–67,102], fusion proteins [57], peptides [68], complement fragments [58], and other blood proteins [66,103].

Table 2 summarizes the quantitation assays that will be described in this section. Although they mostly aim at quantifying antibodies, similar assays could be used to quantify any number of proteins, as long as an interacting partner (or an antibody) is available to play the role of the immobilized ligand. In addition to these, biotherapeutics quality assessment probes will also be discussed. They come in the form of either a kinetic analysis or a qualitative yes/no binding experiment.

Table 2. Quantitation assays of the biotherapeutics reported in this section. The concentration ranges used to build a calibration curve are also reported.

Target Analyte	Ligand	Range (μg/mL)	SPR Instrument	Reference
mAb	Antibody	0.8–50	Biacore T200	[66]
Albumin	Antibody	2–200		
Transferrin	Antibody	0.2–20		
IgA	Antibody	0.4–50		
IgG	Antibody	2–200		
IgG1	Antibody	2.8–90		
IgG2	Antibody	1.8–60		
IgG3	Antibody	0.4–13		
IgG4	Antibody	0.2–5		
Infliximab (IFX, IgG mAb)	Antigen (TNF-α)	0.5–8	Bio-Rad ProteOn	[69]
Anti-IFX antibody	Antibody (IFX)	5–40		
Anti-GFP antibody (total)	Protein A/G	0.03–2	SPR 2/4 (Bruker Daltonics SPR)	[65]
Anti-GFP antibody (bioactive)	Antigen (GFP)	0.03–2		
Anti-PSMA IgG mAb	Antigen (PSMA)	1.35–30	Biacore 3000	[67]
Trastuzumab (IgG mAb)	FcγRI	0.03–3.75	Biacore T100	[55]
Anti-TNF-α antibody	Antigen (TNF-α)	0.02–360	Biacore T200	[37]

3.1. SPR for the Early Development of Biotherapeutics

As the first SPR-based biosensors became available in the early 1990s, they were used to characterize biomolecule interactions and rapidly played an instrumental role in accelerating therapeutic antibody screening and epitope binning experiments [104]. In combination with technologies such as phage display, which enable the creation of large libraries of candidates based on bacteriophage expression, SPR helped identify candidates with the most therapeutic potential [105,106]. These strategies led to the rise of one of the best-selling therapeutic monoclonal antibodies (mAbs), HUMIRA®, commercialized in 2003, and continue to be an essential part in discovery research for new therapeutics against cancer or diseases like hepatitis, HIV, and Alzheimer's [104,107,108]. Kinetic analyses performed by SPR not only allow discriminating between good candidates based on their target specificity but, also, to better understand and estimate the target occupancy and residence time of the therapy [109]. These parameters have become increasingly important in lead optimization in order to minimize off-target effects and ensure therapeutic efficacy by target engagement.

Several studies have described practical approaches based on SPR to study the off-target binding of candidates by serum proteins. This can significantly influence their pharmacokinetic profile and propensity to induce side effects. Frostell et al. showed early on that it is possible to efficiently evaluate the binding of plasma proteins to drug candidates by immobilizing plasma proteins on an SPR surface. Their results correlated with other methods such as ELISA [110]. Gonzales et al. compared the immunogenicity of potential antibody variants against a protein, TAG-72, expressed by several kinds of carcinomas [111]. Ritter et al. used SPR to measure the antibody response in the serum of patients that were treated with humanized anti-A33, an antibody that targets colon cancer [112]. Today, therapeutic drug and immunogenicity monitoring (TDIM) is more and more implemented to guide therapy with biologics by taking into account personalized drug responses to make informed decisions on the course of treatment. Moreover, SPR was shown to be more cost-effective than ELISA for serum concentration analyses, as SPR protocols tend to be simpler and faster [69]. Therefore, with the advantages of low sample volume consumption and real-time and label-free analysis, SPR has rapidly become recognized as a powerful tool and has been extensively used for the early assessment of target specificity, binding stability, and expression levels, which is essential data for the selection of potential drugs [104].

As the market of biotherapeutics has grown over the years, the complexity of their path to commercialization pushed regulatory agencies and industrials to adapt the development workflow of drugs and add the evaluation of more developability aspects such as post-translational modifications, conformation, aggregation, solution stability, and pharmacokinetic properties upstream of the development to reduce the risk of failure downstream as much as possible [104]. The role of SPR also evolved to meet those new requirements, and novel approaches were developed to rapidly characterize and predict a variety of biotherapeutic attributes. For example, antibody clearance can be partly estimated by its interactions with the neonatal Fc receptor FcRn. In fact, FcRn is responsible for recycling antibodies captured in endosomes by binding to their Fc region and leading them back to the cell surface [113,114]. By studying the antibody-FcRn interactions with SPR, it is possible to identify characteristics or residues of the antibody that are favorable to recycling by FcRn and, thus, increase the antibody half-life [115]. Until recently, SPR-based biosensors and their applications were limited by their low throughput. Recent advances, which will be covered in Section 7, have unlocked the full potential of SPR monitoring at every stage of biotherapeutics development and even during production and purification by increasing the throughput.

3.2. SPR for the Quantification of Biotherapeutics

Therapeutic monoclonal antibodies represent a rapidly growing global market in the biopharmaceutical field estimated to reach USD \$300 billion by 2025 [1]. In a recently published study, two SPR assays were suggested to probe the total and bioactive mAb concentrations from a culture broth [65]. Protein A/G was used as the ligand to detect the total concentration, whereas the mAb-specific antigen (here, green fluorescent protein (GFP)) was used to detect only the bioactive antibodies that are properly folded, such that they may play their biological role. Calibration was performed using samples of known concentrations. Various GFP immobilization strategies were compared, and a combination of His-tag capturing and amine coupling stood out for allowing the fast online quantitation of bioactive anti-GFP antibodies. His-tag capturing enables oriented immobilization, whereas amine coupling offers reusability of the sensor surface as the ligand molecules are covalently bound to the dextran layer of the sensor chip. A limit of detection of 1.8 ng/mL was achieved using the adapted immobilization method, which was shown to be similar to that of an ELISA procedure [65].

SPR has also shown to be a capable alternative to ELISA in quantifying proteins in diluted serum samples. Frostell et al. [66] developed SPR assays for the specific quantitation of eight plasma proteins (IgA, IgGs, albumin, and transferrin) and a recombinant mAb from samples taken during production in Chinese Hamster Ovary (CHO) cell cultures. Interestingly, these authors suggested not to perform any reference when conducting a concentration analysis, as they argue that an unmodified surface is not a representative negative control of the active surface when the latter contains large quantities of ligand molecules such as antibodies. They proposed to use a surface on which an irrelevant antibody is bound as a possible alternative but chose not to do so, as adequation to the active surface remained uncertain. The proposed SPR assays were shown to be robust, precise, and faster than an ELISA. In a notable recent study, Infliximab (IFX), a monoclonal antibody against TNF-α, and antibodies against IFX were quantified in human serum samples [69]. In this study, calibration was performed by spiking the serum samples with known concentrations of analytes. Nonspecific contributions by serum proteins binding to the surface were referenced out by injecting the serum on a mock surface (for assays with TNF-α as the ligand to quantify IFX) or a surface harboring a set level of a nonspecific IgG (for assays with IFX as the ligand to quantify anti-IFX antibodies). The calibration and reference protocols used when treating complex media are particularly relevant to the field of bioprocess monitoring, and such studies show the potential of the SPR technique in treating complex matrices.

Harnessing a SPR biosensor to a bioreactor allowed at-line quantitation of the product during its production via cell culture [67,68]. A system of peristaltic pumps enabled automatic sampling at a fixed time interval, with automated dilution of the cell media. The only purification steps between the biosensor and the 3.5-L bioreactor were a decantation column and a filter, which removed cells and cellular debris that could clog the SPR microfluidic channels. The whole process was fully automated. Figure 3 shows a schematic of this experimental setup. The quantitation of anti-PSMA antibodies was performed by immobilizing PSMA (prostate-specific membrane antigen) on the sensor chip [67]. The concentrations measured at-line were in agreement with the concentrations measured offline with SPR and with quantitative Western blotting followed by densitometry analyses. Calibration was automatically repeated every 12h using samples of known concentrations. More work needs to be done to render SPR biosensing more apt to the treatment of samples being automatically harvested, but these two studies showed an interesting proof of concept.

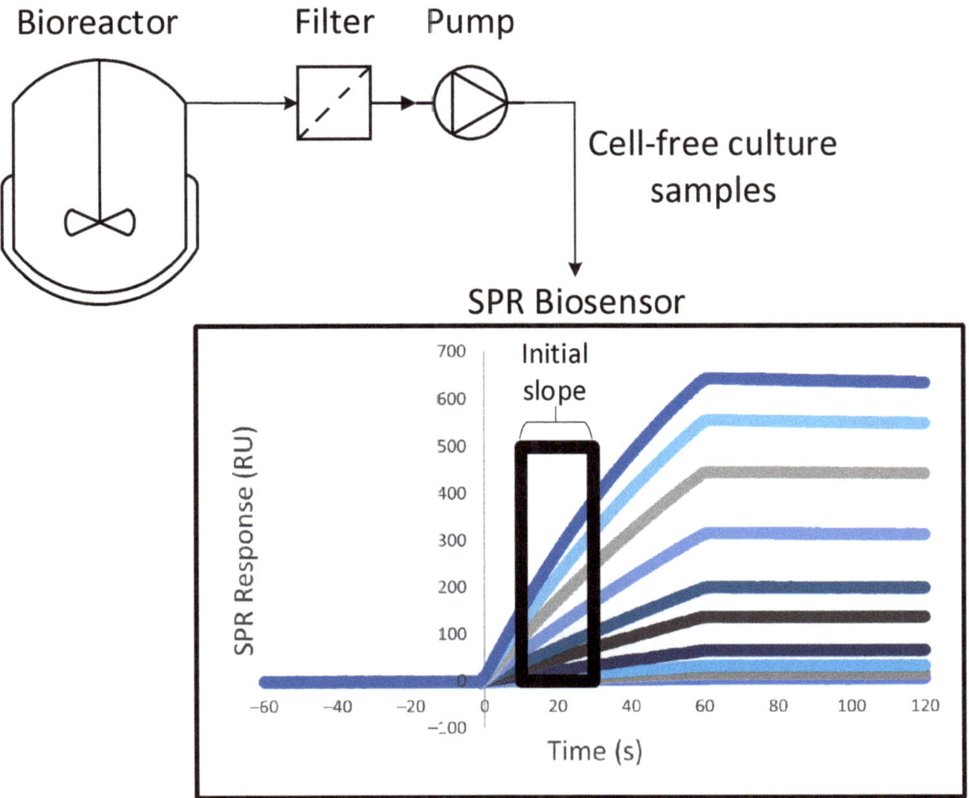

Figure 3. Schematic of the experimental setup suggested by Chavane et al. [67] for an at-line concentration analysis via SPR.

3.3. SPR for the Safety and Quality Assessment of Biotherapeutics

Certain product safety probes can be performed with SPR. As an example, an SPR assay allowing the monitoring of complement activation by nanoparticles was recently detailed [58]. Complement activation is undesirable in this context, as it causes an inflammatory reaction aimed at eliminating the nanoparticles when they are administered intravenously as therapeutic agents. Emulsion-based nanoparticles were incubated in human serum before performing SPR experiments. A total of four SPR surfaces were used. The first one displayed a polyclonal antibody against C3 fragments, while the second contained a monoclonal antibody specific to C3a. Complement activation was detected if either one of these surfaces exhibited a non-null SPR response when injecting the serum samples. The remaining two were reference surfaces: one unmodified and one with an irrelevant antibody. The SPR results were similar to those obtained via immunoelectrophoresis, whereas an ELISA approach resulted in high nonspecific background noise. Removing the nanoparticles prior to the experiment biased the detection of C3a, as it seemed to adsorb on the surface of the nanoparticles, potentially leading to false negatives. C3a could be adequately detected when samples containing the nanoparticles were tested [58].

SPR has also been used to investigate post-translational modifications of proteins. Post-translational modifications may affect the efficacy and safety of biotherapeutics and can therefore be classified as critical quality attributes in a Quality by Design framework. Monoclonal antibody N-glycosylation has been studied abundantly with SPR, as it affects binding to immune system effector cells, by measuring the binding of differently glycosylated mAb samples either to Fc receptors [51–54] or to lectins [56]. In another study, phosphorylation (another post-translational modification) was investigated by immobilizing tumor suppressor P53 on a sensor surface. Kinases were then injected on the surface, causing P53 phosphorylation. As phosphorylation affects P53 binding to the murine double minute 2 (MDM2) protein, the injection of MDM2 enabled the probing of phosphorylation [103].

Most of the current methods to probe the protein quality are ill-suited for online utilization. In contrast, a monitoring tool such as a SPR biosensor could allow the product quality to be considered in the very early stages of bioprocess development, such that it becomes an integral part of the control strategy.

3.4. Sequential SPR Assays

Sequential SPR assays enable both the quantitation of antibodies and the measure of their binding kinetics with Fc receptors in a single sensorgram. A large concentration of antigens is immobilized on the sensor surface. The assay starts with the injection of the antibodies, which bind to the immobilized antigens via their Fab region, enabling quantitation by measuring the slope of the recorded signal. This is followed by an injection of Fc receptors, which bind the antigen-bound antibodies via their Fc region. If the assay is well-calibrated, kinetic evaluation is possible with the recorded signal resulting from the Fc receptor injections. The sequence may be reversed if one aims to evaluate the antigen-binding kinetics. Such a sequential assay was used to quantify Trastuzumab in a culture supernatant by immobilizing Fcγ receptors and then measuring its kinetics to its antigen (HER2) [55]. On the same note, biotinylated TNF-α was immobilized on a SPR surface, enabling the quantitation of an anti-TNF-α antibody. This was followed either by an injection of the TNF-α receptor to verify TNF-α blocking by the antibodies or by an injection of FcγR [37]. In another study, antibodies from crude hybridoma were captured on the SPR surface via an anti-Fc antibody, allowing their quantitation and subsequent antigen-binding kinetic evaluation via a second injection [102]. The first injection (quantitation) may be performed with species in a complex medium (cell lysate), whereas kinetic studies (second injection) are generally performed in a defined buffer for more accurate measurements. Sequential assays could prove a powerful tool for the introduction of SPR biosensors in bioprocess monitoring, as they alleviate the need for purification, multiple SPR surfaces, and regeneration steps by combining the quantitation and quality assessment probes.

4. SPR Applications to Vaccine Production Monitoring

Several research groups have investigated the potential of SPR-based biosensors in quantifying vaccine production. The gold standard in this field has long been the single radial immunodiffusion (SRID) assay, but it requires strain-specific reference antisera and corresponding antigens, which may take up to 6 months to obtain by sheep immunization [40]. ELISA-based assays can be used as robust alternatives, but they remain time-consuming. For this reason, there is a strong interest in methods capable of performing vaccine quantitation without a reference serum, which could supplement SRID to speed up vaccine development. Furthermore, the elaboration of better vaccine potency assays is promoted by regulatory agencies, such as the World Health Organization (WHO) [116].

SPR-based assays have proven to be an efficient alternative to SRID. Low hands-on time and the total analysis time (~10 min for SPR versus 2 to 3 days for SRID [41]) are probably the two main advantages of the technique in a monitoring and/or development context, enabling the rapid measurements required for proper decision-making and control during production and/or purification. The small sample volume (~100 µL) required for the analysis is another notable advantage. Research groups have achieved low detection limits and appreciable quantitation ranges either by changing the assay format or by robust reference and ligand immobilization strategies, rendering monitoring possible from the early stages of production, be it in eggs or in a cell culture, to the end of the purification process. Proofs of concept for the use of SPR in virus quantification have mainly been performed on influenza, but other viruses have been studied with SPR in a diagnostic framework [20,59–61]. For all these reasons, we anticipate a growing role of SPR biosensors in vaccine manufacturing processes in the coming years.

Most research groups aim to quantify vaccine antigen levels [38–40,42,117], while others seek to estimate the viral particle concentration [41,118]. Other than that, vaccines can also be made of viral-like particles (VLPs). Briefly, VLPs are structures that mimic the conformation of viral particles but lack the viral genome [119]. VLPs can play a role as nanovaccines or drug nanocarriers [120]. The methods typically used to quantify VLPs are the same as those employed to quantify viruses. This complicates VLP production monitoring, as those methods are ill-suited to online or at-line measurements [121]. Adenoviruses have also recently been investigated as vaccine vehicles and can be quantified via SPR [122]. Table 3 summarizes the quantitation assays that will be discussed in this section.

Table 3. Quantitation assays of the vaccines or viral particles reported in this section. The concentrations used to build a calibration curve are also reported. The limit of quantitation (LOQ) and the limit of detection (LOD) are given when available. The limit of quantitation is the smallest concentration that can be precisely identified, whereas the limit of detection is the smallest concentration that can be qualitatively detected (yes/no detection).

Target Analyte	Ligand	Range	LOQ	LOD	SPR Instrument	Reference
Strains of Influenza HA and Vaccine HA	Synthetic glycans with α-2,3 or α-2,6 sialic acid conformation	0.33–30 µg/mL			Bio-Rad ProteOn	[40]
Strains of Influenza HA	Biantennary glycan with terminal α-2,6 sialic acid	0.01–2 µg/mL	0.01–0.02 µg/mL		Biacore T200	[42]
HA from viral serum and Vaccine HA	Recombinant HA (Inhibition assay with Anti-HA antibody)	1–15 µg/mL	1 µg/mL	0.5 µg/mL	Biacore T100	[39]
Adenovirus	Antibody against anti-adenovirus antibody (Inhibition assay with anti-adenovirus antibody)	10–5000 PFU/mL		10 PFU/mL	SPRI-Lab+ (Horiba Scientific, Edison, USA)	[118]
Influenza HA	6′-sialyllactose-conjugate with ovalbumin as carrier	10–100 µg/mL			Biacore 2000 & Biacore X	[38]
Vaccine HA	Fetuin glycoprotein bearing both α-2,3 and α-2,6 linkages	0.03–20 µg/mL	0.1 µg/mL	0.03 µg/mL	SPR-2 (Sierra Sensors)	[41]
Cell culture-derived whole influenza virus		~10^5–10^7 PFU/mL	5.3×10^5 PFU/mL	1.8×10^5 PFU/mL		

4.1. SPR for Quantification of Vaccines Preparations

The most used model system is the influenza virus, which can be detected via hemagglutinin (HA), a homotrimeric glycoprotein found on the membrane of the virus. As such, an HA binding partner is used as the immobilized ligand. For the selection of an appropriate ligand, aiming to replicate the infection process of influenza is a sound strategy. Infection occurs when sialic receptors on HA bind terminal sialic acid residues on cell surface glycoproteins [123,124]. Therefore, most research groups use sialic acids or conjugates containing a sialic acid moiety as the ligand. Lectins can also be used [38], as HA exhibits carbohydrate sites on its surface. Note that such SPR assays will detect viral particles whether they are infectious or not, as long as they display HA [41].

Influenza strains recognize sialic acid residues that are linked to galactose either in an α-2,3 (avian strain) or α-2,6 (human strain) conformation. Khurana and colleagues [40] used synthetic glycans containing either of these conformations to quantitate different strains of influenza HA and HA contents of the vaccine preparations of several manufacturers, showing excellent agreement with the SRID methods. The biotinylated ligands were immobilized on a biosensor surface with grafted NeutrAvidin, and regeneration was possible. Doses ranging from 0.33 to 30 µg HA per mL were used for calibration purposes, and the initial slope of the signal was recorded. Biotinylated glycans were also used to study the kinetics and affinity of several influenza strains to multivalent glycans with either α-2,3 or α-2,6 linkage [117]. The ability to measure interaction kinetics is absent in most other virus detection methods, including SRID.

A SPR assay based on a biantennary glycan with terminal α-2,6 sialic acid residues that strongly bind human influenza virus strain HAs from 1999 to 2017 was introduced by Bruce-Staskal and colleagues [42]. An asialoglycan was immobilized on the reference surface. As their SPR assay emulates a biologically relevant binding event, the authors argued it can distinguish between native and denatured HA conformations. The total binding levels recorded with their assay were impacted by stresses (temperature and pH) applied to the sample. As such, the proposed SPR assay can indirectly measure degradation and/or unfolding and, hence, be used to assess the half-life and stability of a vaccine preparation [42].

An inhibition assay was suggested to increase the sensitivity of SPR-based methods [39]. An anti-HA antibody was added in excess to influenza vaccine preparations a sufficiently long time before the SPR experiment to allow the reaching of a steady state. At a steady state, the remaining free antibodies (which had not bound to any HA) were titrated with recombinant HA immobilized at the biosensor surface. One could correlate the recorded SPR signal to the solution concentration of viral particles [72]. Of interest, a lower concentration of viral particles meant more free antibodies in the solution available for binding on the SPR surface and, thus, a higher SPR response. This allowed for a lower detection limit. A quantitation range of 1–15 µg/mL (via calibration with samples of known concentrations) was achieved for three influenza strains with a detection limit of less than 0.5 µg/L (respectively, 10–30 µg/mL and ~5 µg/mL for SRID, according to these authors) [39]. A similar assay was suggested for the detection of adenoviruses [118]. In that case, a detection range of 10–5000 PFU (plaque-forming units) per mL was achieved.

4.2. SPR for Quantification of Vaccines during Production

Mandenius and colleagues [38] probed four compounds containing a sialic acid moiety and three lectins as potential ligands for continuous HA quantification by SPR. Continuous immunosensing requires a low-affinity ligand with a high dissociation rate, so that a regeneration step is unnecessary. A rapidly dissociating analyte also reaches equilibrium rapidly during the association phase. By continuously injecting pulses of the culture medium on the weak-affinity ligand surface, the concentration can be monitored online by monitoring the equilibrium plateau values, which are a function of the analyte concentration [125]. The retained ligand was a 6′-sialyllactose conjugate with a substitution level of 0.6 mol of ligand per mole of carrier protein (ovalbumin). This ligand was shown to exhibit low affinity (~µM) binding to HA samples coming from egg or cell-based productions with a rapid dissociation (~s^{-1}), which would theoretically enable a fast analysis in an online fashion. The other tested ligands bound HA too strongly for this purpose, possibly due to avidity effects. A reference surface containing only ovalbumin was shown to adequately replicate nonspecific contributions. Human serum albumin was rejected as a protein carrier, as it caused more prominent nonspecific contributions than ovalbumin in this system. These authors showed that SPR responses at equilibrium correlated to the concentration measurements obtained with an SRID assay [38].

Using different SPR surfaces to quantify different influenza strains is suboptimal, especially when analyzing multivalent vaccine preparation. As such, ligands bearing both α-2,3 and α-2,6 linkages, such as fetuin glycoproteins, are of particular interest. A fetuin-based SPR imaging (SPRi; see Section 7) assay was suggested, which offered a higher reproducibility and better quantitation range (0.03–20 µg/mL) than SRID [41]. The assay was tested on influenza vaccine antigens and egg- and cell culture-derived whole influenza viruses. Of interest, viral particle production based on a culture of Madin-Darby Canine Kidney cells (MDCK) was monitored at-line from 1 to 4 days post-infection. Only a 10-min clarification by slow centrifugation was performed between the culture vessel (T-175 flask) and the biosensor. Of interest, multiple culture media caused negligeable nonspecific binding when injected alone on the biosensing surface, which paved the way to the at-line monitoring of multiple cell lines. Amongst them, the EMEM medium was selected, as it is appropriate for MDCK cells. Truncated fetuin (without terminal sialic acids) was observed to be a more reliable control surface ligand than bovine serum albumin, as it exhibited a lower nonspecific binding to HA. For antigen quantitation, calibrations were performed on HA samples of known concentrations, and the initial binding rates were recorded. For viral particle quantitation, either an infectivity assay (PFU) or total particles counting (TRPS) were used as a surrogate assay for concentration determination. The SPR signal initial slope strongly correlated linearly to both measurements. However, plaque-forming units only depend on infectious particles, not total particles exhibiting HA, as does SPR. Infectious viral particles cannot be directly quantified using SPR [41]. The ability to monitor vaccine production at-line is certainly desirable. On top of that, SPR can efficiently supplement SRID even in an offline framework by providing rapid measurements that can help speed development and testing.

5. SPR Applications in Bacteria and Contaminant Monitoring

SPR assays developed in the field of food safety could be adapted to bioprocess monitoring to ensure product safety and detect contamination. Such assays aim to detect bacteria or other pathogens in enrichment broths, in liquids, or in food dilutions [43]. Correspondingly, detecting impurities in cell culture media or a purified product is of interest to ensure the safety of biotherapeutics [44]. Traditional methods for bacteria detection such as colony counting and culture-based techniques require several experimental steps. The same thing can be said of alternative techniques such as ELISA and polymerase chain reaction (PCR), making them lengthy and laborious [45]. The ease of use of SPR and the low analysis time could prove very useful in this field.

Bacteria detection is usually done by immobilizing a bacteria-specific antibody on the biosensing surface. Bacteriophages have also been used [45,126,127]. The sensitivity of direct SPR assays may prove limiting, with a typical lower detection limit of about 10^5 colony-forming units (CFU) per milliliter. Here, sensitivity is quantified by the limit of detection, which is defined as the concentration of bacteria that causes an SPR response equivalent to the average value of the instrument noise (control) plus three times its standard deviation [46]. Table 4 summarizes the quantitation assays discussed in this section.

The sensitivity was improved by selecting an oriented ligand immobilization strategy, by taking advantage of the biotin-NeutrAvidin link to immobilize a biotinylated antibody, or by using protein A/G to capture the antibodies via their Fc region. A detection limit of 10^2 CFU/mL was achieved [45].

The sensitivity can also be improved by performing a sandwich assay [47,48]. A sandwich assay requires a second antibody that recognizes a different binding site on the bacteria membrane. After injecting the bacteria on the immobilized antibodies, the second antibodies are injected. This amplifies the signal, making it so a lower concentration of bacteria produces a detectable SPR response [72]. The detection limit of *Campylobacter jejuni* was enhanced from 8×10^6 CFU/mL to 4×10^4 CFU/mL by switching from a direct to a sandwich scheme. This compared favorably to a commercial ELISA assay (10^6–10^7 CFU/mL) [48]. Bhandari and colleagues [47] suggested incubating the secondary antibody with the bacteria prior to injecting them both rather than injecting them sequentially. Their assay achieved a lower detection limit of *Salmonella Typhimurium* in romaine lettuce samples of 4.7×10^5 CFU/mL, which proved more sensitive than the direct (1.9×10^6 CFU/mL) and typical sandwich (1.6×10^6 CFU/mL) assays [47].

Table 4. Quantitation assays of the bacteria reported in this section. The assay type and the limit of detection (LOD) are also given for each assay. Sandwich assays tend to exhibit a better sensitivity (lower LOD) than direct assays. A subtractive inhibition assay was reported with excellent sensitivity, because it requires detecting an antibody rather than the bacterium itself.

Target Analyte	Ligand	Assay Type	LOD (CFU/mL)	SPR Instrument	Reference
Campylobacter jejuni	Antibody	Direct	8×10^6	SPR-4 (Sierra Sensors)	[48]
		Sandwich	4×10^4		
Campylobacter jejuni	Antibody	Subtractive inhibition	131	SPR-4 (Sierra Sensors)	[49]
Salmonella Typhimurium in romaine lettuce	Antibody	Direct	1.9×10^6	Reichert Dual Channel SR7500DC	[47]
		Sandwich	1.6×10^6		
		Sandwich with prior incubation	4.7×10^5		
Escherichia coli O157:H7 in ground beef and cucumber samples	Wheat germ agglutinin (lectin)	Direct	3×10^3	Biacore 3000	[46]
Listeria monocytogenes in milk	Wheat germ agglutinin (lectin)	Direct	10^4	Biacore 3000	[50]
Listeria monocytogenes	Phage-displayed single-chain antibody	Direct	2×10^6	Nomadics SPR3 (SPREETA3, Texas Instruments)	[127]

Another explanation for the sensitivity issues is the insufficient depth of the evanescent wave (approximately 300 nm compared to the size of a bacterium, which is on the order of 1 μm) [45,128]. Mass accumulation on the SPR surface has a decaying importance on the SPR signal with the distance from the metal surface [33]. As such, using smaller ligands could improve the sensitivity. Antibody Fab fragments or other small binders such as aptamers would be appropriate for this purpose. For more details, the reader is referred to a recent review on the use of aptamer-based biosensors to detect *Pseudomonas aeruginosa* [129]. This also signifies that, for a sandwich assay to amplify the signal, the second antibody must bind the bacteria near the metal surface; otherwise, they are not perceived [48,130].

C. jejuni could be detected at a limit of 131 CFU/mL in chicken samples with a subtractive inhibition assay [49]. Briefly, the sample was incubated with an excess of antibodies. The concentration of the remaining free antibodies after incubation was dependent on the concentration of bacteria. These free antibodies were collected via centrifugation and injected on a SPR surface with an immobilized secondary antibody. As such, this method relied on the detection of an antibody, which was smaller than the penetration depth of the evanescent wave, rather than the bacteria themselves. This enabled a better sensitivity.

Assays using lectins as the ligand were also suggested. Wheat germ agglutinin (WGA) has been used to detect *Escherichia coli* O157:H7 in ground beef and cucumber samples [46] and *Listeria monocytogenes* in milk [50]. Nonspecific binding may affect the sensitivity in food samples, as better detection limits were reported in the buffer. Researchers in this field verify the specificity of their methods by evaluating their cross-reactivity to other bacteria, but little seems to be done to reference the contributions from other components of food samples, such as fibers, vitamins, carbohydrates, etc. Advances in referencing strategies will need to be made to counter their effect on the recorded SPR signal.

Due to the limitations reported in this section, we conclude that the detection of microorganisms via SPR is challenging but feasible, especially with more sophisticated assays. Other limitations include the possibility of clogging the microfluidic channels of the biosensor and the very slow diffusion rate of the bacteria towards the surface [131]. Although some of the reported studies showed statistically significant differences in SPR responses for different bacteria concentrations, pointing towards quantitation, the main goal in this field remains the detection of a particular bacterium species while limiting false positives caused by cross-reactivity to other species.

6. Limitations of SPR Biosensing in the Context of Bioprocess Monitoring

We now describe the major hurdles in the application of SPR for bioprocess monitoring. In a monitoring context, assays enabling rapid measurements with limited or no human handling of the samples are strongly preferred. A kinetic SPR assay requires injecting the analyte solution at different concentrations, which implies the need for diluting samples. A concentration analysis requires either a calibration to be repeated periodically or injections of each sample at two different flow rates. However, the total analysis time remains competitive, as each cycle lasts approximately 10 min, meaning a result in a matter of hours. This is still favorable over ELISA or SRID.

SPR sensor surfaces on which a ligand has been immobilized are reusable, but this often comes with the need for a regeneration step. Indeed, the available immobilized ligand molecule density must remain constant from one cycle to another. As such, all analyte molecules must be removed from the surface after each experiment by injecting a more 'aggressive' regeneration solution, increasing the cycle cost and duration. Selecting the proper regeneration solution can be one of the most challenging steps in designing an SPR assay. This is a major shortcoming of affinity biosensors [132]. Using a low-affinity ligand with a rapid dissociation rate has been suggested in the aim of performing continuous immunosensing [38] without the need for regeneration. However, this assumes the availability of multiple ligands for the analyte of interest, from which a suitable one with a weak affinity can be selected.

When taking a sample from a cell culture, minimal purification needs to be performed to at least remove the cells and cell debris, which could clog the flow cells of microfluidic SPR instruments. Dilution and filtration have both been automated to harness a SPR biosensor at-line from a bioreactor using a filter and a decantation column, along with a system of peristaltic pumps [67,68]. More work needs to be done to make at-line measurements readily available both in research and in the industry. We highlighted sequential assays combining concentration and kinetic analyses [37,55] (discussed in detail in Section 3.4) as an example of a smart protocol aiming to gain more information in less injections and regenerations. We also noted the existence of automated sampling systems with integrated cell removal and liquid handling modules [133]. Those have been used mainly to perform chromatography measurements online from a bioreactor but could potentially be adapted to SPR biosensors in the future. We believe that such integrated instruments will play a bigger role in the PAT context.

Nonspecific binding is another problem that may hamper SPR assays. It occurs when materials accumulate on the sensor surface other than by forming an analyte-ligand interaction. It could be the analyte adsorbing on the biosensor surface. It could also be other compounds present in the injected solution, such as blood proteins (when injecting serum

samples), binding the ligand, and/or the surface itself. SPR detects all mass accumulat
near the surface, whether it comes from a specific binding or not. As such, using a liga
with a high specificity for the analyte is crucial. A robust double-referencing protocol
has been widely adopted, but the question of what to immobilize on the reference surf
if anything, remains. Section 7 will present a brief discussion of the antifouling mater
In concentration analysis assays, nonspecific binding is problematic, because it lowers
detection limits. The sensitivity can be improved by using oriented ligand immobilizat
strategies [15] or by using an inhibition (see Section 4.1) or a sandwich assay format (
Section 5). However, these assays increase the cost in time and materials, as they requ
sequential injections or a secondary antibody [45,72].

Mass transport limitations (MTL) can bias kinetic analyses by reducing the obser
association and dissociation rate constants. MTL occurs when a large quantity of liga
molecules has been immobilized on the sensor surface. This speeds up the formatior
analyte-ligand complexes, which depend on the concentration of analyte at proximity
the surface and the concentration of available ligand. Hence, the availability of anal
molecules near the surface can become limiting during the association phase when diffus
of the analyte towards the surface is slow in comparison to the binding rate. Moreo
rebinding can occur during the dissociation phase. In brief, the analyte molecules that
dissociated from a ligand molecule may rebind to another ligand molecule before t
return to the bulk of the flow cell. Ideally, MTL should be avoided when performin
kinetic analysis, either by reducing the immobilized ligand density or by increasing
analyte injection flow rate. However, MTL can be accounted for when modeling the sig
On that note, an algorithm aiming to select the appropriate model (with or without M
while optimizing the injection times online has been suggested and has been shown
allow shorter experimental times while maintaining acceptable precision [27]. The neec
account for MTL in the model is investigated via a dimensionless number that represe
the ratio of the analyte-ligand complex formation rate to the analyte diffusion rate. N
that a concentration analysis, on the other hand, necessitates at least a partial MTL.

Heterogeneity in the ligand or analytes tends to complexify sensorgrams. This com
cates both kinetic [15] and concentration analyses [33,37]. The quality of the immobiliz
ligand molecule should be as high as possible, and the selected ligand should exh
a high specificity for the analyte under study. Improper folding or aggregation co
cause SPR signals that are difficult to interpretate, and poor specificity could lead to h
nonspecific contributions. This requirement is not as strong for the analyte, as long
denatured analyte molecules do not interact with either the ligand or the bioactive anal
molecules [15]. The analysis of SPR sensorgrams recorded with a heterogeneous systen
briefly discussed in Section 8.

The depth of the evanescent wave can also be limiting, as binding events occurr
beyond the penetration depth are not sensed in SPR. Further, mass accumulation ha
decaying influence on the SPR response with the distance from the metallic surface [33].
this reason, the absolute calibration-free concentration analysis (CFCA) is influenced by
distribution of the analyte in the evanescent field. A form factor can be applied to coun
this effect, but it is usually taken from empirical observations or obtained experimentally
calibrations, which is not ideal [32,33]. Additionally, diffusion of the analyte is hampered
the sensor matrix, which is not taken into account when estimating the diffusion coeffici
in a solution [33]. Strategies involving multiple sequential injections could potentially l
reproducibility. Using smaller ligands, such as aptamers or antibody Fab fragments, co
be a solution [45,128].

7. Recent Developments in SPR Instruments

Designing SPR sensor chips with surfaces capable of being highly selective for
analyte of interest in complex media, such as human serum or cell lysates, remains one
the main challenges of SPR biosensing for monitoring [70,134]. In bioprocess monitori
achieving low limits of detection is necessary to allow the monitoring of the early p

duction and purification, as certain downstream processing steps have a diluting effect. As such, surfaces limiting nonspecific binding and, thus, offering more sensitivity are of interest. The idea is to replace the commercial matrices on which the ligand is typically immobilized with another matrix with better antifouling properties. Interested readers are invited to consult a recent extensive review on antifouling coatings used in SPR in the fields of food safety and diagnostics [70]. Briefly, most commonly used antifouling layers are composed of hydrogel or zwitterionic polymers or they constitute a self-assembled monolayer (SAM) often based on oligo ethylene glycol groups such as polyethylene glycol (PEG). The optimal antifouling material remains highly dependent on the characteristics of the media that is analyzed, and as such, a universal solution seems unlikely. In all cases, special care should be taken to ensure that the thickness of the antifouling material is well below the evanescent wave decay length. Otherwise, ligand molecules may be immobilized outside the detection range of the instrument. We note a very recent PEG-based surface exhibiting excellent antifouling properties when analyzing blood sera with a thickness of approximately 2 nm [71], which is well under the evanescent wave penetration depth (dependent on the incident light wavelength, around 200–300 nm [15]).

Other advances have been made to increase the throughput of SPR campaigns. Those were suggested mainly with drug screening in mind, but they could be appropriate for the simultaneous monitoring of multiple bioreactors or multiple bioprocess steps using different SPR surfaces with different ligands. The simultaneous measurement of multiple detection spots is possible with the surface plasmon resonance imaging (SPRi) technique. Briefly, rather than measuring the light intensity at a single point, a charge-coupled device (CCD) or complementary metal oxide semiconductor (CMOS) camera captures images of all the detection spots at every time step. These images can be analyzed to extract the SPR response at every detection spot. This allows 2D measurements across multiple channels rather than the typical 1D probing of a single flow cell. For more technical details on SPRi, the reader is referred to a review on the recent advances in SPRi sensors [135]. SPRi allows for more throughput by parallelizing experiments, but the sensitivity can be slightly compromised, and sample handling and surface functionalization remain challenging.

Traditional SPR biosensors, such as those offered by Biacore (Cytiva), can take measurements in only four flow cells connected in series, with often only two actually being used. SPRi biosensors enable parallel injections at different SPR detection spots. The ProteOn XPR36 instrument (Bio-Rad Laboratories, Hercules, CA, USA; no longer commercially available) contains an array of six ligand channels crossed perpendicularly with six analyte channels, enabling the SPR measurements of 36 analyte-ligand couples in parallel. Different ligands or ligand concentrations can be immobilized on each of the six ligand channels, and different analytes or analyte concentrations can be injected in the six analyte channels [136]. ProteOn XPR36 was found to have approximately 10 times more throughput than a traditional Biacore T100 biosensor when performing a mAb screening campaign [137]. However, Biacore T100 was deemed to generate data of a higher quality, with more consistency. More recently, the Molecular Affinity Screening System (MASS-1) was introduced by Sierra Sensors (Bruker, Billerica, MA, USA) [138]. It contains eight parallel channels with two detection spots per channel: one active surface and one reference surface, limiting the distance between them. This biosensor measured similar kinetics and affinities compared to a Biacore 4000 biosensor when analyzing a panel of mAbs for differently sized antigens. Of interest, reproducible results could be obtained with very low ligand densities, which is sometimes necessary to avoid mass transport limitations when analyzing an analyte-ligand pair with a thermodynamic constant K_D in the picomolar range. This is proof of the high sensitivity and low measurement noise of the MASS-1 as such low ligand density experiments lead to very weak SPR responses [138]. Moreover, the LSA platform (Carterra, Salt Lake City, UT, USA), which was released in 2018, enables the study of 384 antibodies simultaneously via SPRi [139].

Some work has been done in improving the portability of the SPR technology. This has paved the way to applications in point of care testing and environmental monitoring. In the

context of bioprocess monitoring, limiting the distance between the process equipment a
the monitoring devices is preferable, as it limits the handling of the samples. A porta
optical-fiber-based SPR biosensor has been shown to be able to detect mAb levels be
in the buffer and serum [140]. Further, low-cost SPR biosensors that can be harness
to a smartphone have been suggested. In short, a SPR-sensing surface is linked to
phone's LED (light source) and camera (detector) through a system of fiber optics, a
the samples are pumped in and out of the measurement channel. Such an arrangem
enabled repeatable measurements with an appropriate, but slightly lower, sensitivity wl
compared to a traditional biosensor when measuring IgG-protein A interactions [141].
more details, we point to a review on smartphone-based SPR biosensing [142].

Other developments come from new configurations of SPR biosensors. We n
phase shift-based SPR biosensors that detect a rapid phase change between p- an
polarized light when SPR occurs. This phase change is more abrupt than the change
the reflected light intensity, allowing a greater sensitivity and wider dynamic range. 7
spatial resolution can also be enhanced, facilitating SPRi implementation [73,74]. Suc
biosensor enabled a better sensitivity than with traditional SPR biosensors for influe
antibody biomarker detection [143]. SPR biosensors that detect phase shifts can be ba
on either optical heterodyne, ellipsometry, or interferometry. For more details on th
concepts, interested readers are referred to relevant reviews [73,144].

Multi-parametric SPR (MP-SPR) biosensors record the whole SPR curve (rather t
only the minimal point) to enable bulk effect removal without a reference and to meas
various optical parameters, such as refractive indices of multiple surface layers. Nota
MP-SPR was used to measure the binding affinity of monoclonal antibody Adalimumal
FcγRIIb and FcγRIIIb in crude cell culture samples containing more than 10^6 cells/mL. '
affinity measured in the crude samples was similar to that of the purified samples [1
In another study, liposome nanocarriers were immobilized on the surface, and undilu
human serum was injected. The formation of a protein corona on the surface of
liposomes could be tracked, and different formulations were compared [146].

SPR Studies with Live Cells

MP-SPR also enables the study of live cells. Cells can be seeded on a metallic surf
by using fibronectin as an adhesion promoter. Living cells uptake of drugs [147], nano
ticles [148], and extracellular vesicles [149] has been monitored online in this fashion. '
evanescent wave penetration is not deep enough to be impacted by the contact between
drugs or nanoparticles and the cell membrane. The SPR signal is caused by morpholog
changes and rearrangements of the intracellular materials that follow the binding ev
leading to endocytosis [148]. Alternatively, configurations enabling a deeper penetratio
the evanescent wave, called long-range SPR (LRSPR), have been suggested for these ty
of studies [150,151] and for the detection of bacteria [152]. However, LRSPR has yet tu
commercialized [153]. Live cell responses to stimuli have also been probed with a bioser
combining SPR and impedance measurements by using a comb-shaped electrode [154

Living cells can also been used as the analyte. One can aim to detect cell membr
antigens by injecting the cells on a SPR surface on which a specific antibody is immobili
For example, the interaction between transmembrane TNF-α expressed on the membra
of Jurkat cells and anti-TNF agents immobilized on the sensor surface was studied
SPR [155]. The responses were found to be concentration-dependent and inhibited by
anti-TNF agents. On another note, stopping the flow completely after injecting the c
causes them to sediment, which allows them to be used as the ligand in a subsequ
injection. Quantifying the secretion of cells is also possible, for example, by attach
microwell arrays on the sensor surface. With each well containing a single cell, and n
of the cell's secreted species diffusing towards the SPR surface rather than the bulk,
ability of SPRi instruments to take measurements at multiple detection spots enabl
comparison of the secretions of different cells [156]. For more details, the reader is refer
to a recent review on the advances in the field of SPR cytometry [157].

Developments in the field of SPR cytometry are mainly tied to fundamental research, but they have possible implications in terms of validating the quality and effectiveness of manufactured bioproducts or nanoparticles, either during production or during the development stage.

8. Recent Developments in SPR Data Analysis

Some recent advances in SPR data analysis aim to analyze complex systems with more than one interaction occurring at the sensor surface. For example, injecting solutions containing a mixture of analytes that can bind to a common ligand was proposed as a way to increase the throughput of SPR biosensors in a drug screening framework. For this purpose, analytical methods enabling the extraction of the kinetic parameters of two analytes from sensorgrams recorded by injecting a single mixture of these analytes were suggested, and proof of concept was obtained on a model system [26,28]. Other than increasing the throughput, for which one willingly mixes the different analytes, one could also aim to identify the individual kinetic parameters of various (bio)molecules that are difficult to separate from each other. For instance, the various glycoforms of therapeutic monoclonal antibodies produced in mammalian cell cultures may not be readily separated. However, it is well-known that binding to their receptors is affected by their glycosylation state. Hence, studying this interaction via SPR leads to complex sensorgrams that cannot be properly analyzed with a simple Langmuir model [51–54].

In this mindset, the multi-analyte analysis was extended to the case of N analytes [31]. This time, the aim was to analyze samples of analytes that are difficult to separate rather than to increase the throughput, as N mixtures are required to elucidate N analytes. This framework has only been applied to a model system composed of the enzyme carbonic anhydrase II as the ligand and several of its inhibitors as the analytes. However, possible applications include analyzing solutions of a given protein that is differently folded or glycosylated. A method for composition estimation, a potentially powerful tool in quality control, has also been suggested, but it requires a priori knowledge of the individual kinetic parameters of each analyte in the mixture with the ligand [31]. Other modeling advances enabled the analysis of complex sensorgrams hailing from analyte [29,30] or ligand heterogeneity [85–87]. These advances came with more complex interaction models, meaning more parameters to adjust. Care should be taken to avoid overfitting. Adding parameters will lead to better fits to recorded data, but it may not be significant. Broadly, if the mathematical model has no biological significance, the identified parameters are mostly meaningless.

Further, a similarity score has been proposed to assess how closely two given sensorgrams resemble each other [158]. As two analyte samples of similar quality should produce a similar binding response on a given SPR surface, computing the similarity score could prove an important asset in evaluating batch-to-batch variabilities and in validating a biosimilar by comparing its binding response to that of an approved reference biotherapeutic.

9. Conclusions and Future Perspectives

Surface plasmon resonance-based biosensors have long been established as a premiere tool for probing the interaction behavior between a solution species and a surface-immobilized species. The main advantages of the SPR technique are its ability to detect the interaction online and without a label. SPR is already in use in the pharmaceutical world as an investigative technique, for example, in drug screening campaigns to determine the best drug to bind a specific target amongst an array of candidates. With increasing demands in biotherapeutics and continuously stricter quality assurance directives from regulatory agencies, the frameworks of quality by design and quality by control pushed forward the need for powerful monitoring tools that can be used to acquire knowledge of the various critical parameters influencing the quality of bioproducts. In this context, we reviewed potential uses of the SPR technique as a monitoring tool. Three different types of bioprocesses were put forward: biotherapeutics production, vaccine production,

and contaminant and bacteria detection in food samples. In all these cases, it seems th
one of the main limitations remains the need for preprocessing of the samples: dilutic
purification, or just simply handling the samples between the sampling site and the SI
instrument. Very few studies have shown the capability for harnessing a SPR biosens
to a bioreactor to automate sample taking, preprocessing, and SPR measurements onli
or at-line [67,68]. We noted, however, that some experimental setups have enabled t
use of other analytical methods at-line of a bioreactor, such as chromatography and ma
spectroscopy [159,160] and in situ Raman spectroscopy [161]. The recent introduction
automated samplers and liquid handlers in integrated systems [133] has paved the way
online SPR, but more work needs to be done on the instruments themselves and relat
experimental protocols. Namely, the need for regeneration of the surface, the dilution
samples, and the removal of nonspecific contributions are all limiting. The portability
the instruments is being worked on, with several options available, but it seems that t
precision and sensitivity may be slightly compromised in such systems.

Most applications of SPR that have been or could be used to monitor bioprocess
online involve estimating the solution concentrations of an analyte by immobilizing
analyte-specific ligand on the biosensing surface. This, of course, enables monitoring
the production, but it can also be used to track the yields of the purification steps. W
evermore advances aiming to increase the sensitivity, SPR could be used to monitor t
product concentration from early production all the way to the end of the downstre:
processing steps. A concentration analysis has been shown to be possible in complex me
such as blood sera and cell lysates, but the same cannot be said about kinetic analy:
To perform a kinetic analysis, a low density of the immobilized ligand is necessary
avoid mass transport limitations, rendering nonspecific contributions to the SPR sig.
more prominent. Performing a kinetic analysis online or at-line within a bioprocess co
enable the monitoring of certain key quality attributes of bioproducts during producti
As an example, the binding kinetics and affinity of monoclonal antibodies to Fc recept
are strongly dependent on their glycosylation profile, a CQA of mAbs [51]. We no
the existence of sequential assays [37,55,102] in which, as an example, antibodies are fi
quantified in complex media via interactions with their immobilized antigen, followed
a second injection, this time of Fc region receptors in a defined buffer, to allow a kin
analysis between the immobilized antibodies and the injected Fc receptors. Such ingenic
protocols, and others in the future, will be necessary to allow kinetic analyses in cru
samples for various types of bioproducts. This would open the way to the utilization
SPR biosensors in quality by design or even quality by control frameworks as proc
analytical technology.

Author Contributions: J.G., C.F.-N., and O.H. were responsible for conceptualizing and editing
manuscript. G.D.C. and Y.D. provided mentorship and valuable comments on the manuscript.
authors have read and agreed to the published version of the manuscript.

Funding: This research received no external funding.

Institutional Review Board Statement: Not applicable.

Informed Consent Statement: Not applicable.

Data Availability Statement: Not applicable.

Acknowledgments: This work was supported by the Natural Sciences and Engineering Resea
Council of Canada (stipend allocated to Catherine Forest-Nault and Jimmy Gaudreault via
NSERC-CREATE PrEEmiuM program). This work was supported by the Trans-MedTech Insti
(NanoBio Technology Platform) and its main funding partner, the Canada First Research Ex
lence Fund.

Conflicts of Interest: The authors declare no conflict of interest.

erences

Lu, R.-M.; Hwang, Y.-C.; Liu, I.J.; Lee, C.-C.; Tsai, H.-Z.; Li, H.-J.; Wu, H.-C. Development of therapeutic antibodies for the treatment of diseases. *J. Biomed. Sci.* **2020**, *27*, 1. [CrossRef]

Gherghescu, I.; Delgado-Charro, M.B. The Biosimilar Landscape: An Overview of Regulatory Approvals by the EMA and FDA. *Pharmaceutics* **2021**, *13*, 48. [CrossRef]

U. S. Food and Drug Administration. *Guidance for Industry: Q8(R2) Pharmaceutical Development*; U. S. Food and Drug Administration: Rockville, MD, USA, 2009.

Yu, L.X.; Amidon, G.; Khan, M.A.; Hoag, S.W.; Polli, J.; Raju, G.K.; Woodcock, J. Understanding pharmaceutical quality by design. *AAPS J* **2014**, *16*, 771–783. [CrossRef]

Grangeia, H.B.; Silva, C.; Simões, S.P.; Reis, M.S. Quality by design in pharmaceutical manufacturing: A systematic review of current status, challenges and future perspectives. *Eur. J. Pharm. Biopharm.* **2020**, *147*, 19–37. [CrossRef]

Yu, L.X.; Kopcha, M. The future of pharmaceutical quality and the path to get there. *Int. J. Pharm.* **2017**, *528*, 354–359. [CrossRef] [PubMed]

Sommeregger, W.; Sissolak, B.; Kandra, K.; von Stosch, M.; Mayer, M.; Striedner, G. Quality by control: Towards model predictive control of mammalian cell culture bioprocesses. *Biotechnol. J.* **2017**, *12*, 1600546. [CrossRef]

Su, Q.; Ganesh, S.; Moreno, M.; Bommireddy, Y.; Gonzalez, M.; Reklaitis, G.V.; Nagy, Z.K. A perspective on Quality-by-Control (QbC) in pharmaceutical continuous manufacturing. *Comput. Chem. Eng.* **2019**, *125*, 216–231. [CrossRef]

U. S. Food and Drug Administration. *Guidance for Industry: PAT-A Framework for Innovative Pharmaceutical Development, Manufacturing, and Quality Assurance*; U. S. Food and Drug Administration: Rockville, MD, USA, 2004.

Simon, L.L.; Pataki, H.; Marosi, G.; Meemken, F.; Hungerbühler, K.; Baiker, A.; Tummala, S.; Glennon, B.; Kuentz, M.; Steele, G.; et al. Assessment of Recent Process Analytical Technology (PAT) Trends: A Multiauthor Review. *Org. Process Res. Dev.* **2015**, *19*, 3–62. [CrossRef]

Helgers, H.; Schmidt, A.; Lohmann, L.J.; Vetter, F.L.; Juckers, A.; Jensch, C.; Mouellef, M.; Zobel-Roos, S.; Strube, J. Towards Autonomous Operation by Advanced Process Control—Process Analytical Technology for Continuous Biologics Antibody Manufacturing. *Processes* **2021**, *9*, 172. [CrossRef]

Kretschmann, E.; Raether, H. Notizen: Radiative Decay of Non Radiative Surface Plasmons Excited by Light. *Z. Nat. A* **1968**, *23*, 2135–2136. [CrossRef]

Kretschmann, E. Decay of non radiative surface plasmons into light on rough silver films. Comparison of experimental and theoretical results. *Opt. Commun.* **1972**, *6*, 185–187. [CrossRef]

Otto, A. Excitation of nonradiative surface plasma waves in silver by the method of frustrated total reflection. *Z. Phys. A Hadron. Nucl.* **1968**, *216*, 398–410. [CrossRef]

De Crescenzo, G.; Boucher, C.; Durocher, Y.; Jolicoeur, M. Kinetic Characterization by Surface Plasmon Resonance-Based Biosensors: Principle and Emerging Trends. *Cell. Mol. Bioeng.* **2008**, *1*, 204–215. [CrossRef]

Homola, J. Present and future of surface plasmon resonance biosensors. *Anal. Bioanal. Chem.* **2003**, *377*, 528–539. [CrossRef] [PubMed]

Couture, M.; Zhao, S.S.; Masson, J.-F. Modern surface plasmon resonance for bioanalytics and biophysics. *Phys. Chem. Chem. Phys.* **2013**, *15*, 11190–11216. [CrossRef]

Wang, D.S.; Fan, S.K. Microfluidic Surface Plasmon Resonance Sensors: From Principles to Point-of-Care Applications. *Sensors* **2016**, *16*, 1175. [CrossRef]

Prabowo, B.A.; Purwidyantri, A.; Liu, K.-C. Surface Plasmon Resonance Optical Sensor: A Review on Light Source Technology. *Biosensors* **2018**, *8*, 80. [CrossRef]

Shrivastav, A.M.; Cvelbar, U.; Abdulhalim, I. A comprehensive review on plasmonic-based biosensors used in viral diagnostics. *Commun. Biol.* **2021**, *4*, 70. [CrossRef]

Guo, X. Surface plasmon resonance based biosensor technique: A review. *J. Biophotonics* **2012**, *5*, 483–501. [CrossRef]

Myszka, D.G. Improving biosensor analysis. *J. Mol. Recognit. JMR* **1999**, *12*, 279–284. [CrossRef]

Önell, A.; Andersson, K. Kinetic determinations of molecular interactions using Biacore—Minimum data requirements for efficient experimental design. *J. Mol. Recognit.* **2005**, *18*, 307–317. [CrossRef] [PubMed]

Karlsson, R.; Katsamba, P.S.; Nordin, H.; Pol, E.; Myszka, D.G. Analyzing a kinetic titration series using affinity biosensors. *Anal. Biochem.* **2006**, *349*, 136–147. [CrossRef] [PubMed]

De Crescenzo, G.; Woodward, L.; Srinivasan, B. Online optimization of surface plasmon resonance-based biosensor experiments for improved throughput and confidence. *J. Mol. Recognit.* **2008**, *21*, 256–266. [CrossRef] [PubMed]

Mehand, M.S.; De Crescenzo, G.; Srinivasan, B. Increasing throughput of surface plasmon resonance-based biosensors by multiple analyte injections. *J. Mol. Recognit. JMR* **2012**, *25*, 208–215. [CrossRef]

Si Mehand, M.; De Crescenzo, G.; Srinivasan, B. On-line kinetic model discrimination for optimized surface plasmon resonance experiments. *J. Mol. Recognit. JMR* **2014**, *27*, 276–284. [CrossRef]

Mehand, M.S.; Srinivasan, B.; De Crescenzo, G. Optimizing Multiple Analyte Injections in Surface Plasmon Resonance Biosensors with Analytes having Different Refractive Index Increments. *Sci. Rep.* **2015**, *5*, 15855. [CrossRef]

Zhang, Y.; Forssén, P.; Fornstedt, T.; Gulliksson, M.; Dai, X. An adaptive regularization algorithm for recovering the rate constant distribution from biosensor data. *Inverse Probl. Sci. Eng.* **2018**, *26*, 1464–1489. [CrossRef]

30. Forssén, P.; Multia, E.; Samuelsson, J.; Andersson, M.; Aastrup, T.; Altun, S.; Wallinder, D.; Wallbing, L.; Liangsupree, T.; Riekk M.-L.; et al. Reliable Strategy for Analysis of Complex Biosensor Data. *Anal. Chem.* **2018**, *90*, 5366–5374. [CrossRef]

31. Gaudreault, J.; Liberelle, B.; Durocher, Y.; Henry, O.; De Crescenzo, G. Determination of the composition of heterogeneous bin solutions by surface plasmon resonance biosensing. *Sci. Rep.* **2021**, *11*, 3685. [CrossRef]

32. Pol, E.; Roos, H.; Markey, F.; Elwinger, F.; Shaw, A.; Karlsson, R. Evaluation of calibration-free concentration analysis provided Biacore™ systems. *Anal. Biochem.* **2016**, *510*, 88–97. [CrossRef]

33. Karlsson, R. Biosensor binding data and its applicability to the determination of active concentration. *Biophys. Rev.* **201€** 347–358. [CrossRef]

34. Karlsson, R.; Roos, H.; Fägerstam, L.; Persson, B. Kinetic and Concentration Analysis Using BIA Technology. *Methods* **199**4 99–110. [CrossRef]

35. Christensen, L.L. Theoretical analysis of protein concentration determination using biosensor technology under condition: partial mass transport limitation. *Anal. Biochem.* **1997**, *249*, 153–164. [CrossRef]

36. Vaisocherová, H.; Faca, V.M.; Taylor, A.D.; Hanash, S.; Jiang, S. Comparative study of SPR and ELISA methods based on analy of CD166/ALCAM levels in cancer and control human sera. *Biosens. Bioelectron.* **2009**, *24*, 2143–2148. [CrossRef]

37. Karlsson, R.; Fridh, V.; Frostell, Å. Surrogate potency assays: Comparison of binding profiles complements dose response cur for unambiguous assessment of relative potencies. *J. Pharm. Anal.* **2018**, *8*, 138–146. [CrossRef]

38. Mandenius, C.F.; Wang, R.; Aldén, A.; Bergström, G.; Thébault, S.; Lutsch, C.; Ohlson, S. Monitoring of influenza vi hemagglutinin in process samples using weak affinity ligands and surface plasmon resonance. *Anal. Chim. Acta* **2008**, *623*, 66 [CrossRef]

39. Nilsson, C.E.; Abbas, S.; Bennemo, M.; Larsson, A.; Hämäläinen, M.D.; Frostell-Karlsson, A. A novel assay for influenza vi quantification using surface plasmon resonance. *Vaccine* **2010**, *28*, 759–766. [CrossRef]

40. Khurana, S.; King, L.R.; Manischewitz, J.; Coyle, E.M.; Golding, H. Novel antibody-independent receptor-binding SPR-ba assay for rapid measurement of influenza vaccine potency. *Vaccine* **2014**, *32*, 2188–2197. [CrossRef]

41. Durous, L.; Julien, T.; Padey, B.; Traversier, A.; Rosa-Calatrava, M.; Blum, L.J.; Marquette, C.A.; Petiot, E. SPRi-based hemaggluti quantitative assay for influenza vaccine production monitoring. *Vaccine* **2019**, *37*, 1614–1621. [CrossRef]

42. Bruce-Staskal, P.J.; Woods, R.M.; Borisov, O.V.; Massare, M.J.; Hahn, T.J. Corrigendum to "Hemagglutinin from multiple diverg influenza A and B viruses bind to a distinct branched, sialylated poly-LacNAc glycan by surface plasmon resonance" [Vacc 38(43) (2020) 6757–6765]. *Vaccine* **2021**, *39*, 1544–1545. [CrossRef]

43. Poltronieri, P.; Mezzolla, V.; Primiceri, E.; Maruccio, G. Biosensors for the Detection of Food Pathogens. *Foods* **2014**, *3*, 511– [CrossRef] [PubMed]

44. Mattiasson, B.; Teeparuksapun, K.; Hedström, M. Immunochemical binding assays for detection and quantification of tr impurities in biotechnological production. *Trends Biotechnol.* **2010**, *28*, 20–27. [CrossRef] [PubMed]

45. Dudak, F.C.; Boyaci, I.H. Rapid and label-free bacteria detection by surface plasmon resonance (SPR) biosensors. *Biotechnc* **2009**, *4*, 1003–1011. [CrossRef] [PubMed]

46. Wang, Y.; Ye, Z.; Si, C.; Ying, Y. Monitoring of Escherichia coli O157:H7 in food samples using lectin based surface plasm resonance biosensor. *Food Chem.* **2013**, *136*, 1303–1308. [CrossRef]

47. Bhandari, D.; Chen, F.-C.; Bridgman, R.C. Detection of Salmonella Typhimurium in Romaine Lettuce Using a Surface Plasm Resonance Biosensor. *Biosensors* **2019**, *9*, 94. [CrossRef]

48. Masdor, N.A.; Altintas, Z.; Tothill, I.E. Surface Plasmon Resonance Immunosensor for the Detection of Campylobacter jej *Chemosensors* **2017**, *5*, 16. [CrossRef]

49. Masdor, N.A.; Altintas, Z.; Shukor, M.Y.; Tothill, I.E. Subtractive inhibition assay for the detection of Campylobacter jejun chicken samples using surface plasmon resonance. *Sci. Rep.* **2019**, *9*, 13642. [CrossRef]

50. Raghu, H.V.; Kumar, N. Rapid Detection of Listeria monocytogenes in Milk by Surface Plasmon Resonance Using Wheat G€ Agglutinin. *Food Anal. Methods* **2020**, *13*, 982–991. [CrossRef]

51. Forest-Nault, C.; Gaudreault, J.; Henry, O.; Durocher, Y.; De Crescenzo, G. On the Use of Surface Plasmon Resonance Biosens to Understand IgG-FcγR Interactions. *Int. J. Mol. Sci.* **2021**, *22*, 6616. [CrossRef]

52. Cambay, F.; Henry, O.; Durocher, Y.; De Crescenzo, G. Impact of N-glycosylation on Fcγ receptor/IgG interactions: Unravel differences with an enhanced surface plasmon resonance biosensor assay based on coiled-coil interactions. *mAbs* **2019**, *11*, 435– [CrossRef]

53. Cambay, F.; Forest-Nault, C.; Dumoulin, L.; Seguin, A.; Henry, O.; Durocher, Y.; De Crescenzo, G. Glycosylation of Fcγ recep influences their interaction with various IgG1 glycoforms. *Mol. Immunol.* **2020**, *121*, 144–158. [CrossRef] [PubMed]

54. Subedi, G.P.; Barb, A.W. The immunoglobulin G1 N-glycan composition affects binding to each low affinity Fc γ receptor. *m* **2016**, *8*, 1512–1524. [CrossRef] [PubMed]

55. Dorion-Thibaudeau, J.; Durocher, Y.; De Crescenzo, G. Quantification and simultaneous evaluation of the bioactivity of antib produced in CHO cell culture-The use of the ectodomain of FcγRI and surface plasmon resonance-based biosensor. *Mol. Immu* **2017**, *82*, 46–49. [CrossRef] [PubMed]

56. Wang, W.; Soriano, B.; Chen, Q. Glycan profiling of proteins using lectin binding by Surface Plasmon Resonance. *Anal. Bioc* **2017**, *538*, 53–63. [CrossRef] [PubMed]

Wang, H.; Shi, J.; Wang, Y.; Cai, K.; Wang, Q.; Hou, X.; Guo, W.; Zhang, F. Development of biosensor-based SPR technology for biological quantification and quality control of pharmaceutical proteins. *J. Pharm. Biomed. Anal.* **2009**, *50*, 1026–1029. [CrossRef] [PubMed]

Coty, J.B.; Noiray, M.; Vauthier, C. Assessment of Complement Activation by Nanoparticles: Development of a SPR Based Method and Comparison with Current High Throughput Methods. *Pharm. Res.* **2018**, *35*, 129. [CrossRef] [PubMed]

Huang, L.; Ding, L.; Zhou, J.; Chen, S.; Chen, F.; Zhao, C.; Xu, J.; Hu, W.; Ji, J.; Xu, H.; et al. One-step rapid quantification of SARS-CoV-2 virus particles via low-cost nanoplasmonic sensors in generic microplate reader and point-of-care device. *Biosens. Bioelectron.* **2021**, *171*, 112685. [CrossRef]

Mauriz, E. Recent Progress in Plasmonic Biosensing Schemes for Virus Detection. *Sensors* **2020**, *20*, 4745. [CrossRef]

Singh, P. Surface Plasmon Resonance: A Boon for Viral Diagnostics. *Ref. Modul. Life Sci.* **2017**. [CrossRef]

Bellassai, N.; D'Agata, R.; Jungbluth, V.; Spoto, G. Surface Plasmon Resonance for Biomarker Detection: Advances in Non-invasive Cancer Diagnosis. *Front. Chem.* **2019**, *7*, 570. [CrossRef]

Souto, D.E.P.; Volpe, J.; Gonçalves, C.d.C.; Ramos, C.H.I.; Kubota, L.T. A brief review on the strategy of developing SPR-based biosensors for application to the diagnosis of neglected tropical diseases. *Talanta* **2019**, *205*, 120122. [CrossRef] [PubMed]

Brulé, T.; Granger, G.; Bukar, N.; Deschênes-Rancourt, C.; Havard, T.; Schmitzer, A.R.; Martel, R.; Masson, J.-F. A field-deployed surface plasmon resonance (SPR) sensor for RDX quantification in environmental waters. *Analyst* **2017**, *142*, 2161–2168. [CrossRef] [PubMed]

Zschätzsch, M.; Ritter, P.; Henseleit, A.; Wiehler, K.; Malik, S.; Bley, T.; Walther, T.; Boschke, E. Monitoring bioactive and total antibody concentrations for continuous process control by surface plasmon resonance spectroscopy. *Eng. Life Sci.* **2019**, *19*, 681–690. [CrossRef] [PubMed]

Frostell, Å.; Mattsson, A.; Eriksson, Å.; Wallby, E.; Kärnhall, J.; Illarionova, N.B.; Estmer Nilsson, C. Nine surface plasmon resonance assays for specific protein quantitation during cell culture and process development. *Anal. Biochem.* **2015**, *477*, 1–9. [CrossRef] [PubMed]

Chavane, N.; Jacquemart, R.; Hoemann, C.D.; Jolicoeur, M.; De Crescenzo, G. At-line quantification of bioactive antibody in bioreactor by surface plasmon resonance using epitope detection. *Anal. Biochem.* **2008**, *378*, 158–165. [CrossRef]

Jacquemart, R.; Chavane, N.; Durocher, Y.; Hoemann, C.; De Crescenzo, G.; Jolicoeur, M. At-line monitoring of bioreactor protein production by surface plasmon resonance. *Biotechnol. Bioeng.* **2008**, *100*, 184–188. [CrossRef]

Beeg, M.; Nobili, A.; Orsini, B.; Rogai, F.; Gilardi, D.; Fiorino, G.; Danese, S.; Salmona, M.; Garattini, S.; Gobbi, M. A Surface Plasmon Resonance-based assay to measure serum concentrations of therapeutic antibodies and anti-drug antibodies. *Sci. Rep.* **2019**, *9*, 2064. [CrossRef]

D'Agata, R.; Bellassai, N.; Jungbluth, V.; Spoto, G. Recent Advances in Antifouling Materials for Surface Plasmon Resonance Biosensing in Clinical Diagnostics and Food Safety. *Polymers* **2021**, *13*, 1929. [CrossRef]

D'Agata, R.; Bellassai, N.; Giuffrida, M.C.; Aura, A.M.; Petri, C.; Kögler, P.; Vecchio, G.; Jonas, U.; Spoto, G. A new ultralow fouling surface for the analysis of human plasma samples with surface plasmon resonance. *Talanta* **2021**, *221*, 121483. [CrossRef]

Schasfoort, R.B.M. Introduction to Surface Plasmon Resonance. In *Handbook of Surface Plasmon Resonance*, 2nd ed.; Schasfoort, R.B.M., Ed.; The Royal Society of Chemistry: London, UK, 2017; pp. 1–26.

Deng, S.; Wang, P.; Yu, X. Phase-Sensitive Surface Plasmon Resonance Sensors: Recent Progress and Future Prospects. *Sensors* **2017**, *17*, 2819. [CrossRef]

Kashif, M.; Bakar, A.A.A.; Arsad, N.; Shaari, S. Development of phase detection schemes based on surface plasmon resonance using interferometry. *Sensors* **2014**, *14*, 15914–15938. [CrossRef] [PubMed]

Stenberg, E.; Persson, B.; Roos, H.; Urbaniczky, C. Quantitative determination of surface concentration of protein with surface plasmon resonance using radiolabeled proteins. *J. Colloid Interface Sci.* **1991**, *143*, 513–526. [CrossRef]

Affinité Instruments. Unleashing Label Free Sensing-Surface Plasmon Resonance for Rapid Testing. Available online: https://www.affiniteinstruments.com/ (accessed on 16 September 2021).

Cytiva. Biacore SPR-Surface Plasmon Resonance Interaction Analysis. Available online: https://www.cytivalifesciences.com/en/us/solutions/protein-research/interaction-analysis-with-biacore-surface-plasmon-resonance-spr (accessed on 16 September 2021).

BioNavis. BioNavis: Enter the world of MP-SPR and find a solution to your research needs! Available online: https://www.bionavis.com/ (accessed on 16 September 2021).

Biosensing Instrument. Biosensing Instrument Offers Bioanalytical Tools to Accelerate Drug Discovery Research. Available online: https://biosensingusa.com/ (accessed on 16 September 2021).

Carterra. High Throughput Antibody Screening and Characterization. Available online: https://carterra-bio.com/ (accessed on 16 September 2021).

Reichert Technologies. Discover the Reichert SPR Difference. Available online: https://www.reichertspr.com/ (accessed on 16 September 2021).

Bruker. Surface Plasmon Resonance for High-Throughput Analysis. Available online: https://www.bruker.com/en/products-and-solutions/surface-plasmon-resonance.html (accessed on 16 September 2021).

Andersson, K.; Hämäläinen, M.; Malmqvist, M. Identification and Optimization of Regeneration Conditions for Affinity-Based Biosensor Assays. A Multivariate Cocktail Approach. *Anal. Chem.* **1999**, *71*, 2475–2481. [CrossRef] [PubMed]

84. Goode, J.A.; Rushworth, J.V.H.; Millner, P.A. Biosensor Regeneration: A Review of Common Techniques and Outcomes. *Langmuir* **2015**, *31*, 6267–6276. [CrossRef] [PubMed]
85. Svitel, J.; Balbo, A.; Mariuzza, R.A.; Gonzales, N.R.; Schuck, P. Combined Affinity and Rate Constant Distributions of Ligand Populations from Experimental Surface Binding Kinetics and Equilibria. *Biophys. J.* **2003**, *84*, 4062–4077. [CrossRef]
86. Svitel, J.; Boukari, H.; Van Ryk, D.; Willson, R.C.; Schuck, P. Probing the Functional Heterogeneity of Surface Binding Sites Analysis of Experimental Binding Traces and the Effect of Mass Transport Limitation. *Biophys. J.* **2007**, *92*, 1742–1758. [CrossRef]
87. Gorshkova, I.I.; Svitel, J.; Razjouyan, F.; Schuck, P. Bayesian Analysis of Heterogeneity in the Distribution of Binding Properties Immobilized Surface Sites. *Langmuir* **2008**, *24*, 11577–11586. [CrossRef]
88. Khalifa, M.B.; Choulier, L.; Lortat-Jacob, H.; Altschuh, D.; Vernet, T. BIACORE Data Processing: An Evaluation of the Global Fitting Procedure. *Anal. Biochem.* **2001**, *293*, 194–203. [CrossRef]
89. De Crescenzo, G.; Grothe, S.; Zwaagstra, J.; Tsang, M.; O'Connor-McCourt, M.D. Real-time monitoring of the interactions transforming growth factor-beta (TGF-beta) isoforms with latency-associated protein and the ectodomains of the TGF-beta type and III receptors reveals different kinetic models and stoichiometries of binding. *J. Biol. Chem.* **2001**, *276*, 29632–29643. [CrossRef]
90. Sprague, E.R.; Martin, W.L.; Bjorkman, P.J. pH dependence and stoichiometry of binding to the Fc region of IgG by the herpes simplex virus Fc receptor gE-gI. *J. Biol. Chem.* **2004**, *279*, 14184–14193. [CrossRef]
91. Giannetti, A.M.; Snow, P.M.; Zak, O.; Björkman, P.J. Mechanism for Multiple Ligand Recognition by the Human Transferrin Receptor. *PLoS Biol.* **2003**, *1*, e51. [CrossRef] [PubMed]
92. Müller, K.M.; Arndt, K.M.; Plückthun, A. Model and simulation of multivalent binding to fixed ligands. *Anal. Biochem.* **1998**, *2*, 149–158. [CrossRef] [PubMed]
93. De Crescenzo, G.; Grothe, S.; Lortie, R.; Debanne, M.T.; O'Connor-McCourt, M. Real-Time Kinetic Studies on the Interaction of Transforming Growth Factor α with the Epidermal Growth Factor Receptor Extracellular Domain Reveal a Conformational Change Model. *Biochemistry* **2000**, *39*, 9466–9476. [CrossRef] [PubMed]
94. Futamura, M.; Dhanasekaran, P.; Handa, T.; Phillips, M.C.; Lund-Katz, S.; Saito, H. Two-step mechanism of binding of apolipoprotein E to heparin: Implications for the kinetics of apolipoprotein E-heparan sulfate proteoglycan complex formation on c surfaces. *J. Biol. Chem.* **2005**, *280*, 5414–5422. [CrossRef] [PubMed]
95. Bio-Rad Laboratories. *ProteOn XPR36 Experimental Design and Application Guide*; Bio-Rad Laboratoires: Hercules, CA, USA, 20
96. Goldstein, B.; Coombs, D.; He, X.; Pineda, A.R.; Wofsy, C. The influence of transport on the kinetics of binding to surface receptor Application to cells and BIAcore. *J. Mol. Recognit. JMR* **1999**, *12*, 293–299. [CrossRef]
97. Sjoelander, S.; Urbaniczky, C. Integrated fluid handling system for biomolecular interaction analysis. *Anal. Chem.* **1991**, 2338–2345. [CrossRef]
98. Cytiva. *Accurate Comparability Assessment of a Biosimilar Interferon in Process Development*; Cytiva: Marlborough, MA, USA, 20
99. Hu, T.; Wu, L.; Sun, X.; Su, P.; Yang, Y. Comparative study on quantitation of human myoglobin by both isotope dilution m spectrometry and surface plasmon resonance based on calibration-free analysis. *Anal. Bioanal. Chem.* **2020**, *412*, 2777–27 [CrossRef]
100. Imamura, H.; Honda, S. Calibration-free concentration analysis for an analyte prone to self-association. *Anal. Biochem.* **2017**, *5* 61–64. [CrossRef]
101. Hossler, P.; Khattak, S.F.; Li, Z.J. Optimal and consistent protein glycosylation in mammalian cell culture. *Glycobiology* **2009**, 936–949. [CrossRef]
102. Canziani, G.A.; Klakamp, S.; Myszka, D.G. Kinetic screening of antibodies from crude hybridoma samples using Biacore. *A Biochem.* **2004**, *325*, 301–307. [CrossRef]
103. Wu, L.; He, Y.; Hu, Y.; Lu, H.; Cao, Z.; Yi, X.; Wang, J. Real-time surface plasmon resonance monitoring of site-spec phosphorylation of p53 protein and its interaction with MDM2 protein. *Analyst* **2019**, *144*, 6033–6040. [CrossRef] [PubMed]
104. Cytiva. *Biacore Systems in Discovery and Early-Stage Development of Biotherapeutics Antibodies*; Cytiva: Marlborough, MA, US 2016.
105. Yun, S.; Lee, S.; Park, J.P.; Choo, J.; Lee, E.K. Modification of phage display technique for improved screening of high-affin binding peptides. *J. Biotechnol.* **2019**, *289*, 88–92. [CrossRef]
106. Zhao, A.; Tohidkia, M.R.; Siegel, D.L.; Coukos, G.; Omidi, Y. Phage antibody display libraries: A powerful antibody discov platform for immunotherapy. *Crit. Rev. Biotechnol.* **2016**, *36*, 276–289. [CrossRef] [PubMed]
107. Munke, A.; Persson, J.; Weiffert, T.; De Genst, E.; Meisl, G.; Arosio, P.; Carnerup, A.; Dobson, C.M.; Vendruscolo, M.; Know T.P.J.; et al. Phage display and kinetic selection of antibodies that specifically inhibit amyloid self-replication. *Proc. Natl. Acad. USA* **2017**, *114*, 6444–6449. [CrossRef] [PubMed]
108. Jackson, S.; Lentino, J.; Kopp, J.; Murray, L.; Ellison, W.; Rhee, M.; Shockey, G.; Akella, L.; Erby, K.; Heyward, W.L.; et Immunogenicity of a two-dose investigational hepatitis B vaccine, HBsAg-1018, using a toll-like receptor 9 agonist adjuv compared with a licensed hepatitis B vaccine in adults. *Vaccine* **2018**, *36*, 668–674. [CrossRef] [PubMed]
109. Quinn, J.G.; Pitts, K.E.; Steffek, M.; Mulvihill, M.M. Determination of Affinity and Residence Time of Potent Drug-Tar Complexes by Label-free Biosensing. *J. Med. Chem.* **2018**, *61*, 5154–5161. [CrossRef] [PubMed]
110. Frostell-Karlsson, A.; Remaeus, A.; Roos, H.; Andersson, K.; Borg, P.; Hämäläinen, M.; Karlsson, R. Biosensor analysis of interaction between immobilized human serum albumin and drug compounds for prediction of human serum albumin bind levels. *J. Med. Chem.* **2000**, *43*, 1986–1992. [CrossRef]

Gonzales, N.R.; Schuck, P.; Schlom, J.; Kashmiri, S.V. Surface plasmon resonance-based competition assay to assess the sera reactivity of variants of humanized antibodies. *J. Immunol. Methods* **2002**, *268*, 197–210. [CrossRef]

Ritter, G.; Cohen, L.S.; Williams, C., Jr.; Richards, E.C.; Old, L.J.; Welt, S. Serological analysis of human anti-human antibody responses in colon cancer patients treated with repeated doses of humanized monoclonal antibody A33. *Cancer Res.* **2001**, *61*, 6851–6859.

Wang, W.; Lu, P.; Fang, Y.; Hamuro, L.; Pittman, T.; Carr, B.; Hochman, J.; Prueksaritanont, T. Monoclonal antibodies with identical Fc sequences can bind to FcRn differently with pharmacokinetic consequences. *Drug Metab. Dispos.* **2011**, *39*, 1469–1477. [CrossRef]

Pyzik, M.; Sand, K.M.K.; Hubbard, J.J.; Andersen, J.T.; Sandlie, I.; Blumberg, R.S. The Neonatal Fc Receptor (FcRn): A Misnomer? *Front. Immunol.* **2019**, *10*, 1540. [CrossRef] [PubMed]

Lu, Y.; Vernes, J.M.; Chiang, N.; Ou, Q.; Ding, J.; Adams, C.; Hong, K.; Truong, B.T.; Ng, D.; Shen, A.; et al. Identification of IgG(1) variants with increased affinity to FcγRIIIa and unaltered affinity to FcγRI and FcRn: Comparison of soluble receptor-based and cell-based binding assays. *J. Immunol. Methods* **2011**, *365*, 132–141. [CrossRef] [PubMed]

World Health Organization. *Influenza Vaccine Response during the Start of a Pandemic: Report of a WHO Informal Consultation Held in Geneve, Switzerland, 29 June–1 July 2015*; World Health Organization: Geneva, Switzerland, 2016; Volume 2016.

Suenaga, E.; Mizuno, H.; Penmetcha, K.K. Monitoring influenza hemagglutinin and glycan interactions using surface plasmon resonance. *Biosens. Bioelectron.* **2012**, *32*, 195–201. [CrossRef] [PubMed]

Abadian, P.N.; Yildirim, N.; Gu, A.Z.; Goluch, E.D. SPRi-based adenovirus detection using a surrogate antibody method. *Biosens. Bioelectron.* **2015**, *74*, 808–814. [CrossRef] [PubMed]

Roldão, A.; Mellado, M.C.; Castilho, L.R.; Carrondo, M.J.; Alves, P.M. Virus-like particles in vaccine development. *Expert Rev. Vaccines* **2010**, *9*, 1149–1176. [CrossRef] [PubMed]

Nooraei, S.; Bahrulolum, H.; Hoseini, Z.S.; Katalani, C.; Hajizade, A.; Easton, A.J.; Ahmadian, G. Virus-like particles: Preparation, immunogenicity and their roles as nanovaccines and drug nanocarriers. *J. Nanobiotechnol.* **2021**, *19*, 59. [CrossRef]

Thompson, C.M.; Petiot, E.; Lennaertz, A.; Henry, O.; Kamen, A.A. Analytical technologies for influenza virus-like particle candidate vaccines: Challenges and emerging approaches. *Virol. J.* **2013**, *10*, 141. [CrossRef]

Chang, J. Adenovirus Vectors: Excellent Tools for Vaccine Development. *Immune Netw* **2021**, *21*, e6. [CrossRef]

Glick, G.D.; Toogood, P.L.; Wiley, D.C.; Skehel, J.J.; Knowles, J.R. Ligand recognition by influenza virus. The binding of bivalent sialosides. *J. Biol. Chem.* **1991**, *266*, 23660–23669. [CrossRef]

Connor, R.J.; Kawaoka, Y.; Webster, R.G.; Paulson, J.C. Receptor specificity in human, avian, and equine H2 and H3 influenza virus isolates. *Virology* **1994**, *205*, 17–23. [CrossRef]

Ohlson, S.; Jungar, C.; Strandh, M.; Mandenius, C.-F. Continuous weak-affinity immunosensing. *Trends Biotechnol.* **2000**, *18*, 49–52. [CrossRef]

Tawil, N.; Sacher, E.; Mandeville, R.; Meunier, M. Bacteriophages: Biosensing tools for multi-drug resistant pathogens. *Analyst* **2014**, *139*, 1224–1236. [CrossRef] [PubMed]

Nanduri, V.; Bhunia, A.K.; Tu, S.I.; Paoli, G.C.; Brewster, J.D. SPR biosensor for the detection of L. monocytogenes using phage-displayed antibody. *Biosens. Bioelectron.* **2007**, *23*, 248–252. [CrossRef] [PubMed]

Cho, I.H.; Ku, S. Current Technical Approaches for the Early Detection of Foodborne Pathogens: Challenges and Opportunities. *Int. J. Mol. Sci.* **2017**, *18*, 2078. [CrossRef] [PubMed]

Zheng, X.; Gao, S.; Wu, J.; Hu, X. Recent Advances in Aptamer-Based Biosensors for Detection of Pseudomonas aeruginosa. *Front. Microbiol.* **2020**, *11*, 5229. [CrossRef]

Puttharugsa, C.; Wangkam, T.; Huangkamhang, N.; Gajanandana, O.; Himananto, O.; Sutapun, B.; Amarit, R.; Somboonkaew, A.; Srikhirin, T. Development of surface plasmon resonance imaging for detection of Acidovorax avenae subsp. citrulli (Aac) using specific monoclonal antibody. *Biosens. Bioelectron.* **2011**, *26*, 2341–2346. [CrossRef] [PubMed]

Taylor, A.D.; Ladd, J.; Homola, J.; Jiang, S. Surface Plasmon Resonance (SPR) Sensors for the Detection of Bacterial Pathogens. In *Principles of Bacterial Detection: Biosensors, Recognition Receptors and Microsystems*; Zourob, M., Elwary, S., Turner, A., Eds.; Springer: New York, NY, USA, 2008; pp. 83–108.

Kastenhofer, J.; Rajamanickam, V.; Libiseller-Egger, J.; Spadiut, O. Monitoring and control of *E. coli* cell integrity. *J. Biotechnol.* **2021**, *329*, 1–12. [CrossRef]

Hofer, A.; Kroll, P.; Barmettler, M.; Herwig, C. A Reliable Automated Sampling System for On-Line and Real-Time Monitoring of CHO Cultures. *Processes* **2020**, *8*, 637. [CrossRef]

Jiang, C.; Wang, G.; Hein, R.; Liu, N.; Luo, X.; Davis, J.J. Antifouling Strategies for Selective In Vitro and In Vivo Sensing. *Chem. Rev.* **2020**, *120*, 3852–3889. [CrossRef]

Wang, D.; Loo, J.F.; Chen, J.; Yam, Y.; Chen, S.-C.; He, H.; Kong, S.K.; Ho, H.P. Recent Advances in Surface Plasmon Resonance Imaging Sensors. *Sensors* **2019**, *19*, 1266. [CrossRef]

Bravman, T.; Bronner, V.; Nahshol, O.; Schreiber, G. The ProteOn XPR36™ Array System—High Throughput Kinetic Binding Analysis of Biomolecular Interactions. *Cell. Mol. Bioeng.* **2008**, *1*, 216. [CrossRef]

Yang, D.; Singh, A.; Wu, H.; Kroe-Barrett, R. Determination of High-affinity Antibody-antigen Binding Kinetics Using Four Biosensor Platforms. *J. Vis. Exp.* **2017**, 55659. [CrossRef]

138. Kamat, V.; Rafique, A. Exploring sensitivity & throughput of a parallel flow SPRi biosensor for characterization of antibo antigen interaction. *Anal. Biochem.* **2017**, *525*, 8–22. [CrossRef] [PubMed]

139. Carterra. Speeding Antibody Screening for Drug Development. Available online: https://www.nature.com/articles/d42473--00177-x (accessed on 16 September 2021).

140. Zeni, L.; Perri, C.; Cennamo, N.; Arcadio, F.; D'Agostino, G.; Salmona, M.; Beeg, M.; Gobbi, M. A portable optical-fibre-ba surface plasmon resonance biosensor for the detection of therapeutic antibodies in human serum. *Sci. Rep.* **2020**, *10*, 11 [CrossRef] [PubMed]

141. Liu, Y.; Liu, Q.; Chen, S.; Cheng, F.; Wang, H.; Peng, W. Surface Plasmon Resonance Biosensor Based on Smart Phone Platfor *Sci. Rep.* **2015**, *5*, 12864. [CrossRef] [PubMed]

142. Lertvachirapaiboon, C.; Baba, A.; Shinbo, K.; Kato, K. A smartphone-based surface plasmon resonance platform. *Anal. Meth* **2018**, *10*, 4732–4740. [CrossRef]

143. Wong, C.L.; Chua, M.; Mittman, H.; Choo, L.X.; Lim, H.Q.; Olivo, M. A Phase-Intensity Surface Plasmon Resonance Biosensor Avian Influenza A (H5N1) Detection. *Sensors* **2017**, *17*, 2363. [CrossRef]

144. Huang, Y.H.; Ho, H.P.; Wu, S.Y.; Kong, S.K. Detecting Phase Shifts in Surface Plasmon Resonance: A Review. *Adv. Opt. Tech* **2012**, *2012*, 471957. [CrossRef]

145. Kuncova-Kallio, J.; Järvinen, A. *Comparison of MP-SPR Navi^{TM} Instruments to Biacore^{TM} in Protein Research*; WP_601.809 BioNavis: Tampere, Finland, 2018.

146. Kari, O.K.; Rojalin, T.; Salmaso, S.; Barattin, M.; Jarva, H.; Meri, S.; Yliperttula, M.; Viitala, T.; Urtti, A. Multi-parametric surf plasmon resonance platform for studying liposome-serum interactions and protein corona formation. *Drug Deliv. Transl.* **2017**, *7*, 228–240. [CrossRef]

147. Viitala, T.; Granqvist, N.; Hallila, S.; Raviña, M.; Yliperttula, M. Elucidating the signal responses of multi-parametric surf plasmon resonance living cell sensing: A comparison between optical modeling and drug-MDCKII cell interaction measureme *PLoS ONE* **2013**, *8*, e72192. [CrossRef]

148. Suutari, T.; Silen, T.; Karaman, D.S.E.; Saari, H.; Desai, D.; Kerkelä, E.; Laitinen, S.; Hanzlikova, M.; Rosenholm, J.M.; Ylipertt M.; et al. Real-Time Label-Free Monitoring of Nanoparticle Cell Uptake. *Small* **2016**, *12*, 6289–6300. [CrossRef] [PubMed]

149. Koponen, A.; Kerkelä, E.; Rojalin, T.; Lázaro-Ibáñez, E.; Suutari, T.; Saari, H.O.; Siljander, P.; Yliperttula, M.; Laitinen, S.; Viital Label-free characterization and real-time monitoring of cell uptake of extracellular vesicles. *Biosens. Bioelectron.* **2020**, *168*, 112 [CrossRef] [PubMed]

150. Chabot, V.; Miron, Y.; Grandbois, M.; Charette, P.G. Long range surface plasmon resonance for increased sensitivity in living biosensing through greater probing depth. *Sens. Actuators B Chem.* **2012**, *174*, 94–101. [CrossRef]

151. Vala, M.; Robelek, R.; Bocková, M.; Wegener, J.; Homola, J. Real-time label-free monitoring of the cellular response to osm stress using conventional and long-range surface plasmons. *Biosens. Bioelectron.* **2013**, *40*, 417–421. [CrossRef] [PubMed]

152. Vala, M.; Etheridge, S.; Roach, J.A.; Homola, J. Long-range surface plasmons for sensitive detection of bacterial analytes. *S Actuators B Chem.* **2009**, *139*, 59–63. [CrossRef]

153. Jing, J.-Y.; Wang, Q.; Zhao, W.-M.; Wang, B.-T. Long-range surface plasmon resonance and its sensing applications: A review. *Lasers Eng.* **2019**, *112*, 103–118. [CrossRef]

154. Yanase, Y.; Yoshizaki, K.; Kimura, K.; Kawaguchi, T.; Hide, M.; Uno, S. Development of SPR Imaging-Impedance Sensor Multi-Parametric Living Cell Analysis. *Sensors* **2019**, *19*, 2067. [CrossRef]

155. Ogura, T.; Tanaka, Y.; Toyoda, H. Whole cell-based surface plasmon resonance measurement to assess binding of anti-TNF age to transmembrane target. *Anal. Biochem.* **2016**, *508*, 73–77. [CrossRef]

156. Abali, F.; Stevens, M.; Tibbe, A.G.J.; Terstappen, L.; van der Velde, P.N.; Schasfoort, R.B.M. Isolation of single cells for pro therapeutics using microwell selection and Surface Plasmon Resonance imaging. *Anal. Biochem.* **2017**, *531*, 45–47. [CrossRef]

157. Schasfoort, R.B.M.; Abali, F.; Stojanovic, I.; Vidarsson, G.; Terstappen, L. Trends in SPR Cytometry: Advances in Label-F Detection of Cell Parameters. *Biosensors* **2018**, *8*, 102. [CrossRef]

158. Karlsson, R.; Pol, E.; Frostell, Å. Comparison of surface plasmon resonance binding curves for characterization of pro interactions and analysis of screening data. *Anal. Biochem.* **2016**, *502*, 53–63. [CrossRef] [PubMed]

159. Tharmalingam, T.; Wu, C.H.; Callahan, S.; Goudar, T.G. A framework for real-time glycosylation monitoring (RT-GM mammalian cell culture. *Biotechnol. Bioeng.* **2015**, *112*, 1146–1154. [CrossRef]

160. Mou, Z.-L.; Qi, X.-N.; Liu, R.-L.; Zhang, J.; Zhang, Z.-Q. Three-dimensional cell bioreactor coupled with high performance liq chromatography–mass spectrometry for the affinity screening of bioactive components from herb medicine. *J. Chromatogr. A 2* *1243*, 33–38. [CrossRef] [PubMed]

161. Li, M.-Y.; Ebel, B.; Paris, C.; Chauchard, F.; Guedon, E.; Marc, A. Real-time monitoring of antibody glycosylation site occupa by in situ Raman spectroscopy during bioreactor CHO cell cultures. *Biotechnol. Prog.* **2018**, *34*, 486–493. [CrossRef] [PubMed

MDPI

St. Alban-Anlage 66

4052 Basel

Switzerland

Tel. +41 61 683 77 34

Fax +41 61 302 89 18

www.mdpi.com

Processes Editorial Office

E-mail: processes@mdpi.com

www.mdpi.com/journal/processes